Sustainability Sciences in Asia and Africa

Sustainable Agriculture and Food Security

Series Editor

Rajeev K. Varshney, Centre for Crop and Food Innovation, Food Futures Institute, Murdoch University, Perth, WA, Australia

Sustainability as a scientific endeavor requires broad understanding of the interconnections in our global environment. It is therefore positioned at the crossroads of research in natural sciences, technology and human sciences. This book series supports the global efforts by providing timely coverage of the progress, opportunities, and challenges towards achieving sustainable development goals (SDGs) in Asia and Africa. It fosters the research in transdisciplinary academic fields spanning across agriculture and global food security, health and diseases, access to clean water and affordable energy, climate resilience and effects of global climate change on biodiversity, water and land. This series is planned to narrate the stories of scientific development and progression towards SDGs from the regions of Asia and Africa. The overall aim of the series is to provide a comprehensive publication platform and act as a knowledge engine in the growth of sustainability sciences with a special focus on developing nations. The series constitute of four sub-series each focusing on a specific theme closely linked to a set of SDGs. Each sub-series is leaders and experts in agricultural sciences, environmental sciences, biomedicine and medical sciences respectively. Books in this series are useful for environmentalists, biotechnologists, plant scientists, medical practitioners, conservation biologists, policy makers, NGOs and scientific bodies working in the areas of sustainability sciences in the developing regions.

This book series support the global efforts towards sustainability by providing timely coverage of the progress, opportunities, and challenges of sustainable food production and consumption in Asia and Africa. The series narrates the success stories and research endeavors from the regions of Africa and Asia on issues relating to SDG 2: Zero hunger. It fosters the research in transdisciplinary academic fields spanning across sustainable agriculture systems and practices, post- harvest and food supply chains. It will also focus on breeding programs for resilient crops, efficiency in crop cycle, various factors of food security, as well as improving nutrition and curbing hunger and malnutrition. The focus of the series is to provide a comprehensive publication platform and act as a knowledge engine in the growth of sustainability sciences with a special focus on developing nations. The series will publish mainly edited volumes but some authored volumes. These volumes will have chapters from eminent personalities in their area of research from different parts of the world.

Mahalingam Govindaraj
Editor

Breeding Biofortified Millets for Nutritional Security

Springer

Editor
Mahalingam Govindaraj
Alliance of Bioversity International and the International
Centre for Tropical Agriculture (CIAT)
Cali, Colombia

ISSN 2730-6771 ISSN 2730-678X (electronic)
Sustainability Sciences in Asia and Africa
ISSN 2730-6798 ISSN 2730-6801 (electronic)
Sustainable Agriculture and Food Security
ISBN 978-981-95-9826-7 ISBN 978-981-95-9827-4 (eBook)
https://doi.org/10.1007/978-981-95-9827-4

This Springer imprint is published by the registered company Springer Nature Singapore Pte Ltd.
The registered company address is: 152 Beach Road, #21-01/04 Gateway East, Singapore 189721, Singapore

To my beloved parents, whose unwavering values and sacrifices sowed the seeds of curiosity and resilience in me.

To my family, for their endless patience, love, and support, which sustained me during months away from home through the critical time of this journey.

And to my esteemed research partners and colleagues, your shared vision, dedication, and belief in the power of millets have made this work possible.

This book is a dedication to millet farmers and the shared goal of creating a healthier, more sustainable future through traditional grain crops with remarkable potential.

Preface to the Book Series

The UN's second Sustainable Development Goal (SDG) aims to end hunger, achieve food security and improved nutrition, and promote sustainable agriculture by 2030. Yet over 700 million people still suffer from undernourishment, meaning that global food production systems will need to expand their capacities and produce almost twice their existing levels to feed the world's ever-growing population. This poses a significant challenge, and while farmers, researchers, and policymakers the world over have been tireless in their efforts to address the global food security gap, we're a long way off from where we need to be to meet SDG 2: Zero Hunger.

In recent years, the adoption of new research approaches has led to improvements in food access, farmer autonomy, and the stabilisation of agriculture's global carbon footprint. The advent of these approaches, such as genomics-assisted breeding, has dramatically improved crop breeding programmes worldwide, yet the challenge to reduce hunger remains especially pronounced in the Global South.

Furthermore, the agri-food systems in Asia and Africa are severely constrained by the subsistence nature of farming, limited access to new research tools and technologies, and the increasing impacts of environmental pollution, soil degradation, and biodiversity loss.

This book series, "Sustainable Agriculture and Food Security", aims to support global efforts towards sustainability by providing timely coverage of the progress, opportunities, and challenges of sustainable food production and consumption in Asia and Africa. The series narrates the success stories and research endeavours from Africa and Asia on issues related to SDG 2: Zero Hunger. It fosters research in transdisciplinary academic fields spanning sustainable agriculture systems and practices, pre-seed and post-harvest innovations, and food supply chains. The series aims to provide a comprehensive publication platform and serve as a repository of knowledge in sustainability sciences, with particular focus on developing nations.

Our seas, rivers and lakes are a cornerstone of life on Earth, with fisheries and aquaculture providing livelihoods for an estimated 60 million people. Aquaculture is now the fastest-growing food production sector, providing 20 per cent of animal protein to over 3 billion people globally. It is therefore essential that forward-looking research into our aquatic ecosystems is part of the conversation on global food security and sustainability.

This book, "Harnessing Coastal and Marine Resources for Improved Food Security in Africa", edited by Dr Hupenyu A. Mupambwa, Prof Naftal Gabriel, Prof Tamuka Nhiwatiwa, Dr Kenneth P Abasubong and Prof Samuel K. Mafwila, serves as a comprehensive tome detailing Africa's extraordinary and vibrant marine and coastal ecosystems. From the utilisation of seaweed as a renewable resource in Namibia to detailing the impact of marine pollution in Nigeria, this book offers unprecedented insights into coastal and marine resource management in Africa.

I commend the Editors for including multidisciplinary and continent-wide perspectives and congratulate all the authors for their valuable contributions.

I wish to extend my sincere thanks to the Springer staff, particularly Dr. Sanchi Bhimrajka, Editor (Books, Medicine and Life Sciences), and Mr. Nick Melchior, Editorial Director (Books, Medicine and Life Sciences, Asia), for their constant support in the completion of this compendium.

The cooperation received from my supervisors and research leaders, such as Prof Andrew Deeks (Vice Chancellor), Prof Peter Eastwood (Deputy Vice Chancellor-Research and Innovation), Prof Peter Davies (Pro Vice Chancellor), and my colleagues, Anu Chitikineni and Nic Ingram from Murdoch University (Australia), is also gratefully acknowledged. I want to thank my family members – Monika Varshney, Prakhar Varshney and Preksha Varshney for their love and support in discharging my duties as Series Editor.

Rajeev K. Varshney
Director – WA State Agricultural Biotechnology Centre
Director – Centre for Crop & Food Innovation
International Chair – Agriculture & Food Security
Murdoch University, Perth, WA, Australia

Preface

The United Nations declared 2023 as the International Year of Millets (IYM), recognizing the ancient grains' ability to address some of today's most critical global concerns. This book stems from that historic event, going beyond celebration to present a clear blueprint for protecting our collective nutritional future. Millets have traditionally been nutritionally dense, climate adaptable, and crucial to the diets of millions of people in Asia and Africa. The Green Revolution, while increasing food production, inadvertently led to the neglect of millets in favor of water-intensive crops like rice and wheat. The result has been a paradox: a world with more food but widespread hidden hunger, affecting billions, particularly in developing nations. This volume addresses this by focusing on the power of biofortification, which provides a sustainable, cost-effective, and long-term solution by delivering nutrient-rich crop varieties for food directly to those who need it most. This makes millets an indispensable tool for building resilient and sustainable food systems.

Drawing from decades of research and development, particularly from pioneering institutions like the International Crops Research Institute for the Semi-Arid Tropics (ICRISAT) and HarvestPlus program, this book covers the biofortified millet efforts and work areas. It highlights advancements, such as the development of high-iron pearl millet varieties, which have already begun to improve the health outcomes of vulnerable populations. Chapters describe remarkable progress: traditional pearl millet varieties contain an average of 42 mg/kg iron; advanced biofortified cultivars contain 70–75 mg/kg iron and 35–40 mg/kg zinc. Finger millet varieties contain a high amount of calcium (300 mg/100 g) compared to other cereals. Research has focused on enhancing other micronutrients like zinc and protein content to maximize its health benefits. Reported clinical trials have demonstrated the real-world impact. Children consuming biofortified pearl millet absorbed almost three times more iron compared to a control group. This led to a significant reduction of iron deficiency.

Chapters within this volume explore the genetic basis of millet resilience, the methodologies used in conventional and molecular breeding, and the essential role of genomics in accelerating the biofortification process. The book also discusses strategies for overcoming challenges posed by antinutrients like phytates and others. The International Year of Millets 2023 was a call to action, a pivotal moment to raise global awareness and re-prioritize the importance of millets. The momentum generated by the IYM has created a unique opportunity to build on nutrition through breeding successes and drive forward the agenda of nutritional security. This book serves as documentation of the science and strategy required to ensure that millets are mainstreamed for nutrition and reclaim their rightful place in our food baskets. By embracing these ancient grains and leveraging modern science, we can combat hidden hunger and build a healthier, more sustainable world for all.

Cali, Valle, Colombia Mahalingam Govindaraj

Acknowledgments

We extend our deepest gratitude to our esteemed partners, donors, and collaborators for their foundational work in mainstreaming biofortification. HarvestPlus's determined commitment and collaborative model and spirit have been instrumental in translating scientific nutrition research into tangible change in addressing malnutrition in developing countries. The development and delivery of micronutrient-rich crops stand as a testament to partners' efforts, directly improving nutritional security and livelihoods. This volume is an acknowledgment to pioneering vision of biofortification leaders, donors, and partners, without which the strides documented here would not have been possible.

Contents

Series Editor and Editorial Board Members

About the Series Editor

Rajeev K. Varshney FRS FAA
WA State Agricultural Biotechnology Centre;
Centre for Crop and Food Innovation;
International Chair – Agriculture and Food Security,
Food Futures Institute, Murdoch University,
WA, Australia

Professor Rajeev K. Varshney FRS FAA is a globally recognised agricultural scientist with over two decades of leadership in genomics and molecular breeding. He has made a significant contribution to global food and nutrition security by developing genomic resources for "orphan" crops and leveraging these resources to identify genetic loci and candidate genes associated with drought tolerance, pest resistance, and yield improvement in crops widely grown by smallholder farmers in sub-Saharan Africa and India.

He has led major international research programs that have delivered improved, climate-resilient crop varieties to some of the world's most resource-constrained farming systems, contributing to increased farm productivity, enhanced livelihoods, and improved nutritional outcomes.

Currently, Professor Varshney leads several national and international initiatives in wheat, pulse, and horticultural crops in Australia, with a strong focus on enhancing farmer profitability, environmental sustainability, and global food security.

With more than 600 peer-reviewed publications, over 90,000 citations, and an h-index exceeding 150, he is consistently ranked among the world's most highly cited researchers in plant science.

Editorial Board Members

Chike Mba, Deputy Director, Plant Production and Protection Division, Food and Agricultural Organization (FAO), Rome, Italy.

Trilochan Mohapatra, Chairman, Protection of Plant Varieties and Farmers' Rights Authority, Ministry of Agriculture and Farmers Welfare, Government of India, New Delhi, India.

Soumya Swaminathan, Chairperson, MS Swaminathan Research Foundation, Chennai, India.

Wendy Umberger, Chief Executive Officer, Australian Council for International Agricultural Research (ACIAR), Canberra, ACT, Australia.

About the Editor

Mahalingam Govindaraj is an internationally recognized crop scientist with over 15 years of experience in genetics, plant breeding, seed system and biofortification. Currently leading the biofortification traits mainstreaming in CGIAR. He spearheaded the development and release of over 20 biofortified cultivars across Asia and Africa, focusing on iron- and zinc-rich pearl millet, rice and bean that address micronutrient deficiencies in vulnerable populations. His work has significantly influenced national programs in India, Nigeria, and Niger, where biofortified millets are now being mainstreamed into public food and nutrition systems. He developed high-yielding, drought- and heat-resilient millet cultivars tailored for dryland farming systems. He improved the use of diagnostic markers and quick phenotyping methods for micronutrient and stress-resilience trait selection. He was instrumental in scaling zinc wheat and rice in India and currently participates to multi-partner agricultural efforts. He connects research and real-world effect through over 130 peer-reviewed publications and considerable leadership in capacity-building programs. His achievements earned him the 2022 Norman E. Borlaug Award for Field Research and Application, which recognized his scientific prowess and global contributions to food and nutrition security.

Contributors

C. Amasiddha ICAR – Indian Institute of Millets Research, Hyderabad, Telangana, India

I. I. Angarawai International Crops Research Institute for the Semi-Arid Tropics, Kano, Nigeria

M. Aswin Ajay Raj VOC Agricultural College and Research Institute, Tamil Nadu Agricultural University, Thoothukudi, Tamil Nadu, India

C. Backiyalakshmi International Crops Research Institute for the Semi-Arid Tropics (ICRISAT), Hyderabad, Telangana, India

Binu Cherian HarvestPlus Program, International Centre for Tropical Agriculture (CIAT), Patancheru, Telangana, India

M. A. Dawud Lake Chad Research Institute, Maiduguri, Nigeria

S. E. Diwakar Reddy Department of Genetics and Plant Breeding, UAS, Dharwad, Dharwad, Karnataka, India

Y. Dollah HarvestPlus Program, International Food Policy Research Institute, c/o IITA Campus, Kano, Nigeria

K. P. Gaganashree Department of Genetics and Plant Breeding, UAS, Dharwad, Dharwad, Karnataka, India

Manimaran Gajendiran Department of Soil Science and Agricultural Chemistry, Tamil Nadu Agricultural University, Coimbatore, Tamil Nadu, India

K. N. Ganapathy ICAR – Indian Institute of Millets Research, Hyderabad, Telangana, India\

Prakash I. Gangashetty International Crops Research Institute for the Semi-Arid Tropics, Hyderabad, Telangana, India

Bassirou Sani Boubacar Gaoh International Crops Research Institute for the Semi-Arid Tropics, Niamey, Niger

C. M. Gopika College of Agriculture, University of Agricultural Sciences, Bengaluru, Karnataka, India

Mahalingam Govindaraj Alliance of Bioversity International and the International Centre for Tropical Agriculture (CIAT), Cali, Colombia

ICAR – Indian Institute of Millets Research, Hyderabad, Telangana, India

G. Guru Agricultural Research Station (AICRPDA Main Centre), Tamil Nadu Agricultural University, Kovilpatti, Tamil Nadu, India

N. G. Hanamaratti AICRP Sorghum, MARS, UAS Dharwad, Dharwad, Karnataka, India

K. Hariprasanna ICAR – Indian Institute of Millets Research, Hyderabad, Telangana, India

I. A. Hassan HarvestPlus Program, International Food Policy Research Institute, Kano, Nigeria

Zakari Hassane International Crops Research Institute for the Semi-Arid Tropics, Niamey, Niger

M. Hemalatha VOC Agricultural College and Research Institute, Tamil Nadu Agricultural University, Thoothukudi, Tamil Nadu, India

J. Jonah International Crops Research Institute for the Semi-Arid Tropics, Kano, Nigeria

M. Joseph Agricultural Research Station (AICRPDA Main Centre), Tamil Nadu Agricultural University, Kovilpatti, Tamil Nadu, India

T. Kalaimagal Centre for Plant Breeding and Genetics, Tamil Nadu Agricultural University, Coimbatore, Tamil Nadu, India

R. Karthik Department of Genetics and Plant Breeding, UAS, Dharwad, Dharwad, Karnataka, India

Epiphane Lamine International Crops Research Institute for the Semi-Arid Tropics, Niamey, Niger

K. Manonmani Department of Genetics and Plant Breeding, Amrita School of Agricultural Sciences, Amrita Vishwa Vidyapeetham, Coimbatore, Tamil Nadu, India

Ranjitha Marigoudar Department of Genetics and Plant Breeding, UAS, Dharwad, Dharwad, Karnataka, India

Riyazaddin Mohammed International Crops Research Institute for the Semi-Arid Tropics, Niamey, Niger

Mohamed Roshan Abu Firnass Mustaffa Department of Soil Science and Agricultural Chemistry, Tamil Nadu Agricultural University, Coimbatore, Tamil Nadu, India

Sandeep Nanjundappa Global Centre of Excellence on Millets (Shree Anna), ICAR-Indian Institute of Millets Research, Hyderabad, Telangana, India

Nithya Sree Koppula ICAR – Indian Institute of Millets Research, Rajendranagar, Hyderabad, India

Kannan Pandian Centre for Water and Geospatial Studies, Tamil Nadu Agricultural University, Coimbatore, Tamil Nadu, India

Abishek John Prince Department of Soil Science and Agricultural Chemistry, Tamil Nadu Agricultural University, Coimbatore, Tamil Nadu, India

R. Ravikesavan Centre for Plant Breeding and Genetics, Tamil Nadu Agricultural University, Coimbatore, Tamil Nadu, India

M. Riyazaddin International Crops Research Institute for the Semi-Arid Tropics, Niamey, Niger

L. Sampath Centre for Plant Breeding and Genetics, Tamil Nadu Agricultural University, Coimbatore, Tamil Nadu, India

S. Samuthrapandi Centre for Plant Breeding and Genetics, Tamil Nadu Agricultural University, Coimbatore, Tamil Nadu, India

D. S. Shailaja Department of Genetics and Plant Breeding, UAS, Dharwad, Dharwad, Karnataka, India

A. Shanmugam Centre for Plant Breeding and Genetics, Tamil Nadu Agricultural University, Coimbatore, Tamil Nadu, India

D. Sushmitha Centre for Plant Breeding and Genetics, Tamil Nadu Agricultural University, Coimbatore, Tamil Nadu, India

Tara C. Satyavathi Global Centre of Excellence on Millets (Shree Anna), ICAR-Indian Institute of Millets Research, Hyderabad, Telangana, India

Nepolean Thirunavukkarasu Global Centre of Excellence on Millets (Shree Anna), ICAR-Indian Institute of Millets Research, Hyderabad, Telangana, India

Z. G. S. Turaki Lake Chad Research Institute, Maiduguri, Nigeria

Surya Teja Varanasi Department of Soil Science and Agricultural Chemistry, Tamil Nadu Agricultural University, Coimbatore, Tamil Nadu, India

Karnam Venkatesh ICAR – Indian Institute of Millets Research, Hyderabad, Telangana, India

R. Venkateshwarlu ICAR – Indian Institute of Millets Research, Hyderabad, Telangana, India

M. Vetriventhan International Crops Research Institute for the Semi-Arid Tropics (ICRISAT), Hyderabad, Telangana, India

N. Vinitha VOC Agricultural College and Research Institute, Tamil Nadu Agricultural University, Thoothukudi, Tamil Nadu, India

S. Viswabharathy Centre for Plant Breeding and Genetics, Tamil Nadu Agricultural University, Coimbatore, Tamil Nadu, India

Chapter 1
Establishment of Pearl Millet Biofortification Breeding and Milestones in India

Mahalingam Govindaraj

Abstract Pearl millet is a critical cereal staple in arid regions of India, cultivated on approximately eight million hectares, primarily by resource-poor farmers. Addressing iron (Fe) and zinc (Zn) deficiencies affecting over 50% of the rural people, biofortification breeding programs have successfully developed cultivars with enhanced grain Fe, ranging from 65 to 75 mg/kg, a 30–50% increase over conventional varieties. These biofortified cultivars exhibit competitive grain yields of 3.5–4.0 t/ha across heterogeneous environments, with genotype-by-environment interaction contributing less than 15% of micronutrient variance, ensuring genetically stable nutrient expression across diverse agro-ecologies. Marker-assisted selection targeting key quantitative trait loci (QTLs) has accelerated breeding diagnostic marker use, despite the complex polygenic control of Fe accumulation and environmental influences. Adoption of biofortified pearl millet has expanded steadily, reaching an estimated 2.5 million households, highlighting promising diffusion. Agronomic and grain quality traits remain largely unaffected, preserving farmers and consumer acceptability. However, bioavailability constraints linked to anti-nutritional factors persist, underscoring the need for more integrated agronomic and nutritional efficacy interventions. Future breeding strategies integrating genomic selection and high-throughput phenotyping hold potential to enhance selection precision and breeding efficiency to offer competitive breeding products. Overall, biofortified pearl millet cultivars present a cost-effective, scalable, sustainable intervention for reducing micronutrient malnutrition in India, particularly millet-dependent households, contingent upon continued multidisciplinary efforts to overcome genetic, environmental, and socioeconomic challenges for long-term mainstreaming commitments.

M. Govindaraj (✉)
Alliance of Bioversity International and the International Centre for Tropical Agriculture (CIAT), Cali, Colombia
e-mail: m.govindaraj@cgiar.org

M. Govindaraj (ed.), *Breeding Biofortified Millets for Nutritional Security*, Sustainability Sciences in Asia and Africa,
https://doi.org/10.1007/978-981-95-9827-4_1

Keywords Pearl millet · Biofortification · Bioavailability · Iron · Zinc · Micronutrient malnutrition

1.1 Introduction

Pearl millet (*Pennisetum glaucum* (L.) R. Br.) is a climate-resilient cereal predominantly cultivated in the arid and semi-arid regions of India, where it serves as a major source of calories, fodder, and livelihood security for smallholder farmers. Its inherent resistance to drought, high temperatures, and poor soils has cemented its place as a critical "nutri-cereal" in India's agricultural landscape (Yadav and Rai 2013). However, populations heavily dependent on pearl millet often experience micronutrient deficiencies, particularly iron (Fe) and zinc (Zn), contributing to the broader public health challenge of "hidden hunger" (Bouis and Saltzman 2017). Biofortification is a genetic enhancement of crops to improve micronutrient density. This has emerged as a sustainable strategy for addressing widespread Fe and Zn deficiencies. India has been a global leader in pearl millet biofortification thanks to HarvestPlus coordinated efforts involving the International Crops Research Institute for the Semi-Arid Tropics (ICRISAT), ICAR–AICRP on pearl millet, state agricultural universities, and the private sector (Govindaraj et al. 2024; Yadav et al. 2024). These programs use indigenous genetic variability and modern breeding procedures to create varieties rich in Fe and Zn without sacrificing yield or local adaptation.

A landmark development occurred in 2017 when India established minimum micronutrient thresholds of 42 mg kg^{-1} Fe and 32 mg kg^{-1} Zn for all pearl millet varieties proposed for national release, integrating nutrition directly into varietal release criteria (AICRP Pearl Millet 2017; Govindaraj et al. 2019; and Satyavathi et al. 2021). Biofortified varieties such as the open-pollinated variety (OPV) Dhanashakti' and hybrids like AHB1200 (ICMH 1202) have demonstrated both agronomic suitability and proven nutritional benefits in human efficacy trials (Finkelstein et al. 2017; Bouis and Saltzman 2017). These scientific and policy advances have, collectively, enabled the rapid scaling of biofortified cultivars and generated compelling evidence of their impact on human health through feeding trials demonstrating improved iron status and cognitive outcomes. This chapter synthesizes the scientific foundations, breeding approaches, accomplishments, and future directions of India's millet biofortification program with research-for-development support from donors and the government, highlighting its role as a model for nutrition-sensitive crop improvement in India and globally.

1.2 Context and Challenge of Pearl Millet Biofortification

Pearl millet is a hard, warm-season cereal crop, largely intensified in the drought-prone regions of Africa and Asia. This supports 90 million poor smallholder families living in harsh agroecology where prevailing malnutrition is reported. Pearl millet is one of the traditional food crops that they continue to grow for food, nutrition, and income security. Because of its nutritional values (energy, protein, calcium, iron, and zinc) and climate resilience (drought and heat), demand for pearl millet as a food, in addition to its importance as a source of cattle feed, is expected to increase significantly in Asia (India) and Africa (West and Central Africa). Iron and zinc deficiencies affect around 2 billion people worldwide, primarily in India and Africa (Ritchie and Roser 2018; Bouis et al. 2024). Even though government-supported nutrition intervention programs have reduced malnutrition in India over the years, the National Family Health Survey (NFHS 2021) revealed an unacceptably high prevalence of anemia (50%), underweight (35%), and stunting (38%) among children under the age of 5. More than half of women in 20 Indian states are reported to be anemic. Sustainable Development Goals (SDGs) globally aim to significantly reduce malnutrition by 2030, as 12 of the 17 goals deal with nutrition. Biofortification is a proven, economical, and sustainable dietary-based strategy to address the root cause of micronutrient malnutrition.

India is the world's greatest producer of pearl millet, with 7 million ha and 10 million tons (Satyavathi 2024). It is a staple dish in many states of India and Sub-Saharan Africa. It is an important and inexpensive source of iron and zinc for rural poor households, accounting for 19–63% of total iron and 16–56% of total zinc intake from all dietary sources in India's pearl millet-growing regions of Rajasthan, Haryana, Maharashtra, and Gujarat (Parthasarathy Rao et al. 2006). Although pearl millet is an appealing and healthy crop, not all available cultivars contain significant iron and zinc content. The average Fe concentration of 122 commercial cultivars published in India is 42 mg kg^{-1}, with 32 mg kg^{-1} Zn. (Rai et al. 2016). Biofortification research at the International Crops Research Institute for the Semi-Arid Tropics (ICRISAT), supported by HarvestPlus, showed good prospects for developing cultivars with higher levels of these micronutrients. For instance, six-fold variations for Fe density (18–121 mg kg^{-1}) and four-fold variation for Zn density (22–87 mg kg^{-1}) have been found across a large number of germplasm and breeding lines (Rai et al. 2012, 2015). The focus of pearl millet biofortification research in India is on the production of high-yielding, high-Fe hybrids, as all efforts in the public and private sectors are directed toward hybrid development. While OPVs are for Africa. Several studies have found a positive and highly significant relationship between Fe and Zn; the primary goal of pearl millet biofortification research at ICRISAT is to improve Fe density with Zn density as an associated trait.

1.3 Biofortification Research Foundation and Expansion

(a) *Theory of Change and Impact Pathways*

The core pearl millet cultivar breeding targets high grain yield (>10% than control) combined with disease resistance (downy mildew and blast) and medium to maturity duration (75–90 days). Since the inception of the biofortification program at ICRISAT, several research questions were raised on trait priority, genetics, association with grain yield, and breeding methodology, and all these were progressively addressed. Briefly, ICRISAT's pearl millet biofortification breeding program has entered a three-pronged breeding phase.

- The first phase focuses on short-term approaches such as trait genetics, correlation, germplasm screening, and establishing new genetic variability.
- The second phase involves certifying high-iron and zinc breeding lines and hybrid parents from standard breeding programs to create fast-track biofortified varieties/hybrids.
- The third phase focuses on developing high-Fe/Zn breeding lines and hybrid parents, as well as diversifying genetics by mainstreaming micronutrient characteristics in ICRISAT and NRAS breeding programs.

Pearl millet's grain Fe (31–125 mg kg^{-1}) and Zn (35–82 mg kg^{-1}) levels varied greatly among advanced breeding lines, population progenies, hybrid parents, and released or commercial hybrids (Table 1.1). This table highlights clear differences in micronutrient density (Fe and Zn) across various categories of pearl millet genetic materials. Biofortified breeding lines show the greatest improvement, with over 87% of entries containing more than 85 mg kg^{-1}Fe, and 38% exceeding 105 mg kg^{-1}, compared to negligible representation in other categories. In terms of Zn, most biofortified lines fall within the 36–55 mg kg^{-1}range, though none surpass 65 mg kg^{-1},

Table 1.1 Genetic variability for grain iron (Fe) and zinc (Zn) densities in different pearl millet collections

Material	Entry	Percent entry in micronutrient class (mg kg^{-1})							
		Fe density							
		≤ 45	**46–55**	**56–65**	**66–75**	**76–85**	**86–95**	**96–105**	**>105**
Mainstream hybrid-parents	290	24	31	18	10	7	7	3	1
Commercial cultivars	140	56	35	9	1	0	0	0	0
Germplasm accessions	406	11	19	16	19	17	12	5	1
Biofortified breeding lines	514	0	0	0	2	11	22	27	38
		Zn density							
		≤ 35	**36–45**	**46–55**	**56–65**	**66–75**	**76–85**	**86–95**	**>95**
Mainstream hybrid-parents	290	5	47	34	11	2	0	0	0
Commercial cultivars	140	8	76	16	0	0	0	0	0
Germplasm accessions	406	2	16	31	32	17	2	0	0
Biofortified breeding lines	514	8	45	40	7	0	0	0	0

indicating room for further enhancement. Commercial cultivars are the least nutrient-dense, with 91% of entries having ≤55 mg kg^{-1}Fe and none exceeding 55 mg kg^{-1}Zn. Mainstream hybrid parents perform slightly better but still fall short of biofortified standards, while germplasm accessions display wide variability in both Fe and Zn, suggesting their value as donor sources in breeding. Overall, the data underscores the success of biofortification efforts in enhancing Fe content in pearl millet and highlights the potential for further gains in Zn through continued breeding and selection.

The global baseline for iron in pearl millet, regardless of cultivar (hybrid or OPVs), is 47 mg $kg^{-1.}$ with aims of 77 mg kg^{-1} (+30 mg kg^{-1}) to reduce iron shortages by meeting up to 70% of daily needs (www.harvetsplus.org). Furthermore, the biofortification priority index (BPI) shows pearl millet as a prominent target crop for iron and zinc biofortification (Fig. 1.1). The map depicts the global priority

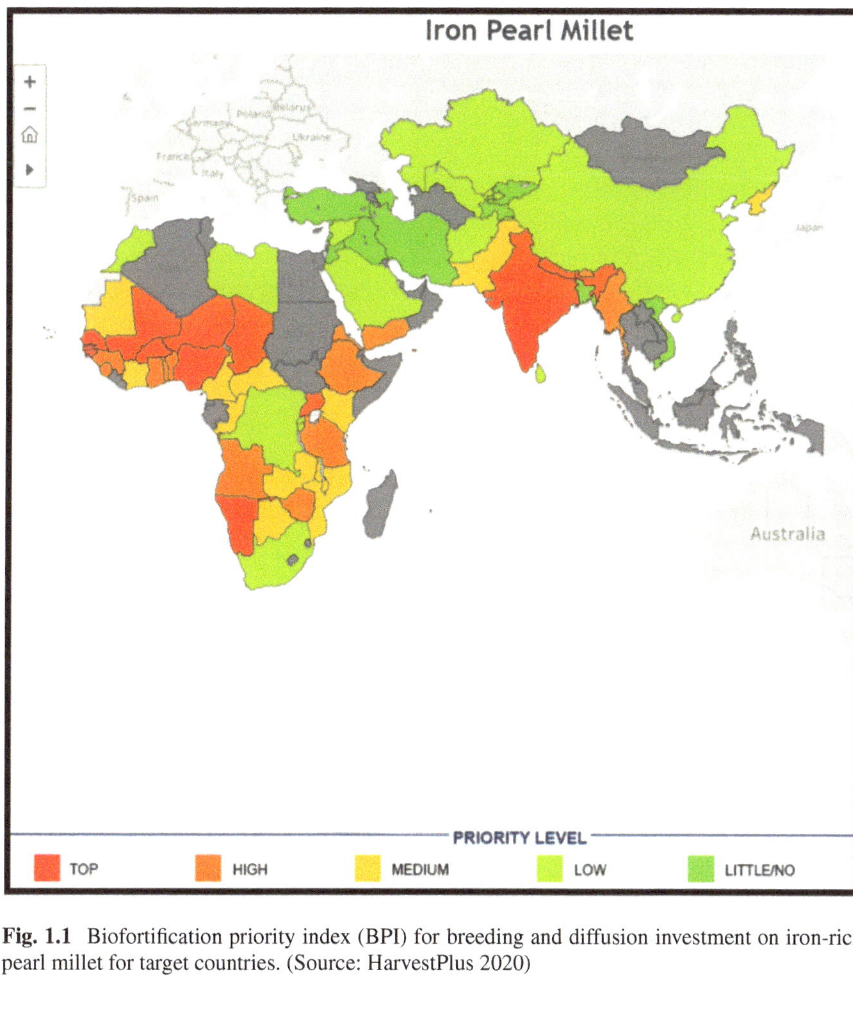

Fig. 1.1 Biofortification priority index (BPI) for breeding and diffusion investment on iron-rich pearl millet for target countries. (Source: HarvestPlus 2020)

levels for the promotion and deployment of iron-biofortified pearl millet, with nations ranked from highest to lowest priority based on iron deficiency prevalence, pearl millet consumption, and production relevance (Bouis and Saltzman 2017; Govindaraj et al. 2020). High-priority countries (orange) include those in the Sahel and Sub-Saharan Africa, as well as sections of South Asia, including Pakistan. Several other countries in Africa and Asia suffer from iron shortage, but pearl millet may play a minor dietary role. Low (light green) and Little/No priority countries (dark green) are locations where pearl millet is not a staple and iron deficiency is less severe (HarvestPlus 2020). Therefore, BPI stresses India and Africa's Sahelian region as crucial intervention zones for biofortified, iron-rich pearl millet (Bouis et al. 2011; Rai et al. 2020). These communities are particularly vulnerable to micronutrient deficiency and have traditionally relied on pearl millet, making them ideal for scaling up biofortified cultivars.

(b) *Disease Screening and Advancement for Nutrient-Dense Lines*

One of the prerequisites of breeding and product advancement with novel traits is to prioritize durable and broad-spectrum major disease resistance. The biofortification program is well equipped with mainstream disease-resistance screening facilities at ICRISAT and NARES locations. This emphasizes the importance of identifying and incorporating stable resistance sources to effectively manage competitive biofortified varieties and hybrids. For instance, screening of 100 and 165 pearl millet lines for downy mildew (*Sclerospora graminicola*) and blast (*Pyricularia grisea*), respectively, revealed clear differences in the distribution of resistant (R), moderately resistant (MR), and susceptible (S) classes, providing useful insights for biofortified variety development (Fig. 1.2a, b).

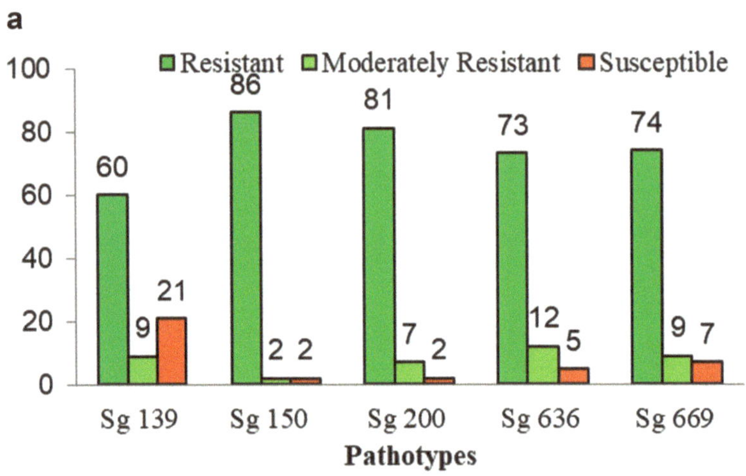

Fig. 1.2 (**a**) Display of downy mildew resistance among pearl millet biofortified breeding lines bred at ICRISAT, Patancheru. (**b**) Display of blast disease resistance among pearl millet biofortified breeding lines bred at ICRISAT, Patancheru

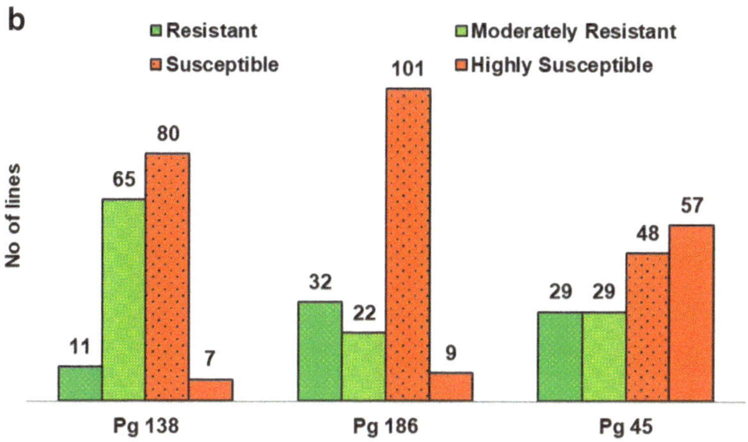

Fig. 1.2 (continued)

For downy mildew, resistance was consistently high across most pathotypes. Pathotypes Sg150 and Sg200 recorded the strongest resistance, with approximately 95–98% of lines resistant (≈45–48 resistant lines out of 50), and up to 3% in MR or S categories. Pathotypes Sg636 and Sg669 showed slightly lower but still strong resistance, with 80% resistant lines (74 lines), while MR and S categories contributed 10–20%. Pathotype Sg139 exhibited the lowest resistance, with only 70% resistant lines (60 lines), indicating relatively higher pathogen virulence (Fig. 1.2a). Overall, downy mildew resistance remained robust (combined mean resistance of 80–90% across pathotypes), with 42 lines being resistant to all the five pathotypes (≤10% DM). In contrast, blast disease displayed substantially lower resistance stability because of the lack of a resistance source in mainstream breeding materials. Screening of 165 breeding lines across the virulent pathotypes showed that approximately 44% of lines were resistant, and 70% are moderately resistant. For instance, 7%, 20%, and 18% of lines were resistant to Pg138, Pg186, and Pg45. First pathotypes have the least resistance, but most of the lines (40%) had moderate resistance, whereas the other two pathotypes had an equal number of resistance (18–20%) and moderate resistance (13–18%) lines for blast (Fig. 1.2b). The resistant lines remained at only ≤20%, and susceptible lines stabilized >50% for three pathotypes. Thus, overall blast resistance across pathotypes is considerably lower than that of downy mildew. These results highlight that while downy mildew resistance remains strong and readily exploitable, blast resistance is weak and unstable over time, challenging national pearl millet breeding programs. For breeding biofortified pearl millet varieties where disease pressure can limit grain quality and micronutrient expression, pathotype-stable downy mildew–resistant lines can be directly integrated, whereas blast resistance will require introgression of durable genes, pyramiding strategies, and multi-environment screening to ensure resilience in the near future.

(c) *Research for Development of Partnerships and Delivery Approaches*

HarvestPlus is a product delivery-oriented program. ICRISAT believes a strong network of diverse partnerships is crucial for measurable delivery in the HarvestPlus-funded program. Public and private breeding sectors in India recognized ICRISAT as a source of pearl millet germplasm and advanced breeding lines. This effort contributed 60% pearl millet cultivars released so far in India, which are from ICRISAT-bred lines. Recognizing the role of Public–Private Partnership (PPP) in enhancing the new trait breeding efficiency, pearl millet biofortification research is well aligned with the NARS and seed companies in India. Twenty-three partners (8 public sector organizations and 15 private seed companies) are engaged, covering all pearl millet growing regions. ICRISAT develops high-iron/zinc pearl millet hybrid-parents as part of the HarvestPlus program mission towards reducing malnutrition. Effective and sustainable biofortification is highly possible through PPP to develop, disseminate, and promote nutrient-dense cultivars. This led to the formation of the Pearl Millet Biofortification Partnership in 2010 under the HarvestPlus-supported Biofortification program at ICRISAT. This challenging partnership exploits harmonizing the existing capacity both at ICRISAT and partners' centers for efficiency, and made a linkage between the mandates of ICRISAT and NARS for quick delivery of biofortified products. This is a research and development partnership and provides a platform for conducting multi-location, On-farm trials, and to breed, test, and release biofortified pearl millet. Therefore, NARS partners have the opportunity to select hybrids of choice for on-farm and farmer trials, followed by seed production and marketing. Hence, products were rapidly delivered to farmers in a fast-track approach and accelerated the adoption of high-iron hybrids. Most of the advanced breeding lines and finished products become international public goods (IPGs). Established X-ray Fluorescence Spectrometry lab at the ICRISAT campus enabled high-throughput screening and equipped breeding program to improve selection efficiency for invisible traits (iron and zinc) (Govindaraj et al. 2016). This screening facility also accesses partners for rapid screening of multilocation trials.

(d) *Product Development and Nutrition Efficiency*

Dhanashakti was the first biofortified pearl millet variety introduced in 2014. It is an early-maturing, open-pollinated variation with the greatest iron content (71 mg kg^{-1}) of any pearl millet cultivar created to date. Dhanashakti is the first high iron biofortified cultivar of any crop variety to enter the public domain in India and has been embraced by thousands of farmers across the country (Table 1.2). This was released in collaboration with Mahatma Phule Krishi Vidyapeeth (MPKV), Maharashtra, and the organized seed production taken by Nirmal Seed Company. According to Nirmal Seeds, the major distributors in India, the peak sale of the first biofortified variety was 200 tons per year. Since an OPV, the target area is limited in India. In addition, state seed corporations (Maharashtra and Karnataka) are also distributing Dhanashakti seeds because of their complementary role in

Table 1.2 Developmental metric of pearl millet biofortification breeding in India

Metrics class	Metrics number
Biofortification partnership	6 State Agricultural University 3 State Seed Corporations 15 Seed Companies.
Breeding material shared	500+ elite high-iron lines to 19 public and 20 private sector partners.
Contribution to national trials	100 Advanced breeding lines 50+ test hybrids
Cultivar released	3 OPV 12 hybrids
Breeder seed production (cumulative)	2500 kg of Dhanashakti 1000 kg each of hybrid ICMH 1202 and ICMH 1203 parents
Seed production and dissemination	Cumulative 4000 MT (2017–2024) 100 MT Karnataka State Seed Corporation 250 MT Maharashtra state seed corporation 1500 MT Nirmal seed Pvt. Ltd. Others 2150 MT.
Annual Area coverage (approx.)	120,000 ha (Maharashtra, Rajasthan, Karnataka, Gujarat, Tamil Nadu)
Household reached (estimated)	2500,000
Bioavailability	Iron: 7.5% (80% of daily iron requirements) Zinc: 25% (100% of daily zinc requirements).

disseminating improved cultivars to farmers. A high iron and high-yielding hybrid ICMH 1201 had 74 mg kg^{-1} Fe density and 3.6 t ha^{-1} of grain yield. Shakti Vardhak Seed Company has proactively undertaken its seed production under the brand name Shakti-1201, and about 25 tons of Truthfully Labeled Seed (TLS) were supplied in 2015 and 2016. ICRISAT recently developed three biofortified hybrids, ICMH 1202, ICMH 1203, and ICMH 1301 (Table 1.2 and Fig. 1.3), officially notified as AHB 1200Fe, HHB299, and Phule Maha Shakti, respectively, in collaboration with agricultural universities of Maharashtra (VNMKV and MPKV) and Haryana (CCSHAU).

Overall, biofortification strategies have resulted in the rapid development and adoption of high-iron pearl millet cultivars containing 70–80 mg kg^{-1} of iron and 35–40 mg kg^{-1} of zinc, which, when consumed 250 g day^{-1}, can meet 80% of the RDA for iron and 100% of the RDA for zinc in men and non-pregnant-non-lactating (NPNL) women. Children require only 150 g per day (Kodkany et al. 2013; Tako et al. 2015). The biofortification program's initial goal was to generate a high-iron pearl millet variety capable of supplying up to 70% of the RDA. Currently, biofortified pearl millet can meet more than 80% of the daily iron requirement. This exceeds the program's aim and contributes to the development of more biofortified cultivars for future releases.

Fig. 1.3 First-wave of biofortified pearl millet hybrids released and notified for commercial cultivation in India

1.4 Critical Partnerships in Dissemination

Biofortified cultivars breeding and delivery systems need strong partnerships and effective partnering practices. Working with NARS breeders in India, ICRISAT has developed the first wave of biofortified cultivars that can provide 25–30% more iron with similar maturity and yield potential and resistance to common diseases. The first wave of products is yet to show impact in the seed market and farm acreage, for which private seed companies should come forward to take over seed dissemination partnerships. Pearl millet commercial hybrid turnover in India is about 5–6 years. Thus, more private players are requested to mainstream micronutrient traits in R4D and to scale up new products. Government seed production investments for biofortified pearl millet via seed corporations and private institutions are highly incentivized. Also, emerging small private seed companies can directly market biofortified cultivars to enter the seed business. Including high-iron pearl millet in public distribution system (PDS) may perhaps increase the biofortified market segment and mainstreaming.

1.5 Impacts of Biofortified Pearl Millet Seed in India

The data show a strong and consistent increase in both IPM seed production and seed distribution in India between 2017 and 2024, demonstrating the expanding reach and impact of biofortified iron pearl millet. Seed production rose from 351 tons in 2017 to 675 tons in 2024, nearly doubling over the eight-year period. Seed distribution followed a similar upward trend, increasing from 271 tons in 2017 to 641 tons in 2024, indicating effective scaling and delivery to millet farming communities (Fig. 1.4). The most significant gains occurred between 2020 and 2022, where seed production climbed from 513 to 685 tons, and distribution increased from 418 to 621 tons. This rapid scale-up reflects heightened demand, strengthened public seed systems, and successful partnerships with the private sector and HarvestPlus projects supporting biofortified crop dissemination. Even in years with slight fluctuations (e.g., 2023), distribution remained high, emphasizing the stability and sustained adoption of IPM seed. During the pandemic, production and distribution were stable, ensuring a high demand of nutrition-rich seed supply. Overall, the progressive increase in both production and distribution highlights the growing national demand for nutrient-rich crops like pearl millet and the expanding role of biofortified iron pearl millet in improving dietary iron intake among rural households, especially in Rajasthan, Maharashtra, and Karnataka. The continued upward trend by 2024 confirms the success of India's millet scaling strategy, besides the 2023 international year of millet, and underscores biofortified millet's contribution to nutrition-sensitive agriculture, farmer livelihoods, and national food system resilience in India.

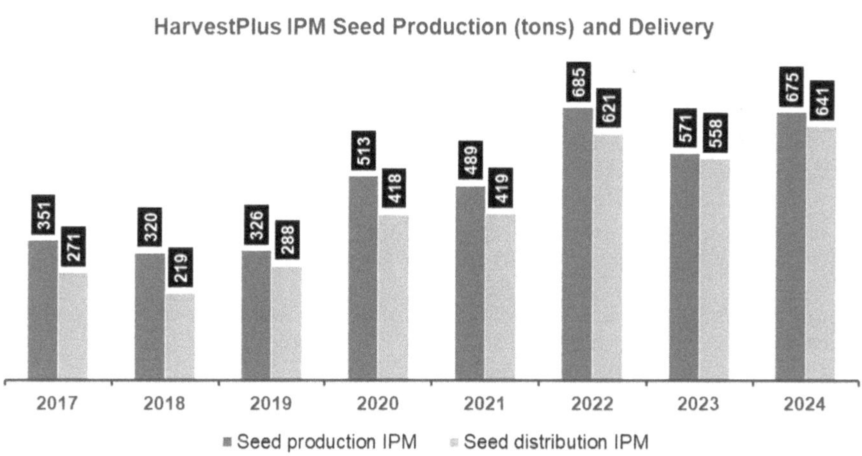

Fig. 1.4 Biofortified pearl millet varieties and hybrid seed production and distribution from 2017 to 2024

1.6 Key Milestones in Pearl Millet Biofortification (India)

Table 1.3 summarizes the evolution of pearl millet biofortification breeding in India, highlighting key scientific, institutional, and policy milestones. Pearl millet biofortification program has progressed through key milestones, beginning with the discovery of a wide spectrum of genetic variability for iron and zinc and supported by quick discovery and understanding their largely additive inheritance, implying that improvement requires accumulation of favourable alleles in both parental lines, an insight critical for hybrid breeding (Govindaraj et al. 2024). The release of Dhanashakti in 2014 marked the world's first biofortified pearl millet variety, followed by the first iron-hybrids like ICMH-1201 that combined high micronutrients with hybrid vigor. The adoption of the national nutritional standards policy in 2017 marks a breakthrough, ensuring micronutrient traits are prioritized alongside yield, accelerating the release and adoption of multiple nutrient-rich cultivars. Human efficacy trials confirmed significant reductions in iron deficiency and cognitive gains (Table 1.3), firmly establishing biofortified pearl millet as a powerful tool for nutrition security for India, and spillover benefits are highly visible in Sub-Saharan Africa. ICRISAT-India's biofortification program has played a crucial role in

Table 1.3 Development of pearl millet biofortification breeding and key milestones in India

Milestone	Details	Reference
Genetic variability identified	Broad natural variation for Fe and Zn established the foundation for biofortification breeding.	Yadav and Rai (2013), Rai et al. (2015)
Genetic architecture defined	Additive inheritance of Fe and Zn highlighted the need for favourable alleles in both parents, guiding hybrid breeding strategy.	Govindaraj et al. (2013, 2016)
Biofortification breeding strategy	Developed fast-track and special breeding (exploring existing variation) approaches for pearl millet breeding to deliver biofortified varieties and hybrids	Govindaraj et al. (2016, 2019)
Release of biofortified variety	First biofortified pearl millet variety *Dhanshakti*; combined high Fe/Zn with good yield and farmer acceptance.	Rai et al. (2014)
Development of biofortified hybrids	High-micronutrient hybrids (e.g., ICMH-1201; 1202) developed to integrate nutritional quality with hybrid vigor and productivity.	Rai et al. (2014), Govindaraj et al. (2016)
Human efficacy evidence	Feeding trials showed reduced iron deficiency and cognitive improvements, confirming a strong public-health impact.	Finkelstein et al. (2017)
National policy--micronutrient standards	India established mandatory Fe (42 mg kg^{-1}) and Zn (32 mg kg^{-1}) thresholds for varietal release, first such policy worldwide.	ICAR–AICRP Pearl Millet (2017), Govindaraj et al. (2019, 2024)
Cultivar expansion and adoption	Multiple biofortified cultivars were released; adoption increased due to farmer demand and nutrition-oriented policy support.	Govindaraj et al. (2024)

supplying breeding materials and products to Africa. Subsequent release of biofortified hybrids and HarvestPlus advocacy in mainstreaming within ICAR and ICRISAT programs demonstrate strong research-based policy integration. These milestones collectively showcase India's global leadership in nutrition-sensitive breeding product dissemination.

1.7 Lessons Learnt, Gaps, and Good Practices at Scale

India's experience with pearl millet biofortification highlights several important lessons for scaling nutrition-sensitive breeding. Strong partnerships between HarvestPlus, ICRISAT, ICAR, state universities, and seed companies proved essential for rapid varietal development and dissemination. Policy supports, especially mandatory Fe and Zn standards, helped integrate nutrition into mainstream breeding. However, gaps remain, including limited seed availability, the need for better genomic tools, low consumer awareness, and post-harvest constraints like flour rancidity. Good practices that enabled scale include multi-location testing, public–private collaboration, nutrition-focused extension, and integration of biofortified millets into government programs. Together, these elements demonstrate an effective model for large-scale biofortification with the following segments to make mainstream biofortification in future varieties.

To Breeders The correlation between the Fe and Zn with other farmers' chosen qualities (maturity, seed size, and color) was not significant, indicating that genetic augmentation of these micronutrients is conceivable without sacrificing these traits. Whereas Fe and Zn levels demonstrated a negative and mainly non-significant connection with grain yield. This could be the result of unidirectional selection, as most relationships reported thus far are in lines/cultivars developed solely for yield (as the target attribute). To break such a linkage, selection in large segregating populations is advised. The genetic architecture (predominantly additive gene action) suggests that both parents in hybrid breeding must be high in micronutrients; this complicates but does not preclude hybrid development. The decision to institutionalize micronutrient thresholds (42 mg/kg Fe/32 mg/kg Zn) in the national varietal release framework is particularly significant; it reflects a shift in breeder priorities, placing nutritional quality on par with yield and agronomic traits. Hence, biofortification can be an integral part of a regular breeding program. Differences in soil Fe and Zn levels across the environment are not reflected in the grain Fe and Zn contents if the soil is above critical limits.

To Partners In India, seed firms dominate the seed market for pearl millet hybrids, with hybrids accounting for nearly 90% of the improved cultivars planted area. As a result, the initial biofortified variety had a limited potential impact. To breed high-yielding and high-iron hybrids, more collaboration and multi-environment data gathering are required. In India, private seed producers can commercialize hybrids as genuinely labeled seed without an official release. This will mutually benefit emerging small seed companies for biofortified cultivars. This is the right time to engage seed and food industries to incorporate biofortified crops into their core business activities. Otherwise, the extent of cultivation, marketing, and consumption of biofortified crops will not be sustained.

To Government While biofortified products begin to reach the seed market, government-supported authorities should put systems in place to certify biofortified seed. For long-term, national cultivar release policy should include the high-iron attribute (standard levels) as a non-negotiable trait. Human feeding trials using high-Fe pearl millet demonstrated significant reductions in iron deficiency and improvements in cognitive performance among school-aged children, validating the public-health value of biofortification. Therefore, including high-iron grains in PDS would be highly rewarding for the market segment, thereby benefiting farmers and consumers.

References

AICRP-Pearl Millet (2017) Proceedings of the 52nd Annual Pearl Millet Workshop All India Coordinated Research Project on Pearl Millet Held at Punjab Agricultural University, Ludhiana (Punjab), April 28–30, 2017

Bouis HE, Saltzman A (2017) Improving nutrition through biofortification: a review of evidence from HarvestPlus, 2003 through 2016. Glob Food Sec 12:49–58. https://doi.org/10.1016/j.gfs.2017.01.009

Bouis HE, Hotz C, McClafferty B, Meenakshi JV, Pfeiffer WH (2011) Biofortification: a new tool to reduce micronutrient malnutrition. Food Nutr Bull 32(1_suppl1):S31–S40. https://doi.org/10.1177/15648265110321S105

Bouis H, Foley J, Lividini K, Jumrani J, Reinke R, Straeten D, Zagado R, Boy E, Brown L, Mudyahoto B, Alioma R, Hussain M, Pfeiffer WH (2024) Biofortification: future challenges for a newly emerging technology to improve nutrition security sustainably. Curr Develop Nutr 8(12):104478

Finkelstein JL, Haas JD, Mehta S (2017) Iron-biofortified pearl millet improves iron status in Indian schoolchildren: a randomized feeding trial. J Nutr 147:1–7

Govindaraj M, Rai KN, Shanmugasundaram P, Dwivedi SL, Sahrawat KL, Muthaiah AR, Rao AS (2013) Combining ability and heterosis for grain iron and zinc densities in pearl millet. Crop Sci 53(2):505–517

Govindaraj M, Rai KN, Pfeiffer WH, Kanatti A, Shivade H (2016) Energy-dispersive X-ray fluorescence spectrometry for cost-effective and rapid screening of pearl millet germplasm and breeding lines for grain iron and zinc density. Commun Soil Sci Plant Anal 47(18):2126–2134

Govindaraj M, Rai KN, Cherian B, Pfeiffer WH, Kanatti A, Shivade H (2019) Breeding biofortified pearl millet varieties and hybrids to enhance millet markets for human nutrition. Agriculture 9(5):1–11

Govindaraj M, Rai KN, Shanmugasundaram P, Dwivedi SL, Sahrawat KL, Muthusamy A, Muli TK (2020) Breeding biofortified pearl millet varieties and hybrids to enhance millet markets for human nutrition. Indian J Genet Plant Breed 80(4):417–429. https://doi.org/10.5958/0975-6906.2020.00058.6

Govindaraj M, Pujar M, Srivastava R, Gupta SK, Pfeiffer WH (2024) Genetic biofortification of pearl millet: trait priority, breeding and genomic progress. In: Pearl millet in the 21st century food-nutrition-climate resilience-improved livelihoods. Springer, Singapore, pp 221–246

HarvestPlus (2020) Biofortification priority index: Iron Pearl Millet Map. CGIAR. https://www.harvestplus.org

ICAR–AICRP on Pearl Millet (2017) Annual Report: Integration of micronutrient criteria in varietal release guidelines. ICAR, New Delhi

Kodkany BS, Bellad RM, Mahantshetti NS, Westcott JE, Krebs NF, Kemp JF, Hambidge KM (2013) Biofortification of pearl millet with iron and zinc in a randomized controlled trial increases absorption of these minerals above physiologic requirements in young children. J Nutr 143:1489–1493

National Family Health Survey (NFHS-5) (2021) 2019–21: International Institute for Population Sciences (IIPS) and ICF; National Family Health Survey (NFHS-5), 2019–21: India: Volume I. pp 671. http://www.rchiips.org/nfhs

Parthasarathy Rao P, Birthal PS, Reddy BVS, Rai KN, Ramesh S (2006) Diagnostics of sorghum and pearl millet grains-based nutrition in India. Int Sorgh Millets Newsl 47:93–96

Rai KN, Govindaraj M, Rao AS (2012) Genetic enhancement of grain iron and zinc content in pearl millet. Qual Assur Safety Crops Foods 4:119–125

Rai KN, Patil HT, Yadav OP, Govindaraj M, Khairwal IS, Cherian B, Rajpurohit BS, Rao AS, Shivade H, Kulkarni MP (2014) Notification of crop varieties and registration of germplasm: Pearl millet variety 'Dhanashakti'. Indian J Genet Plant Breed 74(3):405–406

Rai KN, Velu G, Govindaraj M, Upadhyaya HD, Rao AS, Shivade H, Reddy KN (2015) *Iniadi* pearl millet germplasm as a valuable genetic resource for high grain iron and zinc densities. Plant Genet Resour 13(1):75–82

Rai KN, Yadav OP, Govindaraj M, Pfeiffer WH, Yadav HP, Rajpurohit BS, Patil HT, Kanatti A, Rathore A, Rao AS, Shivade H (2016) Grain iron and zinc densities in released and commercial cultivars of pearl millet. Indian J Agric Sci 86(3):291–296

Rai KN, Govindaraj M, Kumar A, Rao AS (2020) Genetic enhancement of pearl millet for iron and zinc bioavailability. In: Grusak MA, Taylor P, Pfeiffer WH (eds) Biofortification of staple crops. Springer, Cham, pp 87–106. https://doi.org/10.1007/978-3-030-42366-0_5

Ritchie H, Roser M (2018) Micronutrient deficiency. Published online at OurWorldInData.org. Retrieved from https://ourworldindata.org/micronutrient-deficiency

Satyavathi CT (2024) Project coordinator review, 59th Annual Group Meeting, ICAR-AICRP on Pearl Millet, Jodhpur Date: 27–28 June 2024

Satyavathi CT, Ambawat S, Khandelwal V, Govindaraj M, Neeraja CN (2021) Micronutrient Rich Pearl Millet for Nutritionally Secure India. ICAR-All India Coordinated Research Project on Pearl Millet, Jodhpur

Tako E, Reed SM, Budiman J, Hart JJ, Glahn RP (2015) Higher iron pearl millet (*Pennisetum glaucum* L.) provides more absorbable iron that is limited by increased polyphenolic content. Nutr J 14:11

Yadav OP, Rai KN (2013) Genetic improvement of pearl millet in India. Agric Res 2:275–292

Yadav OP, Gupta SK, Govindaraj M, Singh DV, Verma A, Sharma R, Mahala RS, Srivastava SK, Birthal PS (2024) Strategies for enhancing productivity, resilience, nutritional quality, and consumption of pearl millet [Pennisetum glaucum (L.) R. Br.] for food and nutritional security in India. Crop Science (TSI):1–19

Chapter 2
Variability for Grain Nutrient Traits in Finger Millet and Its Interaction with the Environment

C. Backiyalakshmi, K. Manonmani, and M. Vetriventhan

Abstract Finger millet (*Eleusine coracana (L.)* Gaertn.) is one of the important staple food crops, drawing considerable attention owing to its nutritional richness and resilience to challenging growing conditions. The substantial grain nutrients, especially the high calcium content in its grain's positions finger millet as a promising solution for addressing malnutrition particularly calcium deficiencies such as osteoporosis and, contributes to overcoming food security challenges. Nevertheless, the grain nutrient profile of finger millet, like that of other crops, is significantly shaped by factors such as genotype, environment, and their interaction. This chapter underscores the significance of finger millet as a nutritious food source and its potential in combating nutrient deficiencies. It emphasizes the impact of genetic diversity on shaping the grain nutrient traits of finger millet, stressing the importance of efficient characterization and utilization of germplasm to harness desired nutritional traits. The chapter delves into existing studies on grain nutrient composition, exploring the intricate interplay of genetic and environmental factors influencing grain nutrient content, and provides valuable insights for researchers and breeders, guiding them in the development of biofortified finger millet cultivars.

Keywords Finger millet · Malnutrition · Osteoporosis · Germplasm · Grain nutrients

---

C. Backiyalakshmi · M. Vetriventhan (✉)
International Crops Research Institute for the Semi-Arid Tropics (ICRISAT), Hyderabad, Telangana, India
e-mail: Vetriventhan.mani@icrisat.org

K. Manonmani
Department of Genetics and Plant Breeding, Amrita School of Agricultural Sciences, Amrita Vishwa Vidyapeetham, Coimbatore, Tamil Nadu, India

M. Govindaraj (ed.), *Breeding Biofortified Millets for Nutritional Security*, Sustainability Sciences in Asia and Africa,
https://doi.org/10.1007/978-981-95-9827-4_2

2.1 Introduction

Finger millet (*Eleusine coracana (L.)* Gaertn.) holds a crucial role as a staple crop, particularly in regions like Africa and Asia. Finger millet grains are used as food, while the stover is utilized as fodder. The grain nutritional value of finger millet surpasses that of major cereals, as reviewed by various authors (Upadhyaya et al. 2011; Kumar et al. 2016; Gupta et al. 2017; Sood et al. 2019; Vetriventhan et al. 2020). It is also called "future smart food" by FAO (Li and Siddique 2018). In recent years, the importance of finger millet and other small millets in agriculture has grown significantly, especially in semi-arid regions globally. These regions often face food crises, and millets have emerged as an ideal solution for their climatic conditions. Finger millet is highly regarded for its resilience in the face of soil moisture stress, its commendable nutritional value, the ample palatable straw it produces for fodder, and its versatile usability in various forms of consumption. Furthermore, these grains possess excellent storage qualities and can even serve as a source of sustenance during times of famine. While finger millet crop improvement programs have traditionally focused on increasing yield, there is a growing recognition of the need to enhance the nutritional quality as well, given the persistent challenge of malnutrition on a global scale. The nutrient content in finger millet, like that of other crops, is influenced by a multitude of factors, including genetics, environmental conditions, crop management practices, and their intricate interactions. To shed light on the genetic variability present in diverse germplasm collections, the factors impacting nutrient traits, and the genotype-environment interplay, this chapter seeks to facilitate a deeper understanding of nutrient traits within finger millet germplasm, factors influencing nutrient traits, and genotype × environment interaction. This knowledge can subsequently be harnessed for the development and dissemination of biofortified cultivars, further contributing to the fight against malnutrition.

2.2 Potential of Finger Millet to Address Malnutrition and Food Security Challenges

Calcium is the major mineral required in the human diet to address Ca malnutrition in children and Ca deficiency in adults. Calcium is crucial for bone health and preventing osteoporosis. The populations at the highest risk of calcium malnutrition and osteoporosis are young children, pregnant and nursing women, and the elderly osteoporosis is a condition characterized by low bone mass, deterioration of bone tissue, disruption of bone microarchitecture, compromised bone strength, which leads to an increase in the risk of fractures. Calcium also plays a crucial role in regulating several basic body functions, such as muscle contraction and relaxation, nerve impulse transmission, blood coagulation, enzymatic reaction activation, and hormonal secretion stimulation. Osteoporosis causes 8.9 M fractures each year, affecting 500 M people worldwide (https://www.osteoporosis.foundation/

facts-statistics). A systematic review of global dietary Ca intake among adults showed that average national calcium intake ranges from 175 to 1233 mg/day across 74 countries. Many in Asia consume less than 500 mg/day, while Africa/South America averages 400–700 mg/day (Balk et al. 2017).

Finger millet is the richest source of calcium (364 ± 58 mg/100 g), which is many fold higher than rice (raw, brown; 10.91 ± 1.79 mg/100 g), wheat (whole, 39.36 ± 5.65 mg/100 g), maize (8.91 ± 0.61 mg/100 g), and pearl millet (27.6 ± 2.16 mg/100 g) and also three times more than in milk (121 ± 3 mg/100 g in buffalo milk, 118 ± 2.9 mg/100 g in cow milk) (Shobana et al. 2013; Kumar et al. 2016; Gupta et al. 2017; Puranik et al. 2017; Antony Ceasar et al. 2018; Vetriventhan et al. 2020). These could be reasons for recommending the finger millet-based diet to pregnant, lactating women, and growing children (Verma and Patel 2013). The retention rate of calcium from finger millet is 19.6%, which is similar to that of rice at 19.7%. However, finger millet has a higher calcium content; the Ca retention from a finger millet-based diet is 4.4 times higher than a rice-based diet (Anitha et al. 2021). When cooked, the bioavailability of calcium in finger millet is around 28.6%, comparable to milk's 32.1%, and can be further increased through various processing methods (Anitha et al. 2021). These qualities make finger millet an ideal crop to reduce calcium-related health issues globally.

In addition, finger millet also contains high dietary fiber (10–18%), balanced protein (6–13%), minerals (2.5–3.5%), phytates (0.48%), tannins (0.61%), phenolic compounds (0.3–3%), etc. (Devi et al. 2014; Chandra and Chandra 2016). The dietary fiber content of finger millet (10–18%) is comparable to that of wheat and maize, whereas it is higher than in polished and brown rice (Gopalan et al. 2009; Shobana et al. 2013; Rodríguez et al. 2020). The protein content of finger millet is 6% to 13%, which is better balanced with sulphur-containing amino acids, namely, methionine and cystine, as well as lysine, threonine, and valine content, than in rice and other millets (Shobana et al. 2013; Saleh et al. 2013; Sharma et al. 2017; Rodríguez et al. 2020). It is also high in iron content; thus, it has been promoted in Africa to reduce the risk of anemia (Tripathi and Platel 2010; Udeh et al. 2017). It is also high in polyunsaturated fatty acids and contains water-soluble and liposoluble vitamins. Moreover, it contains valuable nutraceuticals like phytates, polyphenols, tannins, trypsin-inhibitory factors, and dietary fiber (Devi et al. 2014; Chandra and Chandra 2016). Beyond its nutritional benefits, finger millet has been associated with potential health advantages, including the potential to mitigate hyperglycemia, hypercholesterolemia, ulcers, and cataracts (Shobana et al. 2009; Muthamilarasan and Prasad 2015). Lipids contribute as a source of fats and essential fatty acids in a diet. Finger millet contained the highest concentration of digalactosyl monoglyceride among foxtail millet and proso millet. Additionally, it contains phosphatidylglycerol, phosphatidylethanolamine, and phosphatidylcholine. It was also observed that finger millet lipid had a higher amount of triacylglycerols (84.8%) when compared to other millets, such as foxtail millet (83.2%) and proso millet (81.2%) (Gupta et al. 2017).

2.3 Factors Influencing Grain Nutrients

Grain yield and nutritional quality depend on many factors, including genetic factors, environmental factors, management factors, and their interactions. Environmental factors such as altitude, temperature, rainfall, slope, aspect of the mountain, etc., and the management factors such as fertilizer dose, spacing, irrigation, etc., can affect plants' morphology, physiology, biochemistry, and nutrient composition. Crops have mechanisms to adapt to such factors by altering their physiological and morphological characteristics (Royer et al. 2005; Qi et al. 2020).

2.3.1 *Genetic Factors*

Grain nutrient composition varies with genotypes, and the genetic characteristics such as maturity duration, plant height, grain yield, seed size, grain colour, etc., cause genetic variability among finger millet germplasm. Thus, it is important to consider these genetic factors and its relationship with grain nutrient content, while breeding for nutrient-dense cultivars. The success of crop improvement programs depends on the germplasm available to breeders to create desirable cultivars. Utilizing variability in germplasm collection is an effective way to improve grain micronutrients in crops. Although selecting micronutrient-dense lines from existing breeding populations and varieties is a good starting point, choosing from germplasm collections has more potential for enhancing crop nutrient content and diversity. Finger millet core and mini core collections (Upadhyaya et al. 2006, 2010), which represent the maximum diversity of the entire collection conserved at ICRISAT genebank, provide better access to germplasm resources and diversity. However, very little is known about the grain nutrient traits of finger millet, despite the enormous variation and nutrient content reported among finger millet germplasm, while a few studies reporting grain nutrient variability are presented in Table 2.1. Evaluation of finger millet germplasm for grain nutrient traits revealed considerable range of variation for protein (0.50–12.4%), Fe (1.37–53.39 mg/100 g), Zn (0.04–3.73 mg/100 g), Ca (50.66–554.0 mg/100 g), Cu (0.18–0.87 mg/100 g), Mg (31.0–217.8 mg/100 g), Mn (4.40–48.43 mg/100 g), P (3.46–380.07 mg/100 g), K (45.0–1427.0 mg/100 g), S (54.8–191.7 mg/100 g), and Na (0.60–41.44 mg/100 g) respectively (Vadivoo et al. 1998; Shashi et al. 2007; Admassu et al. 2009; Upadhyaya et al. 2011; Das et al. 2017; Badigannavar and Ganapathi 2018; Puranik et al. 2020; Fred et al. 2021; Backiyalakshmi et al. 2023). It is also important to note that grain nutrient analysis varies with analytical methods in different labs. These results imply that an ample source of micronutrient content is available on finger millet germplasm, which could be useful in the improvement of grain nutrient content through breeding programs.

Finger millet grains can be grouped into various colors, such as white, dark brown, reddish brown, copper brown, and light brown. Exploring the relationship

Table 2.1 Characterization of finger millet grain nutrient traits

No. of accessions	Fe (mg/100 g)	Zn (mg/100 g)	Ca (mg/100 g)	Protein (%)	Cu (mg/100 g)	Mg (mg/100 g)	Mn (mg/100 g)	P (mg/100 g)	K (mg/100 g)	S (mg/100 g)	Na (mg/100 g)	References
628	1.37–30.04	0.04–3.73	155.5–554.0	0.50–10.10		53.0–217.8	4.40–22.07	7.10–380.3	45.0–1427.0	54.8–191.7		Fred et al. (2021)
36			162.0–487.0 **(320.8)**	6.7–12.4 **(9.7)**								Vadivoo et al. (1998)
622	2.17–6.52 **(2.93)**	1.66–2.5 **(1.99)**	184–489 **(285)**	6.00–11.09 **(7.32)**								Upadhyaya et al. (2011)
48			188.66–324.33 **(235.34)**	5.96–9.33 **(7.50)**								Das et al. (2017)
190	3.025–16.623 **(5.77)**	1.02–2.66 **(1.58)**	223.62–422.55 **(314.81)**	3.85–11.27 **(7.53)**		106.36–179.99 **(133.54)**			266.59–668.83 **(383.15)**		6.25–41.44 **(13.73)**	Puranik et al. (2020)
49	2.00–21.56 **(9.126)**		135–312 **(199.7)**			31.00–139.00 **(77.10)**	0.26–5.0 (3.1)					Badigannavar and Ganapathi (2018)
8	3.60–7.31	1.76–2.31	264–365			66–130		234–292	294–1160		0.60–0.95	Shashi et al. (2007)
314	2.04–4.28 **(3.07)**	1.75–3.44 **(2.568)**	298.7–486.9 **(396.55)**	7.27–10.74 **(8.95)**	0.4–0.87 **(0.6)**	133.25–190.85 **(162.43)**		289.47–345.78 **(313.38)**	309.3–555.1 **(433.2)**	106.1–148.9 **(127.6)**		Backiyalakshmi et al. (2023)
9	4.59–53.39 **(26.90)**	0.97–2.56 **(1.80)**	50.66–319 **(256.52)**	6.26–10.5 **(7.49)**	0.18–0.79 **(0.62)**	78–201 **(168.78)**	17.61–48.43 **(28.35)**	3.46–147 **(121.83)**				Admassu et al. (2009)

Fe Iron, *Zn* Zinc, *Ca* Calcium, Protein, *Cu* Copper, *Mg* Magnesium, *Mn* Manganese, *P* Phosphorus, *K* Potassium, *S* Sulphur, *Na* Sodium. Note: Bold value in the parentheses represent the mean value between grain color and nutrient content is essential, as it can serve as a useful morphological marker for trait identification. A few studies reported that white-seeded types have higher protein content with moderate calcium content than brown-seeded types (Vadivoo et al. 1998; Das et al. 2017). White grain types are reported to have high protein, low fiber, low tannin, and higher consumer acceptability, especially by the nontraditional/urban millet consumer, or to blend with wheat or rice (Rao 1994; Chaudhari et al. 2012; Fred et al. 2021). According to Fred et al. (2021), the concentration of nutrients is associated with grain color, with dark-colored grains having the highest concentration than white color. Among the nutrients investigated by Backiyalakshmi et al. (2023), except for Cu, all other nutrients do not differ significantly between white- and brown-colored seeds, while slightly higher Zn, Ca, P, and K are noticed in white-seeded finger millet, but the differences were nonsignificant. Thus, selection based on seed color may not be the desired parameter for improving grain nutrient content in the finger millet, as reported in pearl millet (Govindaraj et al. 2018). However, breeding opportunities for improving grain nutritional traits in finger millet can be achieved without compromising regional and consumer preferences, including color. Based on inflorescence morphology, finger millet has four different races, namely, *compacta, elongata, plana*, and *vulgaris*. According to Backiyalakshmi et al. (2023), the finger millet races had no significant differences in grain nutrient traits, except for Ca and Cu. Similarly, Vetriventhan and Upadhyaya (2019) reported that grain nutrient content did not differ among races in kodo millet. However, Vetriventhan et al. (2021) found that race *nana* in little millet had higher Fe, Zn, and protein content compared to race *robusta*. Moreover, in proso millet, race *ovatum* had significantly higher Fe, Zn, Ca, and protein content than other races (Vetriventhan and Upadhyaya 2018). The improvement of grain nutritional traits could be accomplished in finger millet cultivars without being deprived of the diversity of inflorescence morphology.

Several studies have investigated the correlation between grain micronutrients, protein content, and grain yield in finger millet (Vadivoo et al. 1998; Upadhyaya et al. 2011; Das et al. 2017; Badigannavar and Ganapathi 2018; Puranik et al. 2020; Fred et al. 2021; Backiyalakshmi et al. 2023). The studies have found that there is a moderate to high correlation between grain micronutrients and protein content, while the correlation between grain yield and nutrient content was weaker, nonsignificant, and negative. It is important to note that the magnitude and strength of correlations depend upon the population used in the study; therefore, they may vary with the materials investigated. If the population is more homogenous, it gives better estimates of correlation (Upadhyaya et al. 2021). Protein content showed a negative genotypic correlation with both calcium content and grain yield, while the correlation between calcium content and yield was not significant (Vadivoo et al. 1998; Fred et al. 2021). Similarly, Das et al. (2017) found a significant negative correlation between protein content and grain yield, with a nonsignificant correlation between calcium content and yield. However, protein content was significantly and positively correlated with calcium content, and micronutrients were positively correlated with each other (Badigannavar and Ganapathi 2018). A positive and significant association was observed among the grain nutrients, and also a nonsignificant correlation between grain yield and calcium was noticed among accessions within landraces, breeding lines, and accessions from Asia, while this correlation was significantly negative among accessions from Africa and in the entire set (Backiyalakshmi et al. 2023). Likewise, seed weight showed a positive significant correlation with copper and potassium, while a negative significant correlation with protein, zinc, sulphur, phosphorus (Backiyalakshmi et al. 2023), and iron (Fred et al. 2021). It suggests that better prospects of combining higher grain nutrients with higher grain yield potential.

2.3.2 *Environmental Factors*

2.3.2.1 Temperature and Its Impact on Nutrient Composition

Heat stress is one of the abiotic stresses that has major implications in the growth and development of crop plants. Besides crop yield, loss in grain quality and alteration in mineral composition were impacted by high temperature, which could have an effect on market revenue and consumer preference. Monjardino et al. (2005) found that heat stress during the early stage of endosperm development had a negative effect on protein concentration in the maize crop. Among the protein fractions, zein and glutelin contents were reduced by 45% and 33% under heat stress. Luitel et al. (2020) reported that different elevations with temperature had an impact on finger millet grain nutrient content. It implies that higher elevation is associated with low temperature, which has a positive effect on finger millet grain nutrient content by 3.13% protein, 0.47% calcium, and 0.33% iron. Low temperatures in higher elevations may stimulate seed protein synthesis and protein remobilization from vegetative organs to grains, thereby increasing seed protein content (Diacono et al. 2012). However, in sorghum, heat stress induces a reduction in grain iron (by 10%), zinc (by 4%), manganese (by 13%), and copper (by 14%), respectively (Impa et al. 2019). This shows that the effects of nutrient accumulation depend on both the type and intensity of environmental factors.

2.3.2.2 CO_2 Effects on Grain Quality

Rising levels of CO_2 in the atmosphere due to climate change are likely to cause imbalances in plant chemistry when compared to preindustrial times. This can lead to an intensification of micronutrient malnutrition, as noted by Loladze (2014), particularly for minerals such as iron, zinc, iodine, and protein (Broberg et al. 2017). Studies have shown that elevated CO_2 concentrations can decrease mineral concentrations in barley grains (by 6.9%), rice grains (by 7.2%), and wheat grains (by 7.6%), and increase the ratio of nonstructural carbohydrates (TNC) to protein in grains and tubers by 6–47%. The reduction in protein mediated by elevated CO_2 was reported to be 15% for barley, 10% for rice, and 10% for wheat (Taub et al. 2008).

In a meta-analysis of the impact of elevated CO_2 on wheat grains, Broberg et al. (2017) found a significant reduction in the concentration of most minerals (such as calcium, cadmium, copper, iron, magnesium, manganese, phosphorus, sulfur, and zinc). Boron (B) and sodium (Na) were not significantly affected, and potassium (K) showed a significant increase of less than 2%. Studies have shown that CO_2-induced reductions in the above-mentioned minerals have been reported in the edible parts of major crops (Loladze 2014; Dietterich et al. 2015). In rice, elevated temperature along with CO_2 also alters grain protein and mineral nutrient composition by 23%, 22%, 30%, and 34% of iron, zinc, copper, and manganese content, respectively (Chaturvedi et al. 2017). In light of climate change, research should concentrate on how CO_2 affects the nutritional qualities of finger millet for breeding nutri-dense cultivars.

2.3.2.3 Effect of Soil Fertility and Nutrient Uptake on Grain Quality

Researchers found a significant relationship between the nutrient composition of soil and grain nutrient content. An increase in organic carbon, nitrogen, and phosphorus (P_2O_5) positively impacts the nutrient composition of finger millet (Luitel et al. 2020). Application of phosphorus fertilizer significantly results in the increase of grain protein content of 11.0% at 25 kg ha^{-1} P than at zero P content recorded the protein content of 4.4% in finger millet. The calcium content significantly differed between P application of 25 kg ha^{-1} P and zero P content (472 mg/100 g and 390 mg/100 g) (Wafula et al. 2018). An 8% increase in crude protein was recorded with 37.5 kg ha^{-1} P compared to zero P application in cowpea (Magani and Kuchinda 2009). Soleymani and Shahrajabian (2013) reported that higher P application led to higher protein content (10.8%) in maize than in control plots, which contained protein content of 9.5%. Higher levels of phosphorus in the soil can also boost grain protein by 5.66% in finger millet, due to its role in protein synthesis and energy provision for nitrogen uptake from soil to grain (Wafula et al. 2016). Application of other major and micronutrient fertilizers on finger millet and its impact on grain nutrient content needs to be studied for a better understanding of the relationship between nutrient fertilizers and grain nutrient qualities.

2.3.2.4 Impact of Water Scarcity on Grain Quality

The water deficit is a common abiotic stress in the dry and semi-dry regions of the world, resulting in severe yield loss. It has been reported that drought stress has a greater impact on the accumulation of starch than nitrogen during the grain filling process in crop plants. This can potentially affect the conversion of sucrose into starch in grains (Panozzo and Eagles 1998), leading to higher grain protein concentration (% protein) in crops, reported in wheat by 12%, 23.8%, and 4% (Houshmand et al. 2014; Kiliç 2010; Singh et al. 2010) and rice (9.5%) (Crusciol et al. 2008). However, the opposite effect of decline in protein content (upto 13.82%) has been observed in some wheat cultivars (Abd El-Kareem and El-Saidy 2011), possibly due to differences in stress levels and plant development status. In rice, the increased grain Ca (10.3%), Mg (31%), Fe (40%), and Zn (44%) concentration under rainfed conditions has been reported (Crusciol et al. 2008). Da Ge et al. (2010) conducted a field study to analyze the nutritional quality of maize grains at various soil moisture levels. The study found that severe drought increased the accumulation of N, Ca, Mg, Cu, and Zn by 12%, 28%, 11%, 18%, and 33%, respectively, when compared to the control. However, a significant decrease in the concentration of P and K by 17% in both minerals was observed during severe drought. This indicates that the effects of drought on mineral concentration are specific to each nutrient. Finger millet is one of the drought-tolerant crops of millets, and water scarcity had a significant effect on finger millet grain yield (Krishna et al. 2021), while no study is available on finger millet investigating the effect of drought stress on grain nutrient composition.

2.3.2.5 Impact of Altitude on Grain Quality

Altitude is reported to have a positive impact on the grain protein, calcium, and iron content in finger millet (Luitel et al. 2020). Specifically, when altitude increased from 365 to 1856 m, there was an average increase of 3.13% protein, 0.47% calcium, and 0.33% iron. Only 2.04% and 1.09% of average grain protein increased as elevation increased from 365 m to 1040 m and 1040 m to 1856 m, respectively. Similarly, the average Ca increased by 0.21% and 0.26% as altitude increased from 365 m to 1040 m, and 1040 m to 1856 m, respectively. The average iron increased by 0.11% and 0.21% from 365 to 1040 m and from 1040 to 1856 m, respectively (Luitel et al. 2020). On the other hand, lower elevations with higher concentrations of CO_2 gas and temperature could be the reason for the reduction in protein composition and grain nutrient contents (Dong et al. 2018). Similar results have been found on other crops, that the higher altitude areas have comparatively higher grain nutrient content than the lower elevation areas (Nadew 2018; Dong et al. 2018).

2.3.2.6 Effect of Geographical Origin on Nutritional Composition

Finger millet has its primary centre of origin in Africa and secondary center of diversity in Asia, respectively. The nutrient content of grain may vary depending on its geographical origin. This information can be used to develop strategies for collecting germplasm that contains specific nutrients. Studies have found that finger millet accessions from Ethiopia have higher calcium content compared to those from Kenya and India (Barbeau and Hilu 1993). According to a study conducted by (Puranik et al. 2020), the calcium content was found to be higher in the Asian subpopulation, while the African population had higher levels of potassium and zinc. Additionally, Backiyalakshmi et al. (2023) reported that the nutrient content in accessions from Africa was significantly higher than that from Asia for zinc, calcium, and magnesium. There was no significant difference in the levels of iron, protein, sulfur, potassium, phosphorus, and copper between the two populations. Nevertheless, nutrient concentrations varied depending on the collection region or origin, and there was no specific region that consistently had higher levels of all nutrients (Fred et al. 2021). Furthermore, there is no substantial evidence to suggest that nutrient content is related to where the grains come from. Finger millet accessions from both Africa and Asia can have high or low nutrient levels (Upadhyaya et al. 2011). It implies that the inclusion of diverse accessions having high nutrient content from African and Asian germplasm could result in better enhancement of grain nutrients in the finger millet biofortification programme.

2.3.3 Management Factors

Crop management practices, such as seeding rate, fertilizer, spacing, source of irrigation, and their interaction with the growing environment, will greatly affect the yield and grain quality of the crop plants. Melash et al. (2019) reported that improper seeding rate can reduce yield by 24% and grain protein by 8.7% on durum wheat. It should be noted that effects vary depending on the growing environment and response of crop varieties to these factors. It is crucial to continuously monitor and re-evaluate basic agronomic practices to improve the performance and sustainability of crop plants (Melash et al. 2023). Finger millet is a versatile crop that can adapt to various agro-climatic conditions. Farmers highly value finger millet as a crop that requires low fertilizer input; it often suffers from low yields, mainly because finger millet is grown largely in marginal lands with poor management practices. Therefore, optimizing nutrient management practices and other factors related to finger millet cultivation is crucial to attaining better yields under marginal local growing conditions.

2.3.4 Interaction among Genotype × Environment × Management (G × E × M)

G × E × M interaction is a phenomenon where phenotypes respond differently to genotypes depending on various environmental factors. For instance, certain genotypes may be more affected than others by a specific aspect of the environment (Kim et al. 2014). Additionally, the order of merit of a series of genotypes may change when measured under different environmental conditions (Falconer et al. 1996). The study of G × E × M is crucial to accurately assess both genetic and environmental factors, as it can obscure these effects (Ottman 1996). Furthermore, it can reveal how inherited traits make some individuals more vulnerable to the negative or positive outcomes of environments. Investigating G × E × M is significant in identifying mechanisms through which specific environmental processes can counteract or worsen genetic risks, thus indicating potential targets for preventive interventions (Leve et al. 2010). Knowledge of G × E × M interaction aids in designing suitable breeding and selection strategies to enhance nutrients in edible portions of grain (Pfeiffer and McClafferty 2007). G × E × M interactions can reduce selection efficiency and accuracy; thus, G × E × M interactions must be studied to introduce new high-yielding genotypes with high stability in different environments.

Finger millet is grown predominantly as a rainfed crop on different soil types and associated differences in the native soil fertility with/without farmers' managed fertility in semi-arid regions. Therefore, it is necessary to assess the stability of macro- and micronutrients and protein-dense finger millet cultivars across different soil types and soil fertility levels typical of the areas to which these cultivars are targeted. Fewer studies have been reported on the finger millet crop related to environmental interaction with grain nutrient content. Luitel et al. (2020) conducted the study relation between finger millet grain nutrients with soil components and altitude. Similarly, Wafula et al. (2018) found that the application of phosphorus had a significant positive impact on finger millet grain nutrient content. Upadhyaya et al. (2011) and Backiyalakshmi et al. (2023) reported the significant G× E interaction in finger millet germplasm for grain nutrient traits across years. Pooled analysis of variance showed significant G× E effects for grain calcium, iron, and zinc contents (Saritha et al. 2018). Significant G× E for micronutrients was reported in other crops, such as pearl millet (Bashir et al. 2014; Anuradha et al. 2017; Singhal et al. 2018), sorghum (Phuke et al. 2017), maize (Prasanna et al. 2011), emmer wheat (*Triticum dicoccoides*) (Gomez-Becerra et al. 2010), and rice (Suwarto, and Nasrullah 2011). These findings suggest that genotype × environment (G×E) interactions are unavoidable for these traits, as in yield and its components. However, the magnitude of significant genetic effects and G×E interactions should be carefully considered during selection and advancement.

2.4 Conclusion

Finger millet is a staple food for millions of people in Asia and Africa. Therefore, adding it to the biofortification target crop breeding investment portfolio will benefit several million. However, global genebank platforms and finger millet improvement programs have paid little attention to grain nutrient traits characterization. Studying the genotype, environment, management, and their interactions for the grain nutrient stability is crucial for developing nutrient-dense cultivars, especially in current and near-future environments. To achieve this, we need to understand how environmental growth conditions, such as elevated temperature, CO_2 and drought, affect the micronutrient content of the finger millet crop. It is also important to explore the genetic diversity and variability of global finger millet germplasm conserved at ICRISAT to discover genotypes that are more nutrtious and resilient to ongoing climate change. Incorporating stable finger millet accessions and mega varieties that are higher in nutrition can reduce the cost of household nutrition interventions and transform the agri-food system sustainably.

References

Suwarto, Nasrullah (2011) Genotype × environment interaction for iron concentration of Rice in Central Java of Indonesia. Rice Sci 18(1):75–78. https://doi.org/10.1016/S1672-6308(11)60011-5

Abd El-Kareem THA, El-Saidy AEA (2011) Evaluation of yield and grain quality of some bread wheat genotypes under normal irrigation and drought stress conditions in calcareous soils. J Biol Sci 11(2):156–164

Admassu S, Teamir M, Alemu D (2009) Chemical composition of local and improved finger millet [Eleusine coracana (L.) Gaetrtin] varieties grown in Ethiopia. Ethiop J Health Sci 19(7):1–8

Anitha S, Givens DI, Botha R, Kane-Potaka J, Sulaiman NLB et al (2021) Calcium from finger millet—a systematic review and meta-analysis on calcium retention, bone resorption, and in vitro bioavailability. Sustainability 13(16):1–14. https://doi.org/10.3390/su13168677

Antony Ceasar S, Maharajan T, Ajeesh Krishna TP, Ramakrishnan M, Victor Roch G et al (2018) Finger millet [Eleusine coracana (L.) Gaertn.] improvement: current status and future interventions of whole genome sequence. Front Plant Sci 9:1054. https://doi.org/10.3389/fpls.2018.01054

Anuradha N, Tara Satyavathi C, Meena MC, Sankar SM, Bharadwaj C et al (2017) Evaluation of pearl millet [Pennisetum glaucum (L.) R. BR.] for grain iron and zinc content in different agro climatic zones of India. Indian J Genet Plant Breed 77(1):65–73. https://doi.org/10.5958/0975-6906.2017.00009.8

Backiyalakshmi C, Babu C, Deshpande S, Govindaraj M, Gupta R et al (2023) Characterization of finger millet global germplasm diversity panel for grain nutrients content for utilization in biofortification breeding. Crop Sci. https://doi.org/10.1002/csc2.21085

Badigannavar A, Ganapathi TR (2018) Genetic variability for mineral nutrients in indigenous germplasm lines of finger millet (Eleusine coracana Gaertn.). J Cereal Sci 84:1–6. https://doi.org/10.1016/j.jcs.2018.09.014

Balk EM, Adam GP, Langberg VN, Earley A, Clark P et al (2017) Global dietary calcium intake among adults: a systematic review. Osteoporos Int 28(12):3315–3324. https://doi.org/10.1007/s00198-017-4230-x

Barbeau WE, Hilu KW (1993) Protein, calcium, iron, and amino acid content of selected wild and domesticated cultivars of finger millet. Plant Foods Hum Nutr 43:97

Bashir EMA, Ali AM, Ali AM, Ismail MI, Parzies HK et al (2014) Patterns of pearl millet genotype-by-environment interaction for yield performance and grain iron (Fe) and zinc (Zn) concentrations in Sudan. Field Crop Res 166:82–91. https://doi.org/10.1016/j.fcr.2014.06.007

Broberg MC, Högy P, Pleijel H (2017) CO2-induced changes in wheat grain composition: meta-analysis and response functions. Agronomy 7(2). https://doi.org/10.3390/agronomy7020032

Chandra D, Chandra S, Pallavi, Sharma AK (2016) Review of finger millet (Eleusine coracana (L.) Gaertn): A power house of health benefiting nutrients. Food Sci Human Wellness 5(3):149–155. https://doi.org/10.1016/j.fshw.2016.05.004

Chaturvedi AK, Bahuguna RN, Pal M, Shah D, Maurya S et al (2017) Elevated CO2 and heat stress interactions affect grain yield, quality and mineral nutrient composition in rice under field conditions. Field Crop Res 206:149–157. https://doi.org/10.1016/j.fcr.2017.02.018

Chaudhari PP, Patel DA, Saravaiya SN, Patel MC (2012) GN-5: a new high yielding white grain finger millet variety. Int J Plant Sci (Muzaffarnagar) 7(1):51–53

Crusciol CAC, Arf O, Soratto PR, Mateus PG (2008) Grain quality of upland rice cultivars in response to cropping systems in the Brazilian tropical savanna. Sci Agric 65:468–473

Da Ge T, Sui FG, Nie S, Sun NB, Xiao H et al (2010) Differential responses of yield and selected nutritional compositions to drought stress in summer maize grains. J Plant Nutr 33(12):1811–1818. https://doi.org/10.1080/01904167.2010.503829

Das R, Pandravada SR, Harikrishnan PJ (2017) Estimation of grain nutrient and their association with grain yield in finger millet [Eleusine coracana (L.) Gaertn]. Plant Arch 17(1):241–246

Devi PB, Vijayabharathi R, Sathyabama S, Malleshi NG, Priyadarisini VB (2014) Health benefits of finger millet (Eleusine coracana L.) polyphenols and dietary fiber: a review. J Food Sci Technol 51(6):1021–1040. https://doi.org/10.1007/s13197-011-0584-9

Diacono M, Castrignanò A, Troccoli A, De Benedetto D, Basso B et al (2012) Spatial and temporal variability of wheat grain yield and quality in a Mediterranean environment: a multivariate geostatistical approach. Field Crop Res 131:49–62. https://doi.org/10.1016/j.fcr.2012.03.004

Dietterich LH, Zanobetti A, Kloog I, Huybers P, Leakey ADB et al (2015) Impacts of elevated atmospheric CO2 on nutrient content of important food crops. Sci Data 2. https://doi.org/10.1038/sdata.2015.36

Dong J, Gruda N, Lam SK, Li X, Duan Z (2018) Effects of elevated CO_2 on nutritional quality of vegetables: a review. Front Plant Sci 9. https://doi.org/10.3389/fpls.2018.00924

Falconer DS, Mackay TFC, Frankham R (1996) Introduction to quantitative genetics, 4th Edition, Longman Scientific and Technical, Harlow

Fred OH, Patrick S, Joseph K, Anil K, Abhishek R et al (2021) Characterization of finger millet germplasm for mineral contents: prospects for breeding. J Cereals Oilseeds 12(1):33–44. https://doi.org/10.5897/jco2020.0222

Gomez-Becerra HF, Erdem H, Yazici A, Tutus Y, Torun B et al (2010) Grain concentrations of protein and mineral nutrients in a large collection of spelt wheat grown under different environments. J Cereal Sci 52(3):342–349. https://doi.org/10.1016/j.jcs.2010.05.003

Gopalan C, Rama Sastri BV, Balasubramanian SC (2009) Nutritive value of Indian foods. National Institute of Nutrition, Indian Council of Medical Research, Hyderabad

Govindaraj M, Rao AS, Shivade H, Rai KN (2018) Effect of grain colour on iron and zinc density in pearl millet. Indian J Genet Plant Breed 78(2):247–251. https://doi.org/10.5958/0975-6906.2018.00032.9

Gupta SM, Arora S, Mirza N, Pande A, Lata C et al (2017) Finger millet: a "certain" crop for an "uncertain" future and a solution to food insecurity and hidden hunger under stressful environments. Front Plant Sci 8:643. https://doi.org/10.3389/fpls.2017.00643

Houshmand S, Arzani A, Mirmohammadi-Maibody SAM (2014) Effects of salinity and drought stress on grain quality of durum wheat. Commun Soil Sci Plant Anal 45(3):297–308. https://doi.org/10.1080/00103624.2013.861911

Impa SM, Perumal R, Bean SR, John Sunoj VS, Jagadish SVK (2019) Water deficit and heat stress induced alterations in grain physico-chemical characteristics and micronutrient composition in field grown grain sorghum. J Cereal Sci 86:124–131. https://doi.org/10.1016/j.jcs.2019.01.013

Kiliç H (2010) The effect of drought stress on grain yield, yield components and some quality traits of durum wheat (Triticum turgidum ssp. durum) cultivars. Notulae Botanicae Horti Agrobotanici Cluj-Napoca. https://doi.org/10.15835/nbha3814274

Kim J, Lee T, Lee HJ, Kim H (2014) Genotype-environment interactions for quantitative traits in Korea Associated Resource (KARE) cohorts. BMC Genet 15. https://doi.org/10.1186/1471-2156-15-18

Krishna SS, Reddy YAN, Kumar RLR (2021) Assessment of traits for grain yield under drought in finger millet. Plant Physiol Rep 26(1):84–94. https://doi.org/10.1007/s40502-020-00561-1

Kumar A, Metwal M, Kaur S, Gupta AK, Puranik S et al (2016) Nutraceutical value of finger millet [Eleusine coracana (L.) Gaertn.], and their improvement using omics approaches. Front Plant Sci 7:934. https://doi.org/10.3389/fpls.2016.00934

Leve LD, Kerr DCR, Shaw D, Ge X, Neiderhiser JM et al (2010) Infant pathways to externalizing behavior: evidence of genotype × environment interaction. Child Dev 81(1):340–356. https://doi.org/10.1111/j.1467-8624.2009.01398.x

Li X, Siddique KHM (2018) Future smart food – rediscovering hidden treasures of neglected and underutilized species for zero hunger in Asia, Bangkok

Loladze I (2014) Hidden shift of the ionome of plants exposed to elevated CO_2 depletes minerals at the base of human nutrition. elife 2014(3). https://doi.org/10.7554/eLife.02245

Luitel DR, Siwakoti M, Jha PK (2020) Nutrients in finger millet and soil at different elevation gradients in Central Nepal. CABI Agric Biosci:1–10. https://doi.org/10.1186/s43170-020-00018-3

Magani IE, Kuchinda C (2009) Effect of phosphorus fertilizer on growth, yield and crude protein content of cowpea (Vigna unguiculata [L.] Walp) in Nigeria. J Appl Biosci 23:1387–1393. www.biosciences.elewa.orgon

Melash AA, Mengistu DK, Aberra DA, Tsegay A (2019) The influence of seeding rate and micronutrients foliar application on grain yield and quality traits and micronutrients of durum wheat. J Cereal Sci 85:221–227. https://doi.org/10.1016/j.jcs.2018.08.005

Melash AA, Bogale AA, Mengstu SG, Aberra DA, Tsegay A et al (2023) Sustainable management practices for durum wheat production: analyzing specific agronomic interventions on productivity, grain micronutrient content, and quality. Heliyon 9(8). https://doi.org/10.1016/j.heliyon.2023.e18733

Monjardino P, Smith AG, Jones RJ (2005) Heat stress effects on protein accumulation of maize endosperm. Crop Sci 45(4):1203–1210. https://doi.org/10.2135/cropsci2003.0122

Muthamilarasan M, Prasad M (2015) Advances in Setaria genomics for genetic improvement of cereals and bioenergy grasses. Theor Appl Genet 128:1–14. https://doi.org/10.1007/s00122-014-2399-3

Nadew BB (2018) Effects of climatic and agronomic factors on yield and quality of bread wheat (Triticum aestivum L.) seed: a review on selected factors. Adv Crop Sci Technol 06(02). https://doi.org/10.4172/2329-8863.1000356

Ottman R (1996) Gene-environment interaction: definitions and study designs. Prev Med 25(6):764–770

Panozzo JF, Eagles HA (1998) Cultivar and environmental effects on quality charaters in wheat. I Starch. Aust J Agric Res 49(5):757–766. https://doi.org/10.1071/A97161

Pfeiffer WH, McClafferty B (2007) HarvestPlus: breeding crops for better nutrition. Crop Sci. https://doi.org/10.2135/cropsci2007.09.0020ipbs

Phuke RM, Anuradha K, Radhika K, Jabeen F, Anuradha G et al (2017) Genetic Variability, Genotype × Environment Interaction, Correlation, and GGE Biplot Analysis for Grain Iron and Zinc Concentration and Other Agronomic Traits in RIL Population of Sorghum (Sorghum bicolor L. Moench). Front Plant Sci 8(712):1–13. https://doi.org/10.3389/fpls.2017.00712

Prasanna BM, Mazumdar S, Chakraborti M, Hossain F, Manjaiah KM et al (2011) Genetic variability and genotype × environment interactions for kernel Fe and Zn concentrations in maize (Zea mays) genotypes. The Indian Journal of Agricultural Sciences 81(8):704–711. https://www.researchgate.net/publication/266356311

Puranik S, Kam J, Sahu PP, Yadav R, Srivastava RK et al (2017) Harnessing finger millet to combat calcium deficiency in humans: challenges and prospects. Front Plant Sci 8:1311. https://doi.org/10.3389/fpls.2017.01311

Puranik S, Sahu PP, Beynon S, Srivastava RK, Sehgal D et al (2020) Genome-wide association mapping and comparative genomics identifies genomic regions governing grain nutritional traits in finger millet (Eleusine coracana L. Gaertn.) Societal Impact Statement. Plants People Planet 2:649–662. https://doi.org/10.1002/ppp3.10120

Qi J, Liu W, Jiao T, Hamblin A, Li Y (2020) Variation in morphological and physiological characteristics of wild Elymus nutans ecotypes from different altitudes in the northeastern Tibetan plateau. J Sens. https://doi.org/10.1155/2020/2869030

Rao PU (1994) Evaluation of protein quality of brown and white ragi (Eleusine coracana) before and after malting. Food Chem 51(4):433–436. https://doi.org/10.1016/0308-8146(94)90198-8

Rodríguez JP, Rahman H, Thushar S, Singh RK (2020) Healthy and resilient cereals and pseudo-cereals for marginal agriculture: molecular advances for improving nutrient bioavailability. Front Genet 11:1–29. https://doi.org/10.3389/fgene.2020.00049

Royer DL, Wilf P, Janesko DA, Kowalski EA, Dilcher DL (2005) Correlations of climate and plant ecology to leaf size and shape: potential proxies for the fossil record. Am J Bot 92(7):1141–1151. https://doi.org/10.3732/ajb.92.7.1141

Saleh ASM, Zhang Q, Chen J, Shen Q (2013) Millet grains: nutritional quality, processing, and potential health benefits. Compr Rev Food Sci Food Saf 12(3):281–295. https://doi.org/10.1111/1541-4337.12012

Saritha HS, Ravishankar P, Sunitha NC (2018) Stability of grain nutrient concentrations in white finger millet. Int J Curr Microbiol Appl Sci 7(11):2786–2801. https://doi.org/10.20546/ijcmas.2018.711.321
Sharma D, Jamra G, Singh UM, Sood S, Kumar A (2017) Calcium biofortification : three pronged molecular approaches for dissecting complex trait of calcium nutrition in finger millet (Eleusine coracana) for devising strategies of enrichment of food crops. Front Plant Sci 7(January):1–18. https://doi.org/10.3389/fpls.2016.02028
Shashi BK, Sharan S, Hittalamani S, Shankar AG, Nagarathna TK (2007) Micronutrient composition, Antinutritional factors and bioaccessibility of iron in different finger millet (Eleusine coracana) genotypes. Karnataka J Agric Sci 20(3):583–585
Shobana S, Sreerama YN, Malleshi NG (2009) Composition and enzyme inhibitory properties of finger millet (Eleusine coracana L.) seed coat phenolics: mode of inhibition of a -glucosidase and pancreatic amylase. Food Chem 115(4):1268–1273. https://doi.org/10.1016/j.foodchem.2009.01.042
Shobana S, Krishnaswamy K, Sudha V, Malleshi NG, Anjana RM et al (2013) Finger millet (Ragi, Eleusine coracana L.). A review of its nutritional properties, processing, and plausible health benefits. In: Advances in food and nutrition research. Academic Press Inc., pp 1–39
Singh S, Gupta AK, Gupta SK, Kaur N (2010) Effect of sowing time on protein quality and starch pasting characteristics in wheat (Triticum aestivum L.) genotypes grown under irrigated and rain-fed conditions. Food Chem 122(3):559–565. https://doi.org/10.1016/j.foodchem.2010.03.010
Singhal T, Satyavathi CT, Kumar A, Sankar SM, Singh SP et al (2018) Genotype × environment interaction and genetic association of grain iron and zinc content with other agronomic traits in RIL population of pearl millet. Crop Pasture Sci 69(11):1092–1102. https://doi.org/10.1071/CP18306
Soleymani A, Shahrajabian MH (2013) The effects of nitrogen fertilizer on ash, nitrate, organic carbon, protein and total yield of forage maize in semi arid region of Iran. www.tjeas.com
Sood S, Joshi DC, Chandra AK, Kumar A (2019) Phenomics and genomics of finger millet: current status and future prospects. Planta 250(3):731–751. https://doi.org/10.1007/s00425-019-03159-6
Taub DR, Miller B, Allen H (2008) Effects of elevated CO2 on the protein concentration of food crops: A meta-analysis. Glob Chang Biol 14(3):565–575. https://doi.org/10.1111/j.1365-2486.2007.01511.x
Tripathi B, Platel K (2010) Finger millet (Eleucine coracana) flour as a vehicle for fortification with zinc. J Trace Elem Med Biol 24(1):46–51. https://doi.org/10.1016/j.jtemb.2009.09.001
Udeh HO, Duodu KG, Israel A, Jideani O (2017) Finger millet bioactive compounds, bioaccessibility, and potential health effects—a review. Czech J Food Sci 35(1):7–17. https://doi.org/10.17221/206/2016-CJFS
Upadhyaya HD, Gowda CLL, Pundir RPS, Reddy VG, Singh S (2006) Development of core subset of finger millet germplasm using geographical origin and data on 14 quantitative traits. Genet Resour Crop Evol 53(4):679–685. https://doi.org/10.1007/s10722-004-3228-3
Upadhyaya HD, Sarma NDRK, Ravishankar CR, Albrecht T, Narasimhudu Y et al (2010) Developing a mini-core collection in finger millet using multilocation data. Crop Sci 50(5):1924–1931. https://doi.org/10.2135/cropsci2009.11.0689
Upadhyaya HD, Ramesh S, Sharma S, Singh SK, Varshney SK et al (2011) Genetic diversity for grain nutrients contents in a core collection of finger millet (Eleusine coracana (L.) Gaertn.) germplasm. Field Crop Res 121(1):42–52. https://doi.org/10.1016/j.fcr.2010.11.017
Upadhyaya HD, Vetriventhan M, Azevedo VCR (2021) Variation for photoperiod and temperature sensitivity in the global mini core collection of sorghum. Front Plant Sci 12:571243. https://doi.org/10.3389/fpls.2021.571243
Vadivoo AS, Joseph R, Ganesan NM (1998) Genetic variability and diversity for protein and calcium contents in finger millet (Eleusine coracana (L.) Gaertn) in relation to grain color. Plant Foods Hum Nutr 52(4):353–364. https://doi.org/10.1023/A:1008074002390

Verma V, Patel S (2013) Value added products from nutri-cereals: finger millet (Eleusine coracana). Emir J Food Agric 25(3):169–176. https://doi.org/10.9755/ejfa.v25i3.10764
Vetriventhan M, Upadhyaya HD (2018) Diversity and trait-specific sources for productivity and nutritional traits in the global proso millet (Panicum miliaceum L.) germplasm collection. Crop J 6(5):451–463. https://doi.org/10.1016/j.cj.2018.04.002
Vetriventhan M, Upadhyaya HD (2019) Variability for productivity and nutritional traits in germplasm of kodo millet, an underutilized nutrient-rich climate smart crop. Crop Sci 59(3):1095–1106. https://doi.org/10.2135/cropsci2018.07.0450
Vetriventhan M, Azevedo VCR, Upadhyaya HD, Nirmalakumari A, Kane-Potaka J et al (2020) Genetic and genomic resources, and breeding for accelerating improvement of small millets: current status and future interventions. Nucleus (India) 1:13. https://doi.org/10.1007/s13237-020-00322-3
Vetriventhan M, Upadhyaya HD, Azevedo VCR, Allan V, Anitha S (2021) Variability and trait-specific accessions for grain yield and nutritional traits in germplasm of little millet (Panicum sumatrense Roth. Ex Roem & Schult.). Crop Sci:2658–2679. https://doi.org/10.1002/csc2.20527
Wafula WN, Korir N, Ojulong H, Siambi M, Gweyi-Onyango JP (2016) Phosphorus influence on plant tissue nitrogen contents and yield attributes of finger millet varieties in semi-arid region of Kenya. Int J Plant Soil Sci 13(3):1–9. https://doi.org/10.9734/ijpss/2016/29901
Wafula WN, Korir NK, Ojulong HF, Siambi M, Gweyi-onyango JP (2018) Protein, calcium, zinc, and iron contents of finger millet grain response to varietal differences and phosphorus application in Kenya. Agronomy 8(28):1–9. https://doi.org/10.3390/agronomy8020024

Chapter 3
Sorghum Nutritional Traits and Its Health Potentials for Nutritional Security

S. Viswabharathy, D. Sushmitha, L. Sampath, S. Samuthrapandi, A. Shanmugam, T. Kalaimagal, and R. Ravikesavan

Abstract Sorghum (*Sorghum bicolor* L. Moench) is a vital cereal crop globally, valued for its adaptability to diverse environmental conditions and nutritional significance. Genotype–environment interaction (GEI) plays a crucial role in shaping the grain nutritional traits of sorghum. This interaction encompasses the dynamic interplay between genetic factors inherent in sorghum genotypes and various environmental factors, including soil type, climate, agronomic practices, and geographical location. Understanding the complex GEI affecting grain nutritional traits in sorghum is essential for optimizing breeding strategies and crop management practices to enhance nutritional quality and overall productivity. The chapter explores how GEI influences key nutritional attributes of sorghum grains, including protein content, amino acid composition, starch quality, dietary fiber content, mineral and vitamin composition, and phytochemical content. Insights into GEI patterns and their implications for sorghum breeding and production are discussed, highlighting the need for integrated approaches that consider both genetic and environmental factors to develop sorghum varieties with improved nutritional profiles and agronomic performance. Further research efforts are warranted to unravel the complexities of GEI in sorghum and exploit this knowledge for sustainable sorghum production and food security across diverse agroecological settings.

Keywords Sorghum · Nutrition · Stability · Iron · Zinc

S. Viswabharathy · D. Sushmitha · L. Sampath · S. Samuthrapandi · A. Shanmugam
T. Kalaimagal (✉) · R. Ravikesavan
Centre for Plant Breeding and Genetics, Tamil Nadu Agricultural University, Coimbatore, Tamil Nadu, India

M. Govindaraj (ed.), *Breeding Biofortified Millets for Nutritional Security*, Sustainability Sciences in Asia and Africa,
https://doi.org/10.1007/978-981-95-9827-4_3

3.1 Introduction

The societal preference for carbohydrate-rich but nutrient-poor foods has driven agricultural production toward starchy staples, exacerbating global malnutrition. Nutritional security aims to ensure universal access to essential nutrients for balanced diets. Nearly half of deaths among children under five are due to undernutrition, and the World Health Organization recognizes micronutrient deficiency as a global health concern (WHO 2023). Traditional methods of micronutrient supplementation and fortification are unsustainable; genetic biofortification offers a promising alternative to enhance staple crops' nutritional content. Ranked fifth for cost-effectiveness, biofortification enhances global micronutrient intake, highlighting the urgent need for the development of improved crop varieties.

Sorghum, scientifically known as *Sorghum bicolor* L., belongs to the carbohydrate-rich *Gramineae* family. Sorghum is the second cheapest source of energy and micronutrients, after pearl millet (Phuke et al. 2017). In key regions like India, sorghum provides 35% of calorie intake and over 50% of micronutrient requirements for low-income groups (Rao et al. 2006). Renowned for its resilience, sorghum offers abundant dietary fiber, protein, and essential minerals. Compared to wheat, sorghum is a nutrient-dense grain with diverse health benefits, positioning it as a promising staple crop. Cultivated in regions such as India, the United States, and various African nations, sorghum serves as a primary food source in Africa and India and is a significant component of livestock feed in developed nations (Tanwar et al. 2023). Sorghum's nutritional value is determined by its chemical composition, which includes anti-nutritional factors like phytate. Breeding involves the deliberate modification of genetic traits in plants or animals to enhance the characteristics desired for improved product quality and quantity, thereby benefiting both human and animal populations.

Enhancing grain nutritional quality through breeding requires genetic diversity in germplasm and understanding the genetic control of micronutrient content. Previous research indicates a positive correlation between iron and zinc, suggesting shared genetic regions or pathways influencing these traits. Phenotypic associations between traits offer insights for simultaneous improvement (Ortiz-Monasterio et al. 2007). The nutritional quality of sorghum grains is not solely determined by genetic factors but also by the intricate interplay between genotype and environment, known as GEI. Understanding how GEI influences grain nutritional traits in sorghum is essential for optimizing breeding strategies, agronomic practices, and food security efforts. The concept of GEI underscores the notion that a genotype's expression of specific traits, such as nutritional content, can vary significantly across different environmental conditions. Factors such as soil quality, climate variability, water availability, and management practices can all impact the nutritional composition of sorghum grains.

Statistical analysis of GEI on grain nutritional traits in sorghum plays a critical role in unravelling the complex relationships between genetic factors, environmental conditions, and nutritional composition. By employing robust statistical methods, researchers can quantify the magnitude and significance of GEI effects, identify superior genotypes with desirable nutritional profiles across diverse environments, and inform breeding strategies and agronomic practices. Statistical analysis of genotype–environment interaction involves a range of techniques, including ANOVA, biplot analysis, AMMI models, mixed models, and Bayesian approaches (Bänziger et al. 2006; Cullis et al. 2010; Velu et al. 2018). Understanding how genetic and environmental factors interact to influence sorghum nutrition can unlock the crop's full potential to contribute to global food security and address nutritional challenges effectively.

3.2 Importance of Biofortification in Sorghum

Micronutrient deficiency affects billions of individuals worldwide, leading to undernourishment and severe health consequences, constituting one of the most pressing global health challenges. Particularly in rural areas of developing nations, these deficiencies contribute to a cycle of poverty (Simnadis et al. 2016). While commercial food fortification is valuable, it alone may not adequately address malnutrition. Micronutrient deficiencies among individuals in developing nations are often attributed to inadequate levels of these essential elements in staple foods, such as rice or wheat in Asia, and maize or sorghum in Africa (Fang et al. 2008; Qureshi et al. 2021). Biofortification offers a promising solution by enhancing the nutritional content of plant-based foods and feeds, providing a cost-effective, sustainable, and enduring method of delivering micronutrients to those in need (Simnadis et al. 2016). Biofortification, the process of enhancing the bioavailable levels of vital nutrients in the edible parts of crop plants through agricultural techniques or genetic selection, offers a potential solution for mitigating malnutrition or hidden hunger (Singh et al. 2016).

In the context of human nutrition, minerals refer to inorganic chemical elements or their dissociated ions necessary for biological or biochemical processes, including electrolyte balance. Excluded from this category are carbon, hydrogen, nitrogen, and oxygen, as they are commonly found in organic molecules. While there are 16 essential minerals, 11 of them are required in such minute amounts or are so abundant in food and water that deficiencies are rare. The remaining five minerals, viz., iodine (I), iron (Fe), zinc (Zn), calcium (Ca), and selenium (Se), are present in limited quantities in many foods, making deficiency more likely with a monotonous diet (Singh et al. 2016). Addressing micronutrient malnutrition has historically posed a significant challenge for humanity, as a substantial portion of 2/third of the global population may lack one or more essential mineral elements (White and Broadley 2009; Stein 2010; Qureshi et al. 2021). Utilizing agricultural strategies like plant breeding and agronomic biofortification to enhance the micronutrient content of staple cereals represents a valuable, cost-effective, and sustainable approach for addressing deficiencies in human populations (Zou et al. 2012; Qureshi et al. 2021; Bouis and Saltzman 2017). This method offers a long-term solution to

Table 3.1 Nutritive value of sorghum

Macronutrient	%	Mineral	µg/g	Vitamins	mg/100 g
Carbohydrate	70–80	Calcium	132	Niacin	2.927
Protein	8–18	Potassium	2874	Riboflavin	0.142
Lipids	1–5	Magnesium	1496	Thiamin	0.237
Dietary fiber	19	Phosphorous	286	Vitamin B-6	0.59
Crude fiber	3			Vitamin E	0.81

Source: Tanwar et al. (2023)

nutritional deficiencies by leveraging natural genetic diversity to cultivate crop plants with higher mineral content (Qureshi et al. 2021). The nutritive value of sorghum is listed in Table 3.1 (Tanwar et al. 2023).

3.3 Nutritional Traits in Sorghum

Sorghum boasts a unique nutritional composition characterized by a balance of macronutrients, micronutrients, and bioactive compounds. This ancient grain has been a dietary staple for centuries, particularly in regions like Africa, India, and the United States. Renowned for its resilience and nutritional richness, sorghum offers a plethora of health benefits, making it an invaluable component of human and animal diets alike. Here's a closer look at its key attributes.

3.3.1 Starch

Starch, a complex carbohydrate composed of glucose units, serves as a vital energy source for humans and many other organisms worldwide. It is abundant in various plant sources, with cereal grains being primary reservoirs of this polysaccharide. The digestibility of starch, crucial for deriving energy from dietary sources, is influenced by multiple factors, including the properties of the endosperm within grains and the methods used for processing. Sorghum, recognized for its nutritional virtues, presents a noteworthy composition of starch, distinct from other cereal grains. Notably, sorghum possesses higher levels of resistant starch (RS) and slowly digestible starch (SDS), components that have garnered attention for their potential health benefits (Poquette et al. 2014; Teferra and Awika 2019). This abundance of RS and SDS not only fosters prolonged satiety but also potentially reduces snacking frequency and overall caloric intake, thereby contributing to a lower glycemic index and diminishing the risk of chronic conditions like type two diabetes and obesity (Anunciação et al. 2017; Teferra and Awika 2019). Although the precise mechanisms underlying sorghum starch's slow digestion are not fully understood, hypotheses point to sorghum polyphenols and endosperm protein properties. This distinctive feature of sorghum-founded diets in managing hyperglycemia positions sorghum among the select "ancient grain" crops prized for their established health benefits, aligning with the preferences of health-conscious consumers (Dunn et al. 2015; Simnadis et al. 2016; Taylor and Awika 2017; Teferra and Awika 2019).

Carbohydrates, especially nitrogen-free extracts like starch and sugar, serve as the primary energy source in sorghum, crucial for the nutrition of livestock and poultry. The carbohydrate content in nitrogen-free extracts within sorghum can vary widely, ranging from 17.4% to 71.2%. Notably, sorghum stalks and husks contain higher crude fiber levels, with respective carbohydrate contents averaging between 23.8% and 26.4% (Bai et al. 2017; Xu 2019). While sorghum's starch content matches that of corn, its starch granules have higher protein content, resulting in lower digestibility compared to corn. Consequently, sorghum's effective energy value is estimated to be 90–95% that of corn. Despite the nutritional assessment of sorghum stalks and husks being inferior to concentrates, they offer abundant sources at lower prices, thereby reducing feed costs (Xu 2019; Birhanu 2021).

3.3.2 Proteins

Sorghum proteins, referred to as kafirins, belong to the prolamins group of grain proteins. They exhibit higher hydrophobicity compared to maize zein protein, rice proteins, and wheat prolamins. Sorghum kafirins consist of three main subclasses grounded on molecular weight: α-, β-, and γ-kafirins, with relative abundances of 80%, 13%, and 7%, respectively (Winn et al. 2009; Teferra and Awika 2019). These kafirin protein bodies are structured such that the more protease-resistant and hydrophobic β- and γ-kafirins encapsulate the more digestible and less hydrophobic α-kafirin component (Belton et al. 2006; Teferra and Awika 2019).

The digestibility of proteins in sorghum is considerably lower, ranging from 40% to 60%, compared to proteins found in other cereal grains. This digestibility decreases even further during cooking, primarily due to the formation of disulfide cross-links within the kafirin proteins (Teferra and Awika 2019). This relatively poor protein quality diminishes sorghum's nutritional value, particularly for populations relying on sorghum as a staple food source. However, sorghum protein demonstrates higher biological value and digestibility than wheat protein (Labuschagne 2018; Xu 2019). Like many other cereal grains, sorghum contains relatively low levels of certain essential amino acids, with lysine being the most deficient (Aboubacar et al. 2001; Duodu et al. 2003; Teferra and Awika 2019).

The protein content in sorghum grain typically falls between 4.4% and 21.1%, slightly exceeding that of corn. Compared to corn, sorghum grains have slightly higher levels of leucine and valine but slightly lower levels of arginine. Sorghum bran contains approximately 10% protein, while fresh sorghum distillers' grains and fresh sorghum vinegar residue have protein contents of 9.3% and 8.5%, respectively. In contrast, the protein content in sorghum stalks and husks is relatively low, at 3.2% and 2.2%, respectively (Labuschagne 2018; Xu 2019).

3.3.3 Vitamins and Minerals

Sorghum grain boasts a significant abundance of β-carotene (the precursor to vitamin A) and tocopherols, serving as a basis of lipid-soluble vitamins such as A, D, and K, as well as water-soluble vitamins like thiamin, riboflavin, and pyridoxine (Martino et al. 2012). Minerals such as magnesium, zinc, iron, copper, calcium, potassium, and phosphorus are richly present in the sorghum grain pericarp, aleurone layer, and germ (Glew et al. 1997; Anglani 1998). These essential vitamins and minerals are primarily found in the aleurone and germ, while lipids are concentrated in the germ. Therefore, removing the outer layers of sorghum grain during processing results in the loss of crucial nutrients.

In terms of mineral content, sorghum's calcium and phosphorus levels are comparable to those of corn, with phosphorus primarily present as phytate phosphorus, comprising approximately 40–70%. Vitamins B1 and B6 contents match those of corn, while pantothenic acid, niacin, and biotin levels surpass those of corn; however, the absorption rates of niacin and biotin are low. Sorghum grains contain approximately 1.4 mg of thiamine (vitamin B1), 0.7 mg of riboflavin (vitamin B2), and 6 mg of niacin per kilogram (Xu 2019; Khoddami et al. 2023). Carotene is also present in sorghum grains, stems, and leaves, particularly when used as green forage or silage. Tannins, water-soluble polyphenol compounds, are primarily present in the seed coat and pericarp of sorghum grains. These tannins impart a strong bitter taste and can hinder food digestion and absorption, potentially causing constipation. To mitigate the adverse effects of tannins, thorough cleaning and milling of sorghum grains are recommended. Soaking and boiling the grains in water can improve taste and reduce their impact on the body (Xu 2019; Khoddami et al. 2023).

3.3.4 Dietary Fiber

Dietary fiber is an essential component of sorghum (3–8%), contributing to its nutritional value and potential health benefits. Sorghum contains both soluble dietary fiber (SDF) and insoluble dietary fiber (IDF), which play important roles in digestive health, blood sugar regulation, and cholesterol management. Sorghum dietary fiber is primarily composed of insoluble dietary fiber (IDF), making up 95% of the total fiber content, predominantly found in the bran. Soluble dietary fiber (SDF) in sorghum is minimal, constituting about 5% of the total fiber, with similar levels in bran and whole sorghum powder but significantly lower in decorticated flour. Extended milling reduces SDF content in decorticated sorghum flour. IDF, being less soluble, faces challenges in food processing and competes with other insoluble dietary fibers. High production costs, limited applications, and underdeveloped preparation techniques hinder sorghum dietary fiber utilization and research. Total dietary fiber, IDF, SDF, and β-glucan are inversely related to sorghum flour digestibility, resulting in lower estimated glycemic index values (Moraes et al. 2015).

Sorghum dietary fiber exhibits water retention and swelling properties, with water-holding capacity of 7–8 ml/g, expansion force of 7–8 g/g, and fatty acid absorption capacity of 2–3 g/g for unsaturated and 5–6 g/g for saturated fatty acids and 6–7 g/g for cholesterol. These properties are influenced by temperature, pH, and ionic concentration. Increased temperature enhances water-holding capacity and expansibility of IDF, while pH affects hydrogen bond formation and ion interactions, influencing fiber characteristics (Olayinka et al. 2011).

3.3.5 *Gluten-Free*

Gluten, extracted from flour through washing, comprises about 65% water and 75–86% protein on a dry basis, with the remaining portion consisting of carbohydrates and lipids held within its matrix. Its unique characteristics, like extensibility and gas-holding ability, make it favored in various food products.

In recent years, there has been a substantial surge in the demand for gluten-free products, propelled by consumer preferences and health considerations. Gluten-free goods are in significant demand from various countries worldwide. Adverse reactions to gluten, such as celiac disease and wheat allergy, have led to dietary restrictions and the popularity of gluten-free diets.

Removing gluten poses challenges due to its widespread use in the food industry for its technological qualities. Gluten-free bread production faces difficulties in replicating gluten's functionality, resulting in irregular and crumbly textures. Various substitutes such as rice, sorghum, and corn flours and others have been utilized, but replicating gluten's functionality remains a significant challenge (Pineli et al. 2015; Adiamo et al. 2018).

3.3.6 *Phytochemicals and Antioxidants*

Sorghum's phytochemicals and antioxidants, primarily in the bran and endosperm, include vitamins (B, D, E, K), polyphenols, carotenoids, and trace elements (zinc, copper, manganese). Phenolic composites like ferulic acid offer antioxidant, anti-inflammatory, and anticancer effects, while tannins, particularly proanthocyanidins, bolster health benefits (Jideani et al. 2014; Tanwar et al. 2023). Flavonoids such as flavones and flavanones exhibit antioxidant and anti-inflammatory properties. Stilbenes contribute to antioxidant and anti-inflammatory effects, reducing chronic disease risk. Phytosterols and policosanols contribute to cardiovascular health by lowering cholesterol. However, anti-nutritional substances like phytates hinder nutrient absorption (Budhwar et al. 2020; Tanwar et al. 2023). Sorghum extracts inhibit diabetes-related enzymes, enhance glucose metabolism, and lower hyperglycemia risk. Phenolic extracts act as antibacterial agents against pathogens like Salmonella and *E. coli*, enhancing food safety (Irondi et al. 2019; Tanwar et al. 2023).

Bioactive compounds possess anticancer properties by targeting cancer symptoms and reducing oxidative stress. Sorghum extracts prevent triglyceride buildup, aiding weight loss and regulating cholesterol synthesis, reducing cardiovascular disease risk. Phenolic substances inhibit inflammatory chemical production, offering protection against chronic inflammation-related diseases. Sorghum's nutritional benefits extend to potential use in preventing and treating diarrhea, alleviating gastrointestinal symptoms (Tanwar et al. 2023).

3.3.7 Glycemic Index (GI)

Obesity and diabetes pose significant global health challenges. Quickly digested carbohydrates can cause rapid spikes in blood glucose and insulin levels, exacerbating diabetes-related complications. The glycemic index (GI) categorizes carbohydrate-containing foods based on their impact on blood sugar levels, with lower GI foods preferable for stable glucose levels. The Estimated Glycemic Index (EGI) estimates GI based on starch digestibility rates. Low GI foods with higher resistant starch (RS) content slow carbohydrate absorption, preventing blood glucose spikes. RS, undigested in the small intestine, supports colonic bacteria fermentation, producing short-chain fatty acids (SCFA) and lowering colonic pH. RS's properties may condense the risk of colon cancer, obesity, diabetes, and inflammatory bowel disease (Baghurst et al. 1996; Hendrich 2010; Lemlioglu-Austin et al. 2012).

3.3.8 Lipids

Sorghum grains typically contain low lipid levels (1.24–3.07 g/100 g), with a predominance of unsaturated fatty acids (83–88%). Polyunsaturated fatty acids (PUFA) generally outweigh monounsaturated fatty acids (MUFA) in sorghum varieties, with linoleic acid being the most abundant (45.6–51.1%), followed by oleic acid (32.2–42.0%), palmitic acid (12.4–16.0%), and linolenic acid (1.4–2.8%). These lipids, mainly concentrated in the scutellum region of the grain, contribute to sorghum's distinctive fatty acid profile, with linoleic and oleic acids being major components (Zbasnik et al. 2009; de Morais Cardoso et al. 2017; Tanwar et al. 2023).

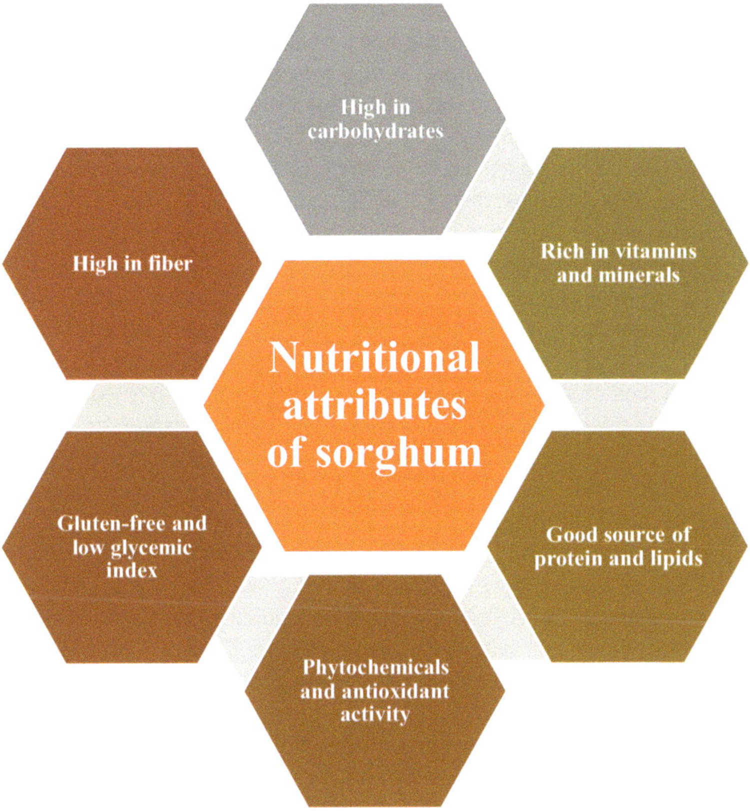

3.4 G × E Interaction in Nutritional Traits

Mg, Mn, Fe, and P contents were reported to be significantly different among the two locations studied, which indicated that the strong influence of environmental factors, viz., precipitation, temperature, soil nutrient composition, and solar radiation, on the concentration of these nutrients in sorghum grains (Motlhaodi et al. 2018). Thus, the selection of sorghum cultivars for higher contents of the above minerals could be done for each location separately. Genotype × environment interaction revealed that the highest G × E interaction was observed for Mn, while Ca, Fe, K, Mg, and P showed high G × E, ranging between 20% and 25% of the total variation (Motlhaodi et al. 2018). The G × E interaction data are important to define environments and select locations for nutritional breeding purposes, especially when targeting simultaneous improvement of grain mineral concentrations such as Mg, Fe, and Zn.

Zinc (Zn) is a vital micronutrient essential for growth and overall health. Addressing micronutrient deficiencies can be achieved through the depletion of biofortified food grains. Sorghum, a staple cereal in arid and semi-arid regions, plays a substantial role in meeting the nutritional needs of vulnerable populations. Zinc is indispensable for numerous bodily functions, including immune system function, genetic material synthesis, and cell integrity maintenance, as it serves as a catalyst for over 100 enzymes. However, genotype × environment interaction (G × E) can lead to inconsistencies in micronutrient expression across different growing conditions (Madhusudhana et al. 2023).

Genotype × environment interaction refers to the variability in genotype performance across different environmental conditions. This interaction occurs when genotypes exhibit varying responses to environmental changes, which can either be cross-over or non-cross-over in nature. Breeders typically prioritize genotypes that demonstrate consistent performance across favourable and unfavorable environments when recommending cultivars. Genotypes displaying cross-over interaction can complicate selection processes in breeding programs. Despite sorghum's adaptability to diverse environments where other cereals struggle, breeders focus on identifying cultivars with stable and optimal performance across assortment of conditions, particularly in terms of yield (Annicchiarico 2002; Mofokeng et al. 2017).

Evaluating the consistency of grain micronutrient expression is decisive for optimizing the use of identified genotypes in breeding initiatives. Successful breeding for grain micronutrient enrichment relies on adequate genetic variability for zinc (Zn) and iron (Fe) content within the germplasm pool, along with comprehensive knowledge of the genetic mechanisms governing micronutrient accumulation in seeds (Hariprasanna et al. 2012).

Several methods are utilized to analyze genotype × environment interaction. These include linear regression (Finlay and Wilkinson 1963; Eberhart and Russell 1966; Freeman and Perkins 1972), additive main effects and multiplicative interaction (AMMI) (Alanís et al. 2010), and genotype by environment (GGE) biplot analysis (Yan et al. 2007; Mofokeng et al. 2017). The AMMI method is effective in distinguishing genotype, environment, and genotype by environment interactions for breeding purposes while minimizing noise.

GGE biplot analysis has been employed to evaluate hybrids for genotype by environment interaction, especially in cases where the AMMI method has limitations (Mofokeng et al. 2017; Yan et al. 2007). GGE biplot analysis can evaluate both genotypes and environments, and it is particularly useful for identifying genotypes adapted to specific environments and analyzing mega-environments (Yan et al. 2007; Mofokeng et al. 2017).

3.5 Conclusion

In conclusion, the genotype–environment interaction (GEI) affecting grain nutritional traits in sorghum represents a dynamic and multifaceted area of research with profound implications for global food security and human health. The intricate interplay between genetic factors and environmental conditions shapes the nutritional composition of sorghum grains, influencing their suitability as a staple food and their potential to address malnutrition. As we navigate the complexities of GEIs in sorghum, it is evident that continued research and innovation are essential to unlock the full potential of this resilient cereal crop. By leveraging advances in genomic technologies, precision breeding techniques, and agronomic practices, we can develop biofortified sorghum varieties that are tailored to meet the nutritional needs of diverse populations and thrive in varying environmental conditions. Ultimately, the study of GEIs in sorghum offers hope for a future where food systems are more resilient, equitable, and nourishing. By harnessing the power of genetics, environmental science, and innovation, we can pave the way toward a healthier and more sustainable food future for generations to come.

3.6 Future Prospects

The genotype–environment interactions (GEIs) affecting grain nutritional traits in sorghum hold significant promise for addressing global malnutrition challenges and improving food security. As research in this field continues to evolve, several key areas are poised for advancement and innovation:

(i) *Precision Breeding for Biofortification*: With advancements in genomic technologies, precision breeding techniques can be further refined to develop sorghum varieties with enhanced nutritional traits, including increased iron, zinc, and protein content. By leveraging genomic data and understanding GEIs, breeders can target specific genomic regions associated with desirable nutritional traits and incorporate them into breeding programs to develop biofortified sorghum varieties.

(ii) *Climate-Resilient Sorghum Varieties*: Climate change poses significant challenges to agricultural productivity, including the nutritional quality of crops. Future research on GEIs can focus on developing climate-resilient sorghum varieties that maintain optimal nutritional traits under changing environmental conditions. By understanding how different genotypes respond to varying climatic factors, breeders can develop sorghum varieties that are resilient to heat, drought, and other environmental stressors while retaining their nutritional value.

(iii) *Integrated Crop Management Strategies*: In addition to genetic approaches, future research can explore integrated crop management strategies that optimize soil health, nutrient availability, and water use efficiency to enhance the nutritional quality of sorghum grains. By integrating agronomic practices such as precision fertilization, irrigation management, and soil amendments with genetic approaches, sorghum production systems can be optimized to produce grains with improved nutritional profiles.

(iv) *Multidisciplinary Collaborations*: Addressing the complex challenges associated with GEIs in sorghum requires multidisciplinary collaborations among plant breeders, agronomists, geneticists, nutritionists, and policymakers. Future research initiatives can foster collaborations across different disciplines to integrate genetic, environmental, and nutritional insights into holistic approaches for improving the nutritional quality of sorghum grains and promoting their adoption at scale.

(v) *Consumer Acceptance and Market Development*: As biofortified sorghum varieties become available, efforts to promote consumer acceptance and market development will be crucial for their successful adoption. Future research can focus on understanding consumer preferences, addressing barriers to adoption, and developing marketing strategies to promote the consumption of biofortified sorghum products. Additionally, partnerships with stakeholders across the food value chain can facilitate the integration of biofortified sorghum into diverse food products and distribution channels.

References

Aboubacar A, Axtell JD, Huang CP, Hamaker BR (2001) A rapid protein digestibility assay for identifying highly digestible sorghum lines. Cereal Chem 78(2):160–165

Adiamo OQ, Fawale OS, Olawoye B (2018) Recent trends in the formulation of gluten-free sorghum products. J Culin Sci Technol 16(4):311–325

Alanís H, Pecina Quintero V, Zavala García F, Montes García N, Gámez Vázquez J, Arcos Cavazos G, Alcalá Salinas L (2010) Finlay and Wilkinson model vs. AMMI model in the analysis of genotype-environment interaction in sorghum. Rev Fitotec Mex 33(2):117–123

Anglani C (1998) Sorghum for human food–a review. Plant Foods Hum Nutr 52:85–95

Annicchiarico P (2002) Genotype x environment interactions: challenges and opportunities for plant breeding and cultivar recommendations. FAO, Rome

Anunciação PC, de Morais Cardoso L, Gomes JVP, Della Lucia CM, Carvalho CWP, Galdeano MC, Pinheiro-Sant'Ana HM (2017) Comparing sorghum and wheat whole grain breakfast cereals: sensorial acceptance and bioactive compound content. Food Chem 221:984–989

Baghurst PA, Baghurst KI, Record SJ (1996) Dietary fiber, non-starch polysaccharides and resistant starch: a review. Food Aust 48:3–35

Bai C, Wang C, Wang P, Zhu Z, Cong L, Li D, Lu X (2017) QTL mapping of agronomically important traits in sorghum (*Sorghum bicolor* L.). Euphytica 213:1–12

Bänziger M, Edmeades GO, Beck D, Bellon M (2006) Breeding for drought and nitrogen stress tolerance in maize: from theory to practice. CIMMYT

Belton PS, Delgadillo I, Halford NG, Shewry PR (2006) Kafirin structure and functionality. J Cereal Sci 44(3):272–286

Birhanu S (2021) Potential benefits of sorghum [*Sorghum bicolor* (L.) Moench] on human health: a review. Int J Food Eng Technol 5(1):8–18

Bouis HE, Saltzman A (2017) Improving nutrition through biofortification: a review of evidence from Harvest Plus, 2003 through 2016. Glob Food Sec 12:49–58

Budhwar S, Sethi K, Chakraborty M (2020) Efficacy of germination and probiotic fermentation on underutilized cereal and millet grains. Food Prod Process Nutr 2:1–17

Cullis B, Smith A, Coombes N (2010) On the design of early generation variety trials with correlated data. J Agric Biol Environ Stat 15(3):334–350

de Morais Cardoso L, Pinheiro SS, Martino HSD, Pinheiro-Sant'Ana HM (2017) Sorghum (Sorghum bicolor L.): nutrients, bioactive compounds, and potential impact on human health. Crit Rev Food Sci Nutr 57(2):372–390

Dunn KL, Yang L, Girard A, Bean S, Awika JM (2015) Interaction of sorghum tannins with wheat proteins and effect on in vitro starch and protein digestibility in a baked product matrix. J Agric Food Chem 63(4):1234–1241

Duodu KG, Taylor JRN, Belton PS, Hamaker BR (2003) Factors affecting sorghum protein digestibility. J Cereal Sci 38(2):117–131

Eberhart ST, Russell WA (1966) Stability parameters for comparing varieties. Crop Sci 6(1):36–40

Fang Y, Wang L, Xin Z, Zhao L, An X, Hu Q (2008) Effect of foliar application of zinc, selenium, and iron fertilizers on nutrients concentration and yield of rice grain in China. J Agric Food Chem 56(6):2079–2084

Finlay KW, Wilkinson GN (1963) The analysis of adaptation in a plant-breeding programme. Aust J Agric Res 14(6):742–754

Freeman GH, Perkins JM (1972) Environmental and genotype-environmental components of variability. VIII. Relations between genotypes grown in different environments and measures of these environments. Heredity. https://doi.org/10.1038/HDY.1971.67

Glew RH, Vander Jagt DJ, Lockett C, Grivetti LE, Smith GC, Pastuszyn A, Millson M (1997) Amino acid, fatty acid, and mineral composition of 24 indigenous plants of Burkina Faso. J Food Compos Anal 10(3):205–217

Hariprasanna K, Agte VL, Patil JV (2012) Genotype× environment interactions for grain micronutrient contents in sorghum [*Sorghum bicolor* (L.) Moench]. Indian J Genet Plant Breed 72(04):429–434

Hendrich S (2010) Battling obesity with resistant starch. Food Technol 64(3):22–30

Irondi EA, Adegoke BM, Effion ES, Oyewo SO, Alamu EO, Boligon AA (2019) Enzymes inhibitory property, antioxidant activity and phenolics profile of raw and roasted red sorghum grains in vitro. Food Sci Human Wellness 8(2):142–148

Jideani AI, Silungwe H, Takalani T, Anyasi TA, Udeh H, Omolola A (2014) Antioxidant-rich natural grain products and human health. In: Oguntibeju O (ed) Antioxidant-antidiabetic agents and human health. InTech Publisher, Rijeka, pp 167–187

Khoddami A, Messina V, Vadabalija Venkata K, Farahnaky A, Blanchard CL, Roberts TH (2023) Sorghum in foods: functionality and potential in innovative products. Crit Rev Food Sci Nutr 63(9):1170–1186

Labuschagne MT (2018) A review of cereal grain proteomics and its potential for sorghum improvement. J Cereal Sci 84:151–158

Lemlioglu-Austin D, Turner ND, McDonough CM, Rooney LW (2012) Effects of sorghum [*Sorghum bicolor* (L.) Moench] crude extracts on starch digestibility, estimated glycemic index (EGI), and resistant starch (RS) contents of porridges. Molecules 17(9):11124–11138

Madhusudhana R, Hariprasanna K, Aruna C, Sajjanar GM, Hanamaratti NG, Sameera S, Tonapi VA (2023) Genetic variability, G× E interaction and stability for iron and zinc content in sorghum grains in advanced breeding lines. J Cereal Sci 110:103653

Martino HSD, Tomaz PA, Moraes EA, Conceição LD, Oliveira DDS, Queiroz VAV, Ribeiro SMR (2012) Chemical characterization and size distribution of sorghum genotypes for human consumption. Rev Inst Adolfo Lutz 71(2):337–344

Mofokeng AM, Shimelis H, Laing M (2017) Breeding strategies to improve sorghum quality. Aust J Crop Sci 11(2):142–148

Moraes ÉA, da Silva Marineli R, Lenquiste SA, Steel CJ, de Menezes CB, Queiroz VAV, Júnior MRM (2015) Sorghum flour fractions: correlations among polysaccharides, phenolic compounds, antioxidant activity and glycemic index. Food Chem 180:116–123

Motlhaodi T, Bryngelsson T, Chite S, Fatih M, Ortiz R, Geleta M (2018) Nutritional variation in sorghum [*Sorghum bicolor* (L.) Moench] accessions from southern Africa revealed by protein and mineral composition. J Cereal Sci 83:123–129

Olayinka OO, Olu-Owolabi BI, Adebowale KO (2011) Effect of succinylation on the physicochemical, rheological, thermal and retrogradation properties of red and white sorghum starches. Food Hydrocoll 25(3):515–520

Ortiz-Monasterio JI, Palacios-Rojas N, Meng E, Pixley K, Trethowan R, Pena RJ (2007) Enhancing the mineral and vitamin content of wheat and maize through plant breeding. J Cereal Sci 46(3):293–307

Phuke RM, Anuradha K, Radhika K, Jabeen F, Anuradha G, Ramesh T, Hariprasanna K, Mehtre SP, Deshpande SP, Anil G, Das RR (2017) Genetic variability, genotype× environment interaction, correlation, and GGE biplot analysis for grain iron and zinc concentration and other agronomic traits in RIL population of sorghum (*Sorghum bicolor* L. Moench). Front Plant Sci 8:712

Pineli LDLDO, Zandonadi RP, Botelho RBA, Oliveira VRD, Figueiredo LFDA (2015) The use of sorghum to produce gluten-free breads: a systematic review. J Adv Nut Hum Metab. https://doi.org/10.14800/janhm.944

Poquette NM, Gu X, Lee SO (2014) Grain sorghum muffin reduces glucose and insulin responses in men. Food Funct 5(5):894–899

Qureshi MT, Ahmad MF, Iqbal N, Waheed H, Hussain S, Brestic M, Noorka IR (2021) Agronomic bio-fortification of iron, zinc and selenium enhance growth, quality and uptake of different sorghum accessions. Plant Soil Environ. https://doi.org/10.17221/137/2021-PSE

Rao PP, Birthal PS, Reddy BV, Rai KN, Ramesh S (2006) Diagnostics of sorghum and pearl millet grains-based nutrition in India. Int Sorghum Millets Newsl 47:93–96

Simnadis TG, Tapsell LC, Beck EJ (2016) Effect of sorghum consumption on health outcomes: a systematic review. Nutr Rev 74(11):690–707

Singh U, Praharaj CS, Chaturvedi SK, Bohra A (2016) Biofortification: introduction, approaches, limitations, and challenges. In: Biofortification of food crops, pp 3–18. https://doi.org/10.1007/978-81-322-2716-8_1

Stein AJ (2010) Global impacts of human mineral malnutrition. Plant Soil 335(1–2):133–154

Tanwar R, Panghal A, Chaudhary G, Kumari A, Chhikara N (2023) Nutritional, phytochemical and functional potential of sorghum: a review. Food Chem Adv 3:100501

Taylor JR, Awika J (2017) Gluten-free ancient grains: cereals, pseudocereals, and legumes: sustainable, nutritious, and health-promoting foods for the 21st century. Woodhead Publishing

Teferra TF, Awika JM (2019) Sorghum as a healthy global food security crop: opportunities and challenges. Cereal Foods World 64(5):1–8

Velu G, Crossa J, Singh RP, Hao Y, Dreisigacker S (2018) Genetic dissection of grain zinc concentration in spring wheat for mainstreaming biofortification in CIMMYT wheat breeding. Sci Rep 8(1):1–12

White PJ, Broadley MR (2009) Biofortification of crops with seven mineral elements often lacking in human diets–iron, zinc, copper, calcium, magnesium, selenium and iodine. New Phytol 182(1):49–84

WHO (2023) Malnutrition. https://www.who.int/news-room/fact-sheets/detail/malnutrition. Accessed 27 Jan 2024

Winn JA, Mason RE, Robbins AL, Rooney WL, Hays DB (2009) QTL mapping of a high protein digestibility trait in *Sorghum bicolor*. Int J Plant Genomics 2009:471853

Xu T (2019) Sorghum. In: Bioactive factors and processing technology for cereal foods, pp 103–135

Yan W, Kang MS, Ma B, Woods S, Cornelius PL (2007) GGE biplot vs. AMMI analysis of genotype-by-environment data. Crop Sci 47(2):643–653

Zbasnik R, Carr T, Weller C, Hwang KT, Wang L, Cuppett S, Schlegel V (2009) Antiproliferation properties of grain sorghum dry distiller's grain lipids in Caco-2 cells. J Agric Food Chem 57(21):10435–10441

Zou CQ, Zhang YQ, Rashid A, Ram H, Savasli E, Arisoy RZ, Cakmak I (2012) Biofortification of wheat with zinc through zinc fertilization in seven countries. Plant Soil 361:119–130

Chapter 4
Sorghum Genomics for Nutritional Traits Improvement

N. G. Hanamaratti, S. E. Diwakar Reddy, R. Karthik, D. S. Shailaja, K. P. Gaganashree, and Ranjitha Marigoudar

Abstract Sorghum (*Sorghum bicolor*) is an important millet crop, particularly in drought affected regions. Enhancing its nutritional profile is vital for addressing global malnutrition challenges. This chapter unveils the present state of sorghum genomic research focused on enhancing key nutritional attributes. Modern genomic studies have identified significant genetic variation in sorghum accessions, revealing loci associated with improved grain nutritional content, including protein, zinc, calcium, and manganese. Researchers are emphasizing the importance of genetic diversity in sorghum breeding programs. Utilizing molecular markers, studies have demonstrated the efficacy in assessing genetic diversity, which is crucial for developing nutritionally enriched sorghum varieties. The combination of genomic tools with traditional breeding methods accelerates the development of sorghum varieties with enhanced nutritional value. These prospects also include the application of advanced gene-editing techniques such as CRISPR–Cas9 to precisely improve nutritional traits. These advancements hold significant promise for addressing malnutrition challenges in resource-limited regions.

Keywords Nutritional genomics · Biofortified sorghum · Functional genomics · Micronutrient

N. G. Hanamaratti (✉)
AICRP Sorghum, MARS, UAS Dharwad, Dharwad, Karnataka, India
e-mail: hanamaratting@uasd.in

S. E. Diwakar Reddy · R. Karthik · D. S. Shailaja · K. P. Gaganashree · R. Marigoudar
Department of Genetics and Plant Breeding, UAS, Dharwad, Dharwad, Karnataka, India

M. Govindaraj (ed.), *Breeding Biofortified Millets for Nutritional Security*, Sustainability Sciences in Asia and Africa,
https://doi.org/10.1007/978-981-95-9827-4_4

4.1 Introduction and Importance of Nutritional Improvement in Sorghum

Millets have traditionally served as staple foods in many developing regions of the world. They are cultivated in nearly 93 countries, with developing nations contributing more than 97% of global millet production and consumption (Meena et al. 2021). Recognizing their importance in ensuring food and nutritional security, the Government of India proposed to the United Nations that the year 2023 be declared the International Year of Millets (IYOM). This initiative aimed to enhance public awareness of millets, promote sustainable production systems, improve grain quality, and stimulate greater investment in research and development for climate-resilient and nutritionally enriched millet crops. On March 5, 2021, the United Nations General Assembly (UNGA), with the endorsement of 72 member countries, officially proclaimed 2023 as the International Year of Millets. Among the diverse millet species, sorghum (*Sorghum bicolor* L. Moench), often referred to as the "great millet," is distinguished by its exceptional adaptability to adverse climatic conditions and its substantial nutritional potential.

Sorghum is a key contributor to global food security, particularly in regions characterized by harsh agro-climatic conditions such as sub-Saharan Africa and large parts of Asia (Amelework et al. 2016). Its inherent tolerance to drought and high temperatures enables reliable cultivation where other cereals frequently fail, making it a primary food and fodder source for millions of people. Globally, sorghum ranks fifth among cereals after wheat, rice, maize, and barley. Its ability to perform well under low rainfall and high-temperature regimes underscores its importance in arid and semi-arid ecosystems, where it can support both human nutrition and livestock production. Despite its agronomic resilience and wide cultivation, sorghum has certain nutritional limitations that restrict its overall contribution to human diets. Compared to major cereals such as wheat and maize, sorghum generally exhibits lower protein content, which could be problematic in regions where dietary protein intake is already inadequate. In addition, sorghum's main storage protein kafirins are deficient in a few essential amino acids, particularly lysine, thereby reducing their biological value (Proietti et al. 2015). The grain is also often low in critical micronutrients such as iron and zinc, whose insufficiencies are associated with serious health concerns in populations that rely heavily on sorghum-based diets. Furthermore, the presence of anti-nutritional compounds, including tannins and phytates in sorghum, impairs nutrient bioavailability and reduces digestibility (Hariprasanna et al. 2015). These constraints highlight the need for focused nutritional improvement in sorghum to fully realize its potential as a climate-resilient and nutritionally valuable cereal crop.

Sorghum is well suited for crop diversification due to its exceptional ability to withstand harsh and resource-limited environments. One of its major advantages is superior water-use efficiency; sorghum requires approximately 332 kilograms of water to produce one kilogram of total dry matter, which is substantially lower than the water requirement of wheat (514 kg), maize (368 kg), and barley (434 kg) (Rao Srinivasa et al. 2014). Studies have indicated that replacing water-intensive crops such as rice with more water-efficient cereals, including sorghum, finger millet, and pearl millet, could reduce India's overall agricultural water use by nearly 33%, while maintaining productivity under drought-prone conditions (Davis et al. 2018). Even though noteworthy efforts have been made to understand variability and inheritance of nutritional traits in sorghum (Hariprasanna et al. 2014a, b; Kamal et al. 2023; Welderufael et al. 2024), and substantial progress has been made through conventional breeding, genetic improvement has remained relatively low. This limitation largely arises because most nutritional and adaptive traits are quantitatively inherited and strongly influenced by environmental factors (Madhusudhana et al. 2023; Welderufael et al. 2024). The integration of molecular breeding approaches with traditional selection methods can significantly enhance selection efficiency and precision, thereby accelerating the development of climate-resilient, high-yielding sorghum cultivars capable of meeting the nutritional demands of a rapidly growing global population (Fig. 4.1).

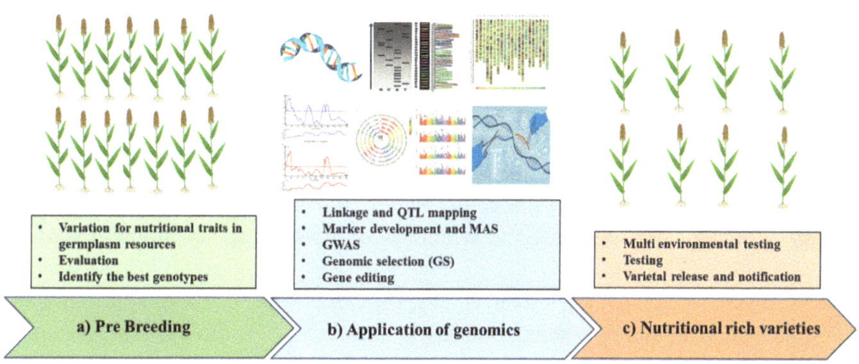

Fig. 4.1 Flow chart illustrating the process of developing nutritional rich sorghum varieties that involves a multi-step, integrated approach. (**a**) "Pre Breeding" – Extensive screening of gene bank accessions and germplasm resources to identify superior genotypes combining enhanced nutritional attributes with desirable yield performance. (**b**) "Application of genomics" – Advanced genomic approaches, like linkage mapping, QTL analysis and GWAS are employed to identify genomic regions controlling key nutritional traits. The identified loci facilitate the development of robust molecular markers, which can be effectively utilized in MAS and genomic selection, and for gene-editing for precise improvement of nutritional quality. (**c**) "Nutritional rich varieties" – The identified nutritionally enriched sorghum genotypes are evaluated in multi-environment trials to assess their stability, adaptability, and agronomic performance, ensuring suitability for varietal release and notification

4.2 Genetic Improvement

Genetic improvement of nutritional traits in sorghum is a complex and time-intensive process that requires harmonized efforts among plant breeders, geneticists, agronomists, and nutrition scientists. The overarching objective is to develop sorghum cultivars that combine enhanced productivity with improved nutritional quality, thereby contributing meaningfully to human health in regions where sorghum forms a major dietary staple. Extensive research has documented substantial genetic variability in sorghum for key nutritional attributes, including protein content, which ranges from approximately 3.5% to 18.79% (Badigannavar et al. 2016; Boyles et al. 2017; Karadi and Kajjidoni 2019; Bodede and Mabelebele 2022), and starch content, reported to vary between 45.1% and 75.7% (Boyles et al. 2017; Sejake et al. 2022). Similarly, widespread variation for grain iron (7-168 mg/kg) and zinc (3-91 mg/kg) concentrations is seen in sorghum germplasm (Table 4.1). Further, for other macro- and micronutrients such as magnesium, calcium, sodium, phosphorus, and potassium (Bodede and Mabelebele 2022; Kamal et al. 2023). In addition, sorghum exhibits diversity for provitamin A content (Cruet-Burgos et al. 2023) and possesses a relatively balanced amino acid composition.

Despite this rich genetic diversity, progress in enhancing grain nutritional quality through traditional breeding has been modest. Targeted efforts to simultaneously increase grain iron and zinc levels have, however, led to the development and release of India's first biofortified sorghum variety, Parbhani Shakti, in 2018 (Ashok Kumar et al. 2022). Nevertheless, traditional breeding approaches are often slow and resource-intensive, principally when handling complex, quantitatively inherited traits that are strongly affected by environmental factors. Moreover, these methods lack the precision required to achieve substantial genetic gains for nutritional traits. As a result, the integration of molecular breeding tools has become progressively crucial to expedite and improve the efficiency of developing nutritionally superior sorghum cultivars.

The sorghum genome is relatively small, with an estimated size of approximately 730 Mb, and is genetically less complex, consisting of 10 diploid chromosomes. The sequencing and public release of a high-quality reference genome based on the elite inbred line BTx623 represented a major milestone and significantly advanced genetic and genomic investigations in sorghum (Paterson et al. 2009). The availability of this reference genome enabled large-scale resequencing of cultivated types and wild relatives, which in turn directed the development of diverse and abundant molecular marker resources. These molecular markers have played a crucial role in the construction of dense linkage maps and the identification of quantitative trait loci (QTLs) and meta-QTLs associated to crucial agronomic traits, as well as tolerance to biotic and abiotic stresses. Such QTL information can be effectively exploited through marker-assisted backcross breeding to develop improved, climate-resilient sorghum cultivars capable of maintaining response under undesirable environmental conditions (Reddy et al. 2014; Wagh et al. 2016; Gorthy et al. 2017; Mwamahonje et al. 2021; Aquib and Nafis 2022). In addition, DNA marker systems

Table 4.1 Summary of extent of genetic variability for iron and zinc in sorghum grains

Grain iron content range (mg/kg)	Grain zinc content range(mg/kg)	Number and types of genetic materials used	References
30–113	11–44	76 farmers' varieties from Benin	Kayodé et al. (2006)
7.7–132.6	15.1–91.3	1394 accessions from core germplasm maintained at the ICRISAT gene bank	Sanjana Reddy et al. (2010)
26–61	21–57	29 selected accessions from the sorghum core collection	Ashok Kumar et al. (2012)
41.17–127.5	13.5–34.67	31 sorghum landrace accessions	Shegro et al. (2012)
28–63	23–55	14 accessions	Ng'uni et al. (2016)
6.3–168.1	3.6–87.0	222 sorghum (49 popular cultivars and parental lines, 26 elite low-amylose content breeding lines, 113 germplasm accessions)	Hariprasanna et al. (2014a, b)
11.0–95.4	11.2–75.8	112 local landraces and varieties	Badigannavar et al. (2016)
25.8–48.9	13.5–42.6	242 sorghum genotypes from mini core collection	Upadhyaya et al. (2016)
43.7–61.2	13.7–23.4	25 sorghum varieties	Ng'uni et al. (2016)
20.6–57.89	13.2–52.45	134 genotypes which includes, mini core collection, red grain type collections, germplasm lines, ABLs, and released varieties	Malaghan and Kajjidoni (2016)
19.54–54.57	16.21–45.78	163 accessions from association panel	Paiva et al. (2017)
10.8–76.4	10.2–58.7	RIL population (296B and PVK801)	Phuke et al. (2017)
18.0–62.2	9.4–43.7	39 exotic lines, 260 indigenous germplasm, 61 cultivars, 25 breeding lines	Hariprasanna et al. (2018)
25–42	18–30	23 accessions from 5 countries in southern Africa	Motlhaodi et al. (2018)
17.3–77.0	2.8–85.0	163 accessions from Asia and Africa	Kamal et al. (2023)

are widely applied to characterize genetic diversity and population structure within germplasm collections, enable accurate cultivar identification, and ensure genetic purity of hybrids and parental lines (Shaikh and Pawar 2021; Enyew et al. 2022; Josia et al. 2021).

Multiple strategies and operational methods have been adopted to expand the nutritional quality of sorghum, with genomics and biotechnology playing an increasingly central role in overcoming existing nutritional constraints. Advances in sorghum genomic research have enabled detailed exploration of its genetic architecture, allowing the identification of genes and molecular markers linked to key nutritional attributes (Hao et al. 2021). Such genetic insights can be effectively utilized to design and develop sorghum cultivars with superior nutritional composition through targeted breeding and selection. This chapter provides an integrated overview of the application of genomics and molecular breeding tools for nutritional enhancement in sorghum. It encompasses a range of approaches, including the use of molecular markers, bulk segregant analysis (BSA), quantitative trait locus (QTL) mapping, marker-assisted backcrossing (MABC), genome-wide association studies (GWAS), genomic selection (GS), and genome-editing technologies. Evaluation of these methodologies, along with their practical applications and achievements, highlights the considerable potential of sorghum genomics in accelerating nutritional improvement. Overall, genomic-enabled breeding offers a robust framework to transform sorghum into not only a climate-resilient crop for marginal environments but also a nutritionally improved and sustainable food source for populations across the globe.

4.3 Development and DNA Marker System

During the early 1990s, restriction fragment length polymorphism (RFLP) markers were widely and successfully employed in sorghum for genetic diversity assessment, linkage mapping, and comparative genome analysis. With advancements in molecular biology, additional marker systems were subsequently adopted, including randomly amplified polymorphic DNA (RAPD) markers (Ayana et al. 2000), amplified fragment length polymorphisms (AFLPs), and simple sequence repeats (SSRs), which were used either individually or in combination to support sorghum genetic improvement programs. Among these, SSR markers gained prominence in plant breeding due to their high genomic abundance, extensive polymorphism, codominant inheritance, and compatibility with high-throughput genotyping platforms. Early efforts to develop SSR markers in sorghum were reported by Brown et al. (1996), who identified 49 SSR loci using approaches such as mining public DNA databases, adapting SSR primers originally developed for maize, and screening sorghum genomic libraries. Subsequently, Taramino et al. (1997) reported 13 sorghum SSR markers, of which ten were recognized through database searches and three through screening of AG-enriched genomic libraries. Further expansion of SSR resources was achieved by mining diverse sequence datasets, including cDNA, expressed sequence tags (ESTs), and unigene collections (Schloss et al. 2002; Arun 2006; Nagaraja Reddy et al. 2012; Zhu et al. 2018).

The availability of the sorghum reference genome further accelerated SSR discovery. Yonemaru et al. (2009) developed 5599 non-redundant SSR markers from whole-genome shotgun sequences of the sorghum inbred line BTx623, representing a wide range of repeat motifs such as (AT/TA)n, (AG/TC)n, (AC/TG)n, and (CG/GC)n. More recently, in silico analysis of the *Sorghum bicolor* reference genome predicted a total of 163,943 SSR loci, of which 130,120 were successfully mapped across the genome using sequence data from 75 accessions, indicating a broad and uniform genomic distribution of SSRs (Baggett et al. 2021). Single-nucleotide polymorphisms (SNPs) have emerged as the marker system of choice for sorghum genotyping due to their high abundance, uniform genome-wide distribution, and suitability for high-throughput and precise genetic analysis. The large-scale discovery of SNPs has been driven by advances in next-generation sequencing technologies, improved bioinformatic pipelines, and the increasing availability of DNA sequence resources, including expressed sequence tags (ESTs) and complete genome assemblies. Using the HaploSNPer platform, Singhal et al. (2011) reported the identification of approximately 77,094 putative SNPs, of which 40,589 were classified as high-confidence variants. Similarly, Nelson et al. (2011) detected about 283,000 SNPs through short-read sequencing of diverse sorghum accessions aligned to the reference genome.

Genotyping-by-sequencing (GbS) approaches have further accelerated SNP discovery in sorghum. Maina et al. (2018) identified 144,299 SNPs in Nigerien sorghum accessions using GbS, while Wondimu et al. (2021) reported 108,107 high-quality SNPs evenly distributed across the genome from 304 Ethiopian sorghum accessions. Whole-genome resequencing has also contributed substantially to variant discovery. Zheng et al. (2011) resequenced two sweet sorghum lines (Keller and E-Tian) and one grain sorghum line (Ji2731), identifying over 1.05 million SNPs along with 99,948 insertion–deletion polymorphisms (InDels). Mace et al. (2013) extended this effort by resequencing 44 sorghum genotypes, including the wild progenitor *Sorghum propinquum*, and uncovered nearly 4.95 million genome-wide SNPs. Comparative resequencing studies have further highlighted extensive genomic variation among elite lines. Evans et al. (2013) identified approximately 1.2 million SNPs and 120,969 InDels between Tx7000 and BTx623, and about 1.6 million SNPs and 152,836 InDels between BTx642 and BTx623. McCormick et al. (2018) reported the discovery of roughly 7.4 million SNPs and 1.9 million InDels by aligning more than 50 resequenced sorghum genomes to the reference assembly. More recently, whole-genome sequencing of 400 accessions from the sorghum association panel (SAP) at an average depth of 38× generated a high-density marker resource comprising nearly 38 million SNPs, InDels, and copy number variants (Boatwright et al. 2022). In parallel, the development of the Sorghum Genome SNP Database (SorGSD) provided access to approximately 62.9 million SNPs identified across 48 sorghum accessions using assembled and annotated *Sorghum bicolor* genome sequences (v2.1) and resequencing data (Luo et al. 2016).

4.4 Genetic Maps

The construction of a genetic or linkage map represents a fundamental prerequisite for genetic analysis of any target trait in sorghum, as it provides the framework for deploying various DNA markers across the genome. The earliest sorghum linkage map was developed in the early 1990s using 36 restriction fragment length polymorphism (RFLP) markers, covering a genetic distance of 283 cM and representing eight of the ten linkage groups (Hulbert et al. 1990). Subsequent attempts to increase marker number and map coverage through the inclusion of additional RFLPs led to improvements in map length and density (Lin et al. 1995; Dufour et al. 1997; Tao et al. 1998), although complete genome representation was still not achieved. Ragab et al. (1994) later constructed a linkage map comprising 15 linkage groups using 38 sorghum and 33 maize genomic DNA probes, spanning a total length of 633 cM. Further progress was made when Subudhi and Nguyen (2000) aligned and integrated five major RFLP-based linkage maps, collectively containing 1036 markers (Boivin et al. 1999; Ragab et al. 1994; Xu et al. 1994), into ten linkage groups. This integrated map not only improved confidence in the accuracy of existing linkage information but also enabled the precise placement and comparison of quantitative trait loci (QTL) markers within defined genomic regions. The widespread adoption of PCR-based marker systems during the 1990s, including RAPDs, AFLPs, and SSRs, substantially enhanced the development of denser sorghum linkage maps. RAPD markers were initially used to construct linkage maps (Tuinstra et al. 1996; Agrama et al. 2002; Knoll et al. 2008); however, concerns regarding their reproducibility across laboratories limited their long-term utility. AFLP markers subsequently gained popularity and were successfully applied to increase marker density and genome coverage (Ritter et al. 2008; Ramu et al. 2009; Shiringani and Friedt 2011). Early SSR-based maps were constrained by the limited availability of microsatellite markers (Taramino et al. 1997; Tao et al. 1998, 2000). This limitation was gradually overcome as additional SSR loci were developed and integrated into existing maps. For example, Kong et al. (2000) and Bhattramakki et al. (2000) incorporated 31 and 113 SSR markers, respectively, into the RFLP-based linkage map reported by Peng et al. (1999). Bhattramakki et al. (2000) subsequently developed a high-density framework map comprising 323 RFLPs and 143 SSRs, which was further expanded by Menz et al. (2002) through the addition of 2926 markers, including AFLPs, SSRs, and RFLPs derived from rice, barley, oat, and maize cDNA and genomic clones. This effort resulted in an average inter-marker distance of approximately 0.5 cM. In a similar vein, Bowers et al. (2003) produced another high-resolution linkage map with an even closer marker spacing of about 0.4 cM using 2512 RFLP loci. Linkage maps developed using EST-derived SSRs (Ramu et al. 2009; Srinivas et al. 2009b) and unigene-based SSR markers (Srinivas et al. 2009a; Nagaraja Reddy et al. 2012) provided valuable resources for comparative genome analysis in sorghum. With continued progress in marker discovery and genotyping technologies, high-throughput marker systems such as Diversity Arrays Technology (DArT) and single-nucleotide polymorphisms (SNPs) increasingly became central to linkage mapping efforts. Linkage maps incorporating both DArT and non-DArT markers, distributed across all ten sorghum chromosomes, were successfully constructed by Mace et al. (2008, 2009).

The advent of next-generation sequencing further enabled the development of ultra-high-density linkage maps. Zou et al. (2012) reported an ultra-dense map based on 3418 bin markers generated through resequencing of 244 RILs derived from the 654 × LTR108 cross. Similarly, Kajiya-Kanegae et al. (2020) constructed a high-resolution linkage map comprising 3710 SNP markers identified via restriction-site-associated DNA sequencing (RAD-seq) in a population of 213 RILs developed from the BTx623 × NOG cross. Using a comparable RAD-seq approach, Jin et al. (2021) generated a dense genetic map for an RIL population obtained from a cross between sorghum (Tx623A) and Sudan grass (Sa). This map consisted of 1065 markers spanning a total genetic distance of 1191.7 cM. Marker distribution varied among chromosomes, with chromosome 2 containing the highest marker density (average interval of 0.88 cM), whereas chromosome 7 exhibited the lowest density (average interval of 1.25 cM). More recently, Cuevas and Vermerris (2022) developed a highly saturated linkage map comprising 33,421 SNP markers obtained by genotyping 205 RILs from the SC1103 × RTx430 cross. In addition to population-specific maps, efforts have been made to generate consensus linkage maps. Satrio et al. (2022) constructed a comprehensive consensus map by integrating publicly available marker data from four different mapping populations. This unified map was based on 3449 non-redundant nucleotide-level polymorphic markers and spanned all ten sorghum chromosomes. The resulting ultra-high-density consensus map covered a total genetic length of 1571.68 cM, with an average inter-marker distance of 0.46 cM, providing a robust framework for QTL mapping and genomic studies in sorghum.

4.5 QTL Mapping

Quantitative trait locus (QTL) mapping has become a powerful and widely used approach in plant breeding for the genetic dissection of complex, polygenic traits. Traits that were previously difficult to improve through conventional breeding can now be partitioned into discrete genomic regions using DNA marker information. This approach allows breeders to identify, monitor, and manipulate multiple interacting loci that collectively govern complex phenotypes (Tanksley 1993). The detection of QTLs associated with economically and agronomically important traits provides a robust basis for efficient genetic improvement through marker-assisted selection (MAS) and facilitates the pyramiding of multiple desirable traits into a single genetic background (Collard et al. 2005; Gupta et al. 2009; Soriano 2020).

In sorghum, QTL mapping studies have identified numerous genomic regions associated with key nutritional attributes. These include QTLs governing crude protein content and protein digestibility (Ayalew et al. 2022; Murray et al. 2008), grain iron and zinc content (Kotla et al. 2016; Kotla et al. 2019), as well as starch and fat content (Murray et al. 2008). In addition, genomic regions influencing β-carotene and total carotenoid accumulation have also been reported (Fernandez et al. 2008). A detailed summary of QTLs identified for nutritionally important traits in sorghum is presented in Table 4.2. The first major QTL influencing protein digestibility was mapped in proximity to the marker *Xtxp11*, exhibiting a LOD score of 3.1 and accounting for approximately 29% of the observed phenotypic variation. A second QTL was detected about 20 cM distal to this region, flanked by the markers *Xtxp88*

Table 4.2 List of QTLs mapped for various nutritionally important traits in Sorghum

Trait	Number of QTLs	Parents	References
Iron	4	296B × PVK 801	Kotla et al. (2019)
Zinc	4	296B × PVK 801	Kotla et al. (2019)
Crude protein content	7	Rio × BTx623	Murray et al. (2008)
	7	RTx430 × SC 35	Ayalew et al. (2022)
	7	BTx642 × BTxARG-1 BTxARG-1 × P850029	Boyles et al. (2017)
Crude fat content	3	Rio × BTx623	Murray et al. (2008)
	8	BTx642 × BTxARG-1 BTxARG-1 × P850029	Boyles et al. (2017)
Starch content	7	Rio × BTx623	Murray et al. (2008)
	5	BTx642 × BTxARG-1 BTxARG-1 × P850029	Boyles et al. (2017)
	10	RTx430 × SC 35	Ayalew et al. (2022)
Amylose	10	BTx642 × BTxARG-1 BTxARG-1 × P850029	Boyles et al. (2017)
	10	RTx430 × SC 35	Ayalew et al. (2022)
β-Carotene content	5	KS115 × Macia	Fernandez et al. (2008)
Lutein content	6	KS115 × Macia	Fernandez et al. (2008)
Zeaxanthin content	2	KS115 × Macia	Fernandez et al. (2008)
Total carotenoids	2	KS115 × Macia	Fernandez et al. (2008)

and *Xtxp329*. This locus showed an LOD score of 2.7 and explained nearly 18% of the phenotypic variance. As opposed to the first QTL, the second locus exerted a positive effect on protein digestibility and is considered to correspond to the HD locus, where the presence of favorable alleles leads to enhanced digestibility (Winn et al. 2009). Using a bulk segregant analysis in an F_2 population obtained from a cross between a normal sorghum line (BTx623) and the high-digestibility mutant P721Q, Massafaro et al. (2016) identified a single major QTL associated with protein digestibility on chromosome 5. This QTL spanned a large genomic interval of approximately 58 Mb and overlapped with the genomic region containing the γ-kafirin gene cluster. F_2 individuals exhibiting high digestibility also displayed the distinctive protein body morphology characteristic of *P721Q*. The researchers inferred that this QTL most likely corresponded to one of the γ-kafirin genes located on chromosome 5; however, precise gene identification was complicated by the high sequence similarity among members of the kafirin gene family. Further dissection of the genetic basis of the *P721Q* mutation was made by Wu et al. (2013) through an integrated strategy combining segregation analysis, DNA and cDNA sequencing, map-based cloning, and heterologous expression studies. The mutation was initially localized to chromosome 5 in the vicinity of a cluster of ten 22-kDa α-kafirin genes. Detailed targeted sequencing revealed the presence of a mutant-specific α-kafirin copy carrying a point mutation that resulted in a G-to-A nucleotide substitution, leading to an alanine-to-threonine amino acid change at position 21 within the endoplasmic reticulum signal peptide. Expression of this mutant α-kafirin in maize produced the distinctive *P721Q*-type protein body structure in the maize endosperm, although protein digestibility itself was not evaluated. Genetic analyses of F_2 and backcross populations further supported the involvement of a single dominant gene controlling the altered endosperm phenotype.

In a broader analysis of grain quality traits, Boyles et al. (2017) employed both association analysis and linkage mapping approaches to identify QTLs controlling crude protein, crude fat, amylose, starch, and gross energy content in sorghum. A single major QTL governing amylose content was detected, explaining between 46% and 61% of the genetic variance, and was located approximately 12 kb from the waxy (*Wx*) gene on chromosome 10, which encodes granule-bound starch synthase. In addition, five QTLs associated with starch content were identified on chromosomes 1, 2, 6, and 10, while seven QTLs for crude protein were distributed across chromosomes 1, 2, 6, 7, and 9. Eight QTLs influencing crude fat content were mapped to chromosomes 1, 2, 3, 4, 5, 6, and 10. Notably, QTLs for starch and crude protein co-localized with a grain number QTL on chromosome 1, and the starch QTL on chromosome 2 was found to overlap with regions controlling crude protein content, indicating potential pleiotropic or linked genetic effects. Micronutrients such as iron (Fe), zinc (Zn), and vitamin A in the form of carotenoids are essential components of the human diet, and deficiencies in these nutrients can result in serious health problems, particularly in developing regions where sorghum serves as a major staple food. Identifying and enhancing the genetic factors governing Fe, Zn, and carotenoids in sorghum grains is therefore a key strategy for enhancing the nutritional value of this crop. To dissect the genetic control of iron and zinc content, Kotla et al. (2019) evaluated an F_6 RIL population comprising 342 lines obtained from a cross between 296B and PVK 801. Using a combination of SSR, DArT, and DArTSeq markers, they detected significant genotype × environment interactions for both micronutrients. A high-density genetic linkage map was developed with 2088 markers (1148 DArTs, 927 DArTSeqs, and 13 SSRs), covering a total genetic distance of 1355.52 cM and an average inter-marker spacing of 0.6 cM. Across multiple environments, eleven individual QTLs and three stable QTLs associated with grain Fe and Zn concentration were identified. Several of these QTLs were linked to putative candidate genes, including *CYP71B34* and *ZFP8* (Zinc Finger Protein 8), which are known to play roles in iron and zinc metabolism in sorghum. Once validated, the associated molecular markers hold strong potential for use in breeding activities aimed at producing sorghum varieties with enhanced grain Fe and Zn content. In a related study focusing on provitamin A, Fernandez et al. (2008) performed QTL mapping for carotenoid concentration and endosperm colour using a RIL population derived from a cross between a yellow-endosperm line ("KS115") and a white-endosperm line ("Macia"). Their genetic map consisted of 112 molecular markers, including nine candidate genes involved in carotenoid biosynthesis. The analysis revealed co-localization of colour-related QTLs with carotenoid QTLs, representing a solid genetic association between these traits. For β-carotene content, five QTLs were detected on chromosomes 1, 2, and 10. Notably, a QTL located on chromosome 2 was stable across environments, exhibited positive additive effects, explained a considerable proportion of phenotypic variation (11.6% and 15.15%), and was associated with a novel phytoene synthase gene (*Psy3*), underscoring its importance for carotenoid enhancement in sorghum.

4.6 Genome-Wide Association Studies (GWAS)

Genome-wide association studies (GWAS), also referred to as association or linkage disequilibrium (LD) mapping, leverage the broad phenotypic variation within a species and the extensive historical recombination that has occurred in natural populations (Raghuvanshi et al. 2021). GWAS provides a complementary and often higher-resolution alternative to conventional QTL mapping for identifying genomic regions linked with complex traits (Wei et al. 2015). This approach primarily examines the associations between single-nucleotide polymorphisms (SNPs) and phenotypic traits of interest. The rapid advancement of high-throughput genotyping platforms has greatly facilitated large-scale SNP discovery, thereby enhancing the effectiveness of GWAS. In contrast to bi-parental QTL mapping, which depends on structured segregating populations, GWAS assesses linkage disequilibrium by genotyping and phenotyping genetically diverse individuals from natural population panels, thereby facilitating the direct detection of trait-associated loci and candidate genes from standing genetic variation.

A major advantage of GWAS is that a single genotyped population can be repeatedly analyzed for multiple traits, maximizing the utility of both genotypic and phenotypic data. GWAS has been broadly applied across economically important crops to elucidate the genetic architecture of essential traits. In rice, GWAS was employed to investigate flowering time, panicle architecture, yellow mottle virus resistance, fertility restoration, and other yield-related traits (Huang et al. 2010; Yano et al. 2016; Li et al. 2020; Cubry et al. 2020). In maize, this method has given insights into genetic variation underlying compositional and pasting properties, stalk biomass, leaf cuticular conductance, and broader physiological traits (Alves et al. 2019; Mazaheri et al. 2019; Lin et al. 2020). Similarly, GWAS has been effectively employed in oilseed crops such as canola to dissect flowering time, and in *Brassica* species to understand the genetic basis of stress tolerance, oil content, and seed quality (Raman et al. 2019; Lu et al. 2019). In sesame, association mapping has been instrumental in identifying loci linked to drought tolerance, vitamin E concentration, yield, and yield-related traits (He et al. 2019; Dossa et al. 2019; Li et al. 2018a, b). In sorghum, GWAS has been broadly applied to study yield-related components (Boyles et al. 2016; Zhang et al. 2023), tolerance to biotic and abiotic stresses (Challa and Neelapu 2018; Osman et al. 2022), and nutritionally important traits, including protein digestibility, antioxidant activity, tannin content, carotenoids, and overall grain quality (Diatta-Holgate et al. 2022; Habyarimana et al. 2019; Cruet-Burgos et al. 2023; Kimani et al. 2020).

Vitamin A deficiency remains a major nutritional challenge, particularly in regions where sorghum constitutes a primary dietary staple. To explain the genetic basis of carotenoid accumulation in sorghum grain, Cruet-Burgos et al. (2020) conducted a GWAS analysis using a diverse sorghum association panel. Their analysis identified 14 SNPs significantly associated with β-carotene content, three of which were in close proximity to a priori candidate genes. In addition, a strong association was noticed on chromosome 2, approximately 75 kb from a putative *CYP97A* hydroxylase gene. For zeaxanthin content, the study detected 38 significant SNPs distributed across 12 genomic regions, including a major association on chromosome 6 near a candidate gene annotated as abscisic acid 8′-hydroxylase 3 (*CYP707A*). Notably, the most prominent zeaxanthin-associated SNP was located within the putative sorghum ortholog of the maize zeaxanthin epoxidase (*ZEP*) gene, indicating that *ZEP* is a key locus regulating carotenoid variation in sorghum. Rhodes et al. (2017) applied GWAS to a diverse sorghum association panel to investigate grain composition traits, including protein, fat, and starch content. They identified several significant SNPs forming association peaks on chromosomes 2 and 4 that were in strong LD with each other and with an a priori candidate gene encoding *alpha-amylase 3*. These outcomes presented molecular evidence for the genetic control underlying variation in grain protein and fat content. Further extending these findings, Cruet-Burgos et al. (2023) quantified lutein, zeaxanthin, β-carotene, and α-carotene in a sorghum diversity panel and performed GWAS, identifying multiple SNPs significantly associated with individual carotenoids across chromosomes 4, 6, and 9. Once again, strong associations near the *ZEP* gene underscored its central role in regulating carotenoid biosynthesis and accumulation in sorghum grain. Kamal et al. (2023) used GWAS to detect genomic regions associated with total kernel weight (TKW), nutritional attributes, and drought-related traits, and demonstrated the feasibility of simultaneous selection for improved grain nutrition and TKW based on co-localization of marker–trait associations. In a GWAS involving 73,512 SNPs, Diatta-Holgate et al. (2022) identified 28 SNPs significantly associated with multiple grain quality traits, including protein digestibility. Several of these associations overlapped with previously reported QTLs, while others represented novel loci contributing to grain quality variation. Among the detected loci, four SNPs showed significant associations with protein digestibility, three of which were in genomic regions previously linked to grain color and tannin-related QTLs. Habyarimana et al. (2019) conducted GWAS for antioxidant traits in sorghum grain and reported 96 significant associations, including major-effect QTLs explaining a substantial proportion of phenotypic variation. Pleiotropic loci with high phenotypic variance effect were identified on chromosomes 1, 4, 9, and 10, providing insights into genes involved in anthocyanin conjugation pathways. Using a high-density dataset of 3,512,517 SNPs, Kimani et al. (2020) identified significant marker–trait associations for tannin content, starch, and amino acid composition in sorghum grain. These SNPs corresponded to both previously reported QTLs and novel candidate genes involved in tannin biosynthesis, starch metabolism, and amino acid pathways (Wu et al. 2012). Earlier, Sukumaran et al. (2015) performed association mapping using 333 SNPs and identified eight significant marker–trait associations related to kernel hardness, starch concentration, and tannin content. Collectively, markers identified through GWAS have proven to be powerful tools for sorghum improvement, facilitating the selection of nutritionally desirable traits and contributing to enhanced tolerance to biotic and abiotic stresses in this important cereal crop.

4.7 Genomic Selection (GS) and Genome Editing

Genomic selection represents a contemporary breeding approach that utilizes genome-wide marker information to forecast the genetic potential of individuals for various traits of interest, including yield, disease resistance, and quality. The marker effect estimates are then employed to assess the genomic estimated breeding values (GEBVs) of different genotypes, which form the basis of selection. This technique has been applied to numerous crops, including sorghum, with the goal of accelerating the breeding process and enhancing the efficiency of selection. In a study by Habyarimana et al. (2019), the effectiveness of four different genomic selection models, GBLUP, Bayesian Ridge Regression (BRR), Bayesian LASSO, and BayesB, was evaluated for four antioxidant traits in sorghum grains: phenols, flavonoids, total antioxidant capacity, and condensed tannins. The research employed genome-wide SNP markers in a novel diversity panel that included *S. bicolor* lines and landraces, as well as RILs derived from crosses between *S. bicolor* and *S. halepense*. The findings revealed that the *S. bicolor* × *S. halepense* lines outperformed the *S. bicolor* lines across all antioxidant traits. Furthermore, the four GS models showed similar performance across the evaluated traits, with prediction accuracies ranging from 0.49 to 0.58. These levels of accuracy are considered sufficiently high to facilitate sorghum breeding programs aimed at developing antioxidant-rich varieties, enabling substantial genetic improvement in a relatively short timeframe and with optimized cost-effectiveness. In another study, Cruet Burgos et al. (2023) calculated GEBVs based on BLUPs for β-carotene, lutein, and zeaxanthin using a dataset of 60,129 SNPs. The very high heritability estimate of 0.78 for β-carotene, along with a prediction accuracy of 0.67, strongly indicated the polygenic nature of carotenoid traits in sorghum. These outcomes advise that genomic selection can serve as an effective strategy for enhancing biofortification efforts in sorghum. Importantly, genomic selection has the potential to transform sorghum breeding by speeding up the development of superior varieties with improved traits. However, its effective application depends on having robust genomic resources, collecting precise phenotypic data, and continuously validating and refining predictive models.

Traditionally, sorghum improvement relied on conventional breeding methods that exploited both the species' inherent genetic diversity and variations introduced through mutagenesis techniques (Hao et al. 2021). The emergence of genome editing, enabled by programmable nucleases, has transformed functional genomics and the creation of genetic diversity for crop breeding. There are four primary technologies for precise genome editing: meganucleases (MegNs, first reported by Puchta et al. 1993), zinc finger nucleases (ZFNs, described by Bibikova et al. 2003 and

Porteus and Baltimore 2003), transcription activator-like effector nucleases (TALENs, by Boch et al. 2009 and Moscou and Bogdanove 2009), and the CRISPR–Cas9 system, elucidated by Jinek et al. 2012 and Cong et al. 2013. All these platforms share the ability to induce targeted double-strand breaks (DSBs) in DNA, which are then repaired by cellular DNA repair pathways (Schmidt et al. 2019). Despite these advancements, the application of genome editing in sorghum has lagged progress that seen in other cereal crops.

In sorghum, researchers have attempted to enhance protein digestibility and overall nutritional quality. The low nutritional value and digestibility of sorghum grain are largely attributed to kafirin storage proteins, particularly β- and γ-kafirins. These proteins possess limited protease cleavage sites and high cysteine content, forming dense disulfide-linked networks that encase the more digestible α- and δ-kafirins within protein bodies. While natural mutations in β-kafirin have occasionally improved grain quality, CRISPR/Cas9-mediated knockout of β-kafirin in the Tx430 background did not result in improved protein digestibility (Massel et al. 2023). Instead, it caused alterations in protein composition and the morphology of the vitreous endosperm, largely due to increased γ-kafirin levels. These findings highlight that peripheral kafirins can exhibit functional redundancy in certain genetic backgrounds, and that the removal of β-kafirin alone may not universally enhance grain protein digestibility.

The enhancement of lysine content and protein digestibility in sorghum using CRISPR/Cas9 has been experimentally demonstrated by Li et al. (2018a, b) and Li et al. (2023). In the study by Li et al. (2018a, b), CRISPR/Cas9 was employed to simultaneously target an entire gene family with the goal of improving sorghum grain quality, which is typically limited by low lysine content and poor digestibility. Kafirins, which account for approximately 70% of total seed protein, are a major determinant of grain quality. Among them, α-kafirins are the most abundant and are encoded by a single gene family consisting of 20 closely related *k1C* gene copies located on chromosome five. To target the entire gene family efficiently, a single guide RNA (sgRNA) was designed to recognize the highly conserved N-terminal endoplasmic reticulum (ER) signal sequence. The CRISPR/Cas9 machinery was introduced into immature embryos of the Tx430 line via Agrobacterium-mediated transformation using a binary vector (pBUN421), with Cas9 expression driven by the maize ubiquitin promoter and the sgRNA by the rice U3 promoter. This process generated 26 T0 transformation events, which were subsequently grown and self-pollinated. Analysis of the T1 and T2 seeds revealed partial opaqueness in the regenerated events carrying the genome-editing construct, reflecting a reduction in the thickness of the vitreous endosperm and serving as an early indicator of successful targeted mutagenesis. Mutation detection was carried out using a T7E1 assay and Illumina MiSeq sequencing, while α-kafirin levels were monitored through SDS–PAGE and Western blot analysis. The study found extensive editing in 25 of the 26 events, with most events showing co-editing of two or more k1C family genes, and one event achieving co-editing across all 12 targeted genes. Flour derived from T2 seeds exhibited improved protein digestibility and lysine content. This work represented the first successful instance of multiplex gene editing in sorghum, achieving an exceptionally high editing efficiency of 92.4%. Notably, the strategy of targeting 20 genes with a single sgRNA, where editing of even one gene qualified the event as edited, contributed to the high observed efficiency. The γ-kafirin protein, encoded by the *K2G* gene on sorghum chromosome 2, was specifically targeted for reduction using CRISPR-/Cas9-mediated gene editing by Li et al. (2023). A guide RNA (sgRNA) was designed to introduce mutations into the coding sequence (CDS) corresponding to the endoplasmic reticulum signal peptide of γ-kafirin. The pK2GsgRNA/Cas9 construct was delivered into sorghum via pollen-mediated transformation. Sequencing of the transformed plants showed that three out of 24 transgenic lines carried mutations in the targeted region. In comparison with wild-type plants, these mutants exhibited a reduction in γ-kafirin content ranging from 12.75% to 19.22%, accompanied by a substantial increase in protein digestibility of raw flour, ranging from 26.91% to 74.31%. Grain weights of the mutants remained similar to the wild type, while the plants demonstrated more vigorous growth, with taller and thicker shoots. This study highlights the potential of genome editing to enhance sorghum protein digestibility and demonstrates how these improvements can be rapidly integrated into breeding programs to develop improved, transgene-free cultivars, thereby benefiting global sorghum production. These findings underscore the broad applicability of cutting-edge genome-editing tools like CRISPR/Cas9 for improving both nutritional and agronomical traits in sorghum.

In conclusion, genomic approaches are pivotal for enhancing nutritional traits in sorghum, providing a robust framework to tackle global challenges associated with food security and human nutrition. Detailed understanding of the sorghum genome and its genetic variation has enabled the identification of key genes and pathways responsible for the synthesis of essential nutrients and the reduction of anti-nutritional compounds. This knowledge has empowered breeders to develop sorghum varieties with enhanced nutritional profiles, making them suitable for human consumption and animal feed. Furthermore, genomics provides insights into the complex interactions between genetics, environment, and nutrition, allowing for the formulation of strategies to improve crop performance under diverse conditions. Advanced genomics tools, including marker-assisted selection and genome editing, have accelerated breeding programs, reducing the time required to develop and release nutritionally improved sorghum cultivars. As a staple crop in many regions, sorghum holds significant potential to combat malnutrition and ensure food security. The application of genomics for nutritional trait improvement not only addresses immediate dietary needs but also contributes to resilience against environmental stressors, promoting sustainable agriculture. Continued research and application of genomic approaches in sorghum breeding will be critical for advancing understanding of the genetic basis of nutritional traits and for developing nutritionally enhanced varieties that support a healthier, food-secure future.

References

Agrama H, Widle G, Reese J, Campbell L, Tuinstra M (2002) Genetic mapping of QTLs associated with greenbug resistance and tolerance in *Sorghum bicolor*. Theor Appl Genet 104(8):1373–1378

Alves ML, Carbas B, Gaspar D, Paulo M, Brites C, Mendes-Moreira P, Brites CM, Malosetti M, Van Eeuwijk F, Vaz Patto MC (2019) Genome-wide association study for kernel composition and flour pasting behavior in wholemeal maize flour. BMC Plant Biol 19(1):1–17

Amelework BA, Shimelis HA, Laing MD, Ayele DG, Tongoona P, Mengistu F (2016) Sorghum production systems and constraints, and coping strategies under drought-prone agro-ecologies of Ethiopia. S Afr J Plant Soil *33*(3):207–217

Aquib A, Nafis S (2022) A meta-analysis of quantitative trait loci associated with stay-green in sorghum. J Trop Crop Sci 9(1):39

Arun SS (2006) In silico EST data mining for elucidation of repeats biology and functional annotation in sorghum (*Sorghum bicolor* (L.) Moench). MSc(Agriculture) Thesis, University of Agricultural Sciences, Dharwad, Karnataka (India)

Ashok Kumar A, Reddy BV, Ramaiah B, Sahrawat KL, Pfeiffer WH (2012) Genetic variability and character association for grain iron and zinc contents in sorghum germplasm accessions and commercial cultivars. Eur J Plant Sci Biotechnol 6(1):1–5

Ashok Kumar A, Mehtre SP, Anil G, Gorthy S, Anuradha K, Phuke RM et al (2022) Biofortifying Sorghum for delivering grain micronutrients in high yielding cultivars with market-preferred traits. In: Biofortification of staple crops. Springer Singapore, Singapore, pp 195–204

Ayalew H, Peiris S, Chiluwal A, Kumar R, Tiwari M, Ostmeyer T, Bean S, Jagadish SK (2022) Stable sorghum grain quality QTL were identified using SC35× RTx430 mapping population. Plant Genome *15*(3):e20227

Ayana A, Bekele E, Bryngelsson T (2000) Genetic variation in wild sorghum (*Sorghum bicolor* ssp. verticilliflorum (L.) Moench) germplasm from Ethiopia assessed by random amplified polymorphic DNA (RAPD). Hereditas 132(3):249–254

Badigannavar A, Girish G, Ramachandran V, Ganapathi TR (2016) Genotypic variation for seed protein and mineral content among post-rainy season-grown sorghum genotypes. Crop J 4(1):61–67

Baggett JP, Tillett RL, Cooper EA, Yerka MK (2021) De novo identification and targeted sequencing of SSRs efficiently fingerprints *Sorghum bicolor* sub-population identity. PLoS One 16(3):248213

Bhattramakki D, Dong J, Chhabra AK, Hart GE (2000) An integrated SSR and RFLP linkage map of *Sorghum bicolor* (L.) Moench. Genome 43(6):988–1002

Bibikova M, Beumer K, Trautman JK, Carroll D (2003) Enhancing gene targeting with designed zinc finger nucleases. Science 300(5620):764–764

Boatwright JL, Sapkota S, Jin H, Schnable JC, Brenton Z, Boyles R, Kresovich S (2022) Sorghum Association Panel whole-genome sequencing establishes cornerstone resource for dissecting genomic diversity. Plant J 111(3):888–904

Boch J, Scholze H, Schornack S, Landgraf A, Hahn S, Kay S, Lahaye T, Nickstadt A, Bonas U (2009) Breaking the code of DNA binding specificity of TAL-type III effectors. Science 326(5959):1509–1512

Bodede O, Mabelebele M (2022) Physical characteristics, nutritional composition and phenolic compounds of some of the sorghum landraces obtained in South Africa. Food Res 6(4):312–328

Boivin K, Deu M, Rami JF, Trouche G, Hamon P (1999) Towards a saturated sorghum map using RFLP and AFLP markers. Theor Appl Genet 98(2):320–328

Bowers JE, Abbey C, Anderson S, Chang C, Draye X, Hoppe AH, Jessup R, Lemke C, Lennington J, Li Z, Lin YR (2003) A high-density genetic recombination map of sequence-tagged sites for sorghum, as a framework for comparative structural and evolutionary genomics of tropical grains and grasses. Genetics 165(1):367–386

Boyles RE, Cooper EA, Myers MT, Brenton Z, Rauh BL, Morris GP, Kresovich S (2016) Genome-wide association studies of grain yield components in diverse sorghum germplasm. Plant Genome 9(2):1–17

Boyles RE, Pfeiffer BK, Cooper EA, Rauh BL, Zielinski KJ, Myers MT, Brenton Z, Rooney WL, Kresovich S (2017) Genetic dissection of sorghum grain quality traits using diverse and segregating populations. Theor Appl Genet 130:697–716

Brown SM, Hopkins MS, Mitchell SE, Senior ML, Wang TY, Duncan RR, Gonzalez-Candelas F, Kresovich S (1996) Multiple methods for the identification of polymorphic simple sequence repeats (SSRs) in sorghum (*Sorghum bicolor* (L.) Moench). Theor Appl Genet 93(1):190–198

Challa S, Neelapu NR (2018) Genome-wide association studies (GWAS) for abiotic stress tolerance in plants. In: Biochemical, physiological and molecular avenues for combating abiotic stress tolerance in plants. Academic, pp 135–150

Collard BC, Jahufer MZZ, Brouwer JB, Pang ECK (2005) An introduction to markers, quantitative trait loci (QTL) mapping and marker-assisted selection for crop improvement: the basic concepts. Euphytica 142(1):169–196

Cong L, Ran FA, Cox D, Lin S, Barretto R, Habib N, Hsu PD, Wu X, Jiang W, Marraffini LA, Zhang F (2013) Multiplex genome engineering using CRISPR/Cas systems. Science 339(6121):819–823

Cruet-Burgos C, Cox S, Ioerger BP, Perumal R, Hu Z, Herald TJ, Bean SR, Rhodes DH (2020) Advancing provitamin A biofortification in sorghum: genome-wide association studies of grain carotenoids in global germplasm. Plant Genome 13(1):e20013

Cruet-Burgos C, Morris GP, Rhodes DH (2023) Characterization of grain carotenoids in global sorghum germplasm to guide genomics-assisted breeding strategies. BMC Plant Biol 23(1):1–17

Cubry P, Pidon H, Ta KN, Tranchant-Dubreuil C, Thuillet AC, Holzinger M, Adam H, Kam H, Chrestin H, Ghesquière A, François O (2020) Genome wide association study pinpoints key agronomic QTLs in African rice Oryza glaberrima. Rice 13(1):1–12

Cuevas HE, Vermerris W (2022) Linkage map construction using limited parental genotypic information. Euphytica 218(5):1–11

Davis KF, Chiarelli DD, Rulli MC, Chhatre A, Richter B, Singh D, DeFries R (2018) Alternative cereals can improve water use and nutrient supply in India. Sci Adv 4(7):eaao1108

Diatta-Holgate E, Hugghis E, Weil C, Faye JM, Danquah A, Diatta C, Tongoona P, Danquah EY, Cisse N, Tuinstra MR (2022) Natural variability for protein digestibility and grain quality traits in a West African sorghum association panel. J Cereal Sci 107:103504

Dossa K, Li D, Zhou R, Yu J, Wang L, Zhang Y, You J, Liu A, Mmadi MA, Fonceka D, Diouf D (2019) The genetic basis of drought tolerance in the high oil crop Sesamum indicum. Plant Biotechnol J 17(9):1788–1803

Dufour P, Deu M, Grivet L, D'Hont A, Paulet F, Bouet A, Lanaud C, Glaszmann JC, Hamon P (1997) Construction of a composite sorghum genome map and comparison with sugarcane, a related complex polyploid. Theor Appl Genet 94(3):409–418

Enyew M, Feyissa T, Carlsson AS, Tesfaye K, Hammenhag C, Geleta M (2022) Genetic diversity and population structure of sorghum [*Sorghum bicolor* (L.) moench] accessions as revealed by single nucleotide polymorphism markers. Front Plant Sci 12:799482

Evans J, McCormick RF, Morishige D, Olson SN, Weers B, Hilley J, Klein P, Rooney W, Mullet J (2013) Extensive variation in the density and distribution of DNA polymorphism in sorghum genomes. PLoS One 8(11):79192

Fernandez MGS, Hamblin MT, Li L, Rooney WL, Tuinstra MR, Kresovich S (2008) Quantitative trait loci analysis of endosperm color and carotenoid content in sorghum grain. Crop Sci 48(5):1732–1743

Geleta N, Labuschagne MT, Viljoen CD (2006) Genetic diversity analysis in sorghum germplasm as estimated by AFLP, SSR and morpho-agronomical markers. Biodivers Conserv 15(10):3251–3265

Gorthy S, Narasu L, Gaddameedi A, Sharma HC, Kotla A, Deshpande SP, Are AK (2017) Introgression of shoot fly (*Atherigona soccata* L. Moench) resistance QTLs into elite post-rainy season sorghum varieties using marker assisted backcrossing (MABC). Front Plant Sci 8:1494

Gupta HS, Agrawal PK, Mahajan V, Bisht GS, Kumar A, Verma P, Srivastava A, Saha S, Babu R, Pant MC, Mani VP (2009) Quality protein maize for nutritional security: rapid development of short duration hybrids through molecular marker assisted breeding. Curr Sci 96:230–237

Habyarimana E, Dall'Agata M, De Franceschi P, Baloch FS (2019) Genome-wide association mapping of total antioxidant capacity, phenols, tannins, and flavonoids in a panel of Sorghum bicolor and S. bicolor× S. halepense populations using multi-locus models. PLoS One 14(12):e0225979

Hao H, Li Z, Leng C, Lu C, Luo H, Liu Y et al (2021) Sorghum breeding in the genomic era: opportunities and challenges. Theor Appl Genet 134(7):1899–1924

Hariprasanna K, Agte V, Elangovan M, Patil JV (2014a) Genetic variability for grain iron and zinc content in cultivars, breeding lines and selected germplasm accessions of sorghum [Sorghum bicolor (L.) Moench]. Indian J Genet Plant Breed 74(01):42–49

Hariprasanna K, Agte V, Patil JV (2014b) Genetic control and heterosis for grain iron and zinc contents in sorghum [Sorghum bicolor (L.) Moench]. Indian J Genet Plant Breed 74(Suppl):638–643

Hariprasanna K, Agte V, Elangovan M, Gite S, Kishore A (2015) Anti-nutritional factors and antioxidant capacity in selected genotypes of sorghum [Sorghum bicolor L.(Moench)]. Int J Agric Sci 7(8):975–3710

Hariprasanna K, Suresh P, Amasiddha B, Deepika C, Dalvi U, Chavan U, Venkateswarlu R (2018) Report of the AICRP on sorghum coordinating team. All India coordinated research project on sorghum, Indian institute of millets research (IIMR): Hyderabad, India

He Q, Xu F, Min MH, Chu SH, Kim KW, Park YJ (2019) Genome-wide association study of vitamin E using genotyping by sequencing in sesame (*Sesamum indicum*). Genes Genomics 41(9):1085–1093

Huang X, Sang T, Zhao Q, Feng Q, Zhao Y, Li C, Zhu C, Lu T, Zhang Z, Li M, Fan D (2010) Genome-wide association studies of 14 agronomic traits in rice landraces. Nat Genet 42(11):961–967

Hufford MB, Xu X, Van Heerwaarden J, Pyhäjärvi T, Chia JM, Cartwright RA, Elshire RJ, Glaubitz JC, Guill KE, Kaeppler SM, Lai J (2012) Comparative population genomics of maize domestication and improvement. Nat Genet 44(7):808–811

Hulbert SH, Richter TE, Axtell JD, Bennetzen JL (1990) Genetic mapping and characterization of sorghum and related crops by means of maize DNA probes. Proc Natl Acad Sci 87(11):4251–4255

Jiao Y, Zhao H, Ren L, Song W, Zeng B, Guo J, Wang B, Liu Z, Chen J, Li W, Zhang M (2012) Genome-wide genetic changes during modern breeding of maize. Nat Genet 44(7):812–815

Jin P, Wang L, Zhao W, Zheng J, Wang YH, Liu Y, Meng R, Dai J, Zhou L, Li J (2021) Construction of high density genetic map and QTL mapping in sorghum× sudangrass. Euphytica 217(8):1–11

Jinek M, Chylinski K, Fonfara I, Hauer M, Doudna JA, Charpentier E (2012) A programmable dual-RNA–guided DNA endonuclease in adaptive bacterial immunity. Science 337(6096):816–821

Josia C, Mashingaidze K, Amelework AB, Kondwakwenda A, Musvosvi C, Sibiya J (2021) SNP-based assessment of genetic purity and diversity in maize hybrid breeding. PLoS One 16(8):e0249505

Kajiya-Kanegae H, Takanashi H, Fujimoto M, Ishimori M, Ohnishi N, Wacera WF, Omollo EA, Kobayashi M, Yano K, Nakano M, Kozuka T (2020) RAD-seq-based high-density linkage map construction and QTL mapping of biomass-related traits in sorghum using the Japanese landrace Takakibi NOG. Plant Cell Physiol 61(7):1262–1272

Kamal NM, Gorafi YSA, Tomemori H, Kim JS, Elhadi GMI, Tsujimoto H (2023) Genetic variation for grain nutritional profile and yield potential in sorghum and the possibility of selection for drought tolerance under irrigated conditions. BMC Genomics 24(1):515

Karadi A, Kajjidoni ST (2019) Genetic variability and diversity for productivity traits and grain quality including nutritional quality traits in selected mini core and promising released varieties of sorghum. J Pharmacogn Phytochem 8(4):2091–2097

Kayodé AP, Linnemann AR, Hounhouigan JD, Nout MJ, van Boekel MA (2006) Genetic and environmental impact on iron, zinc, and phytate in food sorghum grown in Benin. J Agric Food Chem 54(1):256–262

Kimani W, Zhang LM, Wu XY, Hao HQ, Jing HC (2020) Genome-wide association study reveals that different pathways contribute to grain quality variation in sorghum (*Sorghum bicolor*). BMC Genomics 21(1):1–19

Knoll J, Gunaratna N, Ejeta G (2008) QTL analysis of early-season cold tolerance in sorghum. Theor Appl Genet 116(4):577–587

Kong L, Dong J, Hart GE (2000) Characteristics, linkage-map positions, and allelic differentiation of *Sorghum bicolor* (L.) Moench DNA simple-sequence repeats (SSRs). Theor Appl Genet 101(3):438–448

Kotla A, Phuke R, Hariprasanna K, Mehtre SP, Rathore A, Radhika K, Hash CT, Reddy BVS, Patil JV, Jabeen F, Shashikanth D (2016) Identification of QTLs and underlying candidate genes controlling grain Fe and Zn concentration in sorghum [Sorghum bicolor (L). Moench]. International Crops Research Institute for the Semi-Arid Tropics (ICRISAT)

Kotla A, Phuke R, Hariprasanna K, Mehtre SP, Rathore A, Gorthy S, Srivastava RK, Das R, Prakash AB, Radhika K, Hash CT (2019) Identification of QTLs and candidate genes for high grain Fe and Zn concentration in sorghum [Sorghum bicolor (L.) Moench]. J Cereal Sci 90:102850

Li D, Dossa K, Zhang Y, Wei X, Wang L, Zhang Y, Liu A, Zhou R, Zhang X (2018a) GWAS uncovers differential genetic bases for drought and salt tolerances in sesame at the germination stage. Genes 9(2):87

Li A, Jia S, Yobi A, Ge Z, Sato SJ, Zhang C, Angelovici R, Clemente TE, Holding DR (2018b) Editing of an alpha-kafirin gene family increases, digestibility and protein quality in sorghum. Plant Physiol 177(4):1425–1438

Li P, Zhou H, Yang H, Xia D, Liu R, Sun P, Wang Q, Gao G, Zhang Q, Wang G, He Y (2020) Genome-wide association studies reveal the genetic basis of fertility restoration of CMS-WA and CMS-HL in xian/indica and aus accessions of rice (*Oryza sativa* L.). Rice 13(1):1–12

Li X, Liu W, Wang G, Sun SSM, Yuan L, Wang J (2023) Improving digestibility of sorghum proteins by CRISPR/Cas9-based genome editing. Food Energy Secur 13:e506

Lin YR, Schertz KF, Paterson AH (1995) Comparative analysis of QTLs affecting plant height and maturity across the Poaceae, in reference to an interspecific sorghum population. Genetics 141(1):391–411

Lin M, Matschi S, Vasquez M, Chamness J, Kaczmar N, Baseggio M, Miller M, Stewart EL, Qiao P, Scanlon MJ, Molina I (2020) Genome-wide association study for maize leaf cuticular conductance identifies candidate genes involved in the regulation of cuticle development. Genetics 10(5):1671–1683

Lu F, Romay MC, Glaubitz JC, Bradbury PJ, Elshire RJ, Wang T, Li Y, Semagn K, Zhang X, Hernandez AG (2015) High-resolution genetic mapping of maize pan-genome sequence anchors. Nat Commun 6(1):1–8

Lu K, Wei L, Li X, Wang Y, Wu J, Liu M, Zhang C, Chen Z, Xiao Z, Jian H, Cheng F (2019) Whole-genome resequencing reveals Brassica napus origin and genetic loci involved in its improvement. Nat Commun 10(1):1–12

Luo H, Zhao W, Wang Y, Xia Y, Wu X, Zhang L, Tang B, Zhu J, Fang L, Du Z, Bekele WA (2016) SorGSD: a sorghum genome SNP database. Biotechnol Biofuels 9(1):1–9

Mace ES, Xia L, Jordan DR, Halloran K, Parh DK, Huttner E, Wenzl P, Kilian A (2008) DArT markers: diversity analyses and mapping in *Sorghum bicolor*. BMC Genomics 9(1):1–11

Mace ES, Rami JF, Bouchet S, Klein PE, Klein RR, Kilian A, Wenzl P, Xia L, Halloran K, Jordan DR (2009) A consensus genetic map of sorghum that integrates multiple component maps and high-throughput Diversity Array Technology (DArT) markers. BMC Plant Biol 9(1):1–14

Mace ES, Tai S, Gilding EK, Li Y, Prentis PJ, Bian L, Campbell BC, Hu W, Innes DJ, Han X, Cruickshank A (2013) Whole-genome sequencing reveals untapped genetic potential in Africa's indigenous cereal crop sorghum. Nat Commun 4(1):1–9

Madhusudhana R, Hariprasanna K, Aruna C, Sajjanar GM, Hanamaratti NG, Sameera S, Tonapi VA (2023) Genetic variability, G× E interaction and stability for iron and zinc content in sorghum grains in advanced breeding lines. J Cereal Sci 110:103653

Maina F, Bouchet S, Marla SR, Hu Z, Wang J, Mamadou A, Abdou M, Saïdou AA, Morris GP (2018) Population genomics of sorghum (*Sorghum bicolor*) across diverse agroclimatic zones of Niger. Genome 61(4):223–232

Malaghan SV, Kajjidoni ST (2016) Genetic variability for grain micronutrient (iron and zinc) content, plant morphological and other productivity related traits in rabi sorghum [sorghum bicolour sorghum bicolour sorghum bicolour (l.) Moench] (l.) Moench]. Bioscan 11(3):1951–1954

Massafaro M, Thompson A, Tuinstra M, Dilkes B, Weil CF (2016) Mapping the increased protein digestibility trait in the high-lysine sorghum mutant P721Q. Crop Sci 56(5):2647–2651

Massel K, Hintzsche J, Restall J, Kerr ED, Schulz BL, Godwin ID (2023) CRISPR-knockout of β-kafirin in sorghum does not recapitulate the grain quality of natural mutants. Planta 257(1):8

Mazaheri M, Heckwolf M, Vaillancourt B, Gage JL, Burdo B, Heckwolf S, Barry K, Lipzen A, Ribeiro CB, Kono TJ, Kaeppler HF (2019) Genome-wide association analysis of stalk biomass and anatomical traits in maize. BMC Plant Biol 19(1):1–17

McCormick RF, Truong SK, Sreedasyam A, Jenkins J, Shu S, Sims D, Kennedy M, Amirebrahimi M, Weers BD, McKinley B, Mattison A (2018) The *Sorghum bicolor* reference genome: improved assembly, gene annotations, a transcriptome atlas, and signatures of genome organization. Plant J 93(2):338–354

Meena RP, Joshi D, Bisht JK, Kant L (2021) Global scenario of millets cultivation. In: Millets and millet technology, pp 33–50

Menz MA, Klein RR, Mullet JE, Obert JA, Unruh NC, Klein PE (2002) A high-density genetic map of *Sorghum bicolor* (L.) Moench based on 2926 AFLP, RFLP and SSR markers. Plant Mol Biol 48(5):483–499

Moscou MJ, Bogdanove AJ (2009) A simple cipher governs DNA recognition by TAL effectors. Science 326(5959):1501–1501

Motlhaodi T, Bryngelsson T, Chite S, Fatih M, Ortiz R, Geleta M (2018) Nutritional variation in sorghum [Sorghum bicolor (L.) Moench] accessions from southern Africa revealed by protein and mineral composition. J Cereal Sci 83:123–129

Murray SC, Sharma A, Rooney WL, Klein PE, Mullet JE, Mitchell SE, Kresovich S (2008) Genetic improvement of sorghum as a biofuel feedstock: I. QTL for stem sugar and grain non-structural carbohydrates. Crop Sci 48(6):2165–2179

Mwamahonje A, Eleblu JSY, Ofori K, Feyissa T, Deshpande S, Garcia-Oliveira AL, Bohar R, Kigoni M, Tongoona P (2021) Introgression of QTLs for drought tolerance into farmers' preferred sorghum varieties. Agriculture 11(9):883

Nagaraja Reddy R, Madhusudhana R, Murali Mohan S, Chakravarthi DVN, Seetharama N (2012) Characterization, development and mapping of Unigene-derived microsatellite markers in sorghum (*Sorghum bicolor* (L.) Moench). Mol Breed 29(3):543–564

Nelson JC, Wang S, Wu Y, Li X, Antony G, White FF, Yu J (2011) Single-nucleotide polymorphism discovery by high-throughput sequencing in sorghum. BMC Genomics 12(1):1–15

Ng'Uni D, Shargie NG, Andersson S, Van Biljon A, Labuschagne MT (2016) Genetic variation and trait associations of yield, protein and grain micronutrients for identification of promising sorghum varieties. Cereal Res Commun 44(4):681–693

Osman MEFM, Dirar AI, Konozy EHE (2022) Genome-wide screening of lectin putative genes from Sorghum bicolor L., distribution in QTLs and a probable implications of lectins in abiotic stress tolerance. BMC Plant Biol 22(1):1–15

Paiva CL, Queiroz VAV, Simeone MLF, Schaffert RE, de Oliveira AC, da Silva CS (2017) Mineral content of sorghum genotypes and the influence of water stress. Food Chem 214:400–405

Paterson AH, Bowers JE, Bruggmann R, Dubchak I, Grimwood J, Gundlach H, Haberer G, Hellsten U, Mitros T, Poliakov A, Schmutz J (2009) The Sorghum bicolor genome and the diversification of grasses. Nature 457(7229):551–556

Peng Y, Schertz KF, Cartinhour S, Hart GE (1999) Comparative genome mapping of Sorghum bicolor (L.) Moench using an RFLP map constructed in a population of recombinant inbred lines. Plant Breed 118(3):225–235

Phuke RM, Anuradha K, Radhika K, Jabeen F, Anuradha G, Ramesh T Kumar AA (2017) Genetic variability, genotype×environment interaction, correlation, and GGE biplot analysis for grain iron and zinc concentration and other agronomic traits in RIL population of sorghum (Sorghum bicolor L. Moench). Front Plant Sci 8:712

Porteus MH, Baltimore D (2003) Chimeric nucleases stimulate gene targeting in human cells. Science 300(5620):763–763

Proietti I, Frazzoli C, Mantovani A (2015) Exploiting nutritional value of staple foods in the world's semi-arid areas: risks, benefits, challenges and opportunities of sorghum. Healthcare 3(2):172–193

Puchta H, Dujon B, Hohn B (1993) Homologous recombination in plant cells is enhanced by in vivo induction of double strand breaks into DNA by a site-specific endonuclease. Nucleic Acids Res 21(22):5034–5040

Ragab RA, Dronavalli S, Maroof MS, Yu YG (1994) Construction of a sorghum RFLP linkage map using sorghum and maize DNA probes. Genome 37(4):590–594

Raghuvanshi R, Srivastava AK, Verulkar S, Suprasanna P (2021) Unlocking allelic diversity for sustainable development of salinity stress tolerance in rice. Curr Genomics 22(6):393

Raman H, Raman R, Qiu Y, Yadav AS, Sureshkumar S, Borg L, Rohan M, Wheeler D, Owen O, Menz I, Balasubramanian S (2019) GWAS hints at pleiotropic roles for FLOWERING LOCUS T in flowering time and yield-related traits in canola. BMC Genomics 20(1):1–18

Ramu P, Kassahun B, Senthilvel S, Ashok Kumar C, Jayashree B, Folkertsma RT, Reddy LA, Kuruvinashetti MS, Haussmann BIG, Hash CT (2009) Exploiting rice–sorghum synteny for targeted development of EST-SSRs to enrich the sorghum genetic linkage map. Theor Appl Genet 119(7):1193–1204

Rao Srinivasa P, Reddy BV, Nagaraj N, Upadhyaya HD (2014) Sorghum production for diversified uses. Taylor and Francis, pp 1–27

Reddy PS, Reddy BVS, Rao PS (2014) Genotype by sowing date interaction effects on sugar yield components in sweet sorghum (Sorghum bicolor L. Moench). SABRAO J Breed Genet 46(2):241–255

Rhodes DH, Hoffmann JL, Rooney WL, Ramu P, Morris GP, Kresovich S (2014) Genome-wide association study of grain polyphenol concentrations in global sorghum [Sorghum bicolor (L.) Moench] germplasm. J Agric Food Chem 62(45):10916–10927

Rhodes DH, Hoffmann L, Rooney WL, Herald TJ, Bean S, Boyles R, Brenton ZW, Kresovich S (2017) Genetic architecture of kernel composition in global sorghum germplasm. BMC Genomics 18(1):1–8

Ritter KB, Jordan DR, Chapman SC, Godwin ID, Mace ES, Lynne McIntyre C (2008) Identification of QTL for sugar-related traits in a sweet× grain sorghum (*Sorghum bicolor* L. Moench) recombinant inbred population. Mol Breed 22(3):367–384

Sanjana Reddy P, Reddy BV, Ashok Kumar A, Ramesh S, Sahrawat KL, Venkateswara Rao P (2010) Association of grain Fe and Zn contents with agronomic traits in sorghum. Indian J Plant Genet Resour 23:280–284

Satrio RD, Nikmah IA, Fendiyanto MH, Pratami MP, Awwanah M, Sari NIP, Farah N, Nurhadiyanta N (2022) Construction of an ultra-high-density consensus genetic map and analysis of recombination rate variation in *Sorghum bicolor*. Asian J Agric 6(1):47

Schloss S, Mitchell S, White G, Kukatla R, Bowers J, Paterson A, Kresovich S (2002) Characterization of RFLP probe sequences for gene discovery and SSR development in *Sorghum bicolor* (L.) Moench. Theor Appl Genet 105(6):912–920

Schmidt C, Pacher M, Puchta H (2019) DNA break repair in plants and its application for genome engineering. In: Transgenic plants: methods and protocols, pp 237–266

Sejake T, Shargie N, Figlan S, Mofokeng A, Tsilo TJ (2022) Phenotypic variation of sorghum accessions for grain yield and quality traits. Agronomy 12(12):3089

Shaikh TR, Pawar KSG (2021) Evaluation of hybrid purity with their parents in sorghum (*Sorghum bicolor* L. Monech) by using Rapd and SSR markers. Pharm Innov J 10(6):155–159

Shegro A, Shargie NG, van Biljon A, Labuschagne MT (2012) Diversity in starch, protein and mineral composition of sorghum landrace accessions from Ethiopia. J Crop Sci Biotechnol 15(4):275–280

Shiringani AL, Friedt W (2011) QTL for fibre-related traits in grain× sweet sorghum as a tool for the enhancement of sorghum as a biomass crop. Theor Appl Genet 123(6):999–1011

Singhal D, Gupta P, Sharma P, Kashyap N, Anand S, Sharma H (2011) In-silico single nucleotide polymorphisms (SNP) mining of *Sorghum bicolor* genome. Afr J Biotechnol 10(5):580–583

Soriano JM (2020) Molecular marker technology for crop improvement. Agronomy 10(10):1462

Srinivas G, Satish K, Madhusudhana R, Nagaraja Reddy R, Murali Mohan S, Seetharama N (2009a) Identification of quantitative trait loci for agronomically important traits and their association with genic-microsatellite markers in sorghum. Theor Appl Genet 118(8):1439–1454

Srinivas G, Satish K, Madhusudhana R, Seetharama N (2009b) Exploration and mapping of microsatellite markers from subtracted drought stress ESTs in *Sorghum bicolor* (L.) Moench. Theor Appl Genet 118(4):703–717

Srinivasa Rao P, Reddy BV, Nagaraj N, Upadhyaya HD (2014) Sorghum production for diversified uses. In: Genetics, genomics and breeding of sorghum. CRC Press, pp 1–27

Subudhi PK, Nguyen HT (2000) Linkage group alignment of sorghum RFLP maps using a RIL mapping population. Genome 43(2):240–249

Sukumaran S, Dreisigacker S, Lopes M, Chavez P, Reynolds MP (2015) Genome-wide association study for grain yield and related traits in an elite spring wheat population grown in temperate irrigated environments. Theor Appl Genet 128:353–363

Tanksley SD (1993) Mapping polygenes. Annu Rev Genet 27(1):205–233

Tao YZ, Jordan DR, Henzell RG, McIntyre CL (1998) Identification of genomic regions for rust resistance in sorghum. Euphytica 103(3):287–292

Tao YZ, Henzell RG, Jordan DR, Butler DG, Kelly AM, McIntyre CL (2000) Identification of genomic regions associated with stay green in sorghum by testing RILs in multiple environments. Theor Appl Genet 100(8):1225–1232

Taramino G, Tarchini R, Ferrario S, Lee M, Pe ME (1997) Characterization and mapping of simple sequence repeats (SSRs) in *Sorghum bicolor*. Theor Appl Genet 95(1):66–72

Tuinstra MR, Grote EM, Goldsbrough PB, Ejeta G (1996) Identification of quantitative trait loci associated with pre-flowering drought tolerance in sorghum. Crop Sci 36(5):1337–1344

Upadhyaya HD, Dwivedi SL, Singh S, Sahrawat KL, Singh SK (2016) Genetic variation and post flowering drought effects on seed iron and zinc in ICRISAT sorghum mini core collection. Crop Sci 56(1):374–383

Wagh AD, Gahukar SJ, Gangurde SS (2016) Marker assisted selection for shoot fly tolerance in sorghum, Sorghum bicolor. Biosci Biotechnol Res Commun 9(2):179–188

Wei X, Liu K, Zhang Y, Feng Q, Wang L, Zhao Y, Li D, Zhao Q, Zhu X, Zhu X, Li W (2015) Genetic discovery for oil production and quality in sesame. Nat Commun 6(1):1–10

Welderufael S, Abay F, Ayana A, Amede T (2024) Genetic diversity, correlation and genotype× yield× trait (GYT) analysis of grain yield and nutritional quality traits in sorghum (Sorghum bicolor [L.] Moench) genotypes in Tigray, northern Ethiopia. Discover Agric 2(1):4

Winn JA, Mason RE, Robbins AL, Rooney WL, Hays DB (2009) QTL mapping of a high protein digestibility trait in Sorghum bicolor. Int J Plant Genomics 2009(1):471853

Wondimu Z, Dong H, Paterson AH, Worku W, Bantte K (2021) Genetic diversity, population structure, and selection signature in Ethiopian sorghum (*Sorghum bicolor* L.(Moench)) germplasm. G3 11(6):087

Wu Y, Huang Y (2008) Molecular mapping of QTLs for resistance to the greenbug Schizaphis graminum (Rondani) in *Sorghum bicolor* (Moench). Theor Appl Genet 117(1):117–124

Wu Y, Li X, Xiang W, Zhu C, Lin Z, Wu Y et al (2012) Presence of tannins in sorghum grains is conditioned by different natural alleles of Tannin1. Proc Natl Acad Sci 109(26):10281–10286

Wu Y, Yuan L, Guo X, Holding DR, Messing J (2013) Mutation in the seed storage protein kafirin creates a high-value food trait in sorghum. Nat Commun 4(1):2217

Xu GW, Magill CW, Schertz KF, Hart GE (1994) A RFLP linkage map of *Sorghum bicolor* (L.) Moench. Theor Appl Genet 89(2):139–145

Yano K, Yamamoto E, Aya K, Takeuchi H, Lo PC, Hu L, Yamasaki M, Yoshida S, Kitano H, Hirano K, Matsuoka M (2016) Genome-wide association study using whole-genome sequencing rapidly identifies new genes influencing agronomic traits in rice. Nat Genet 48(8):927–934

Yonemaru JI, Ando TSUYU, Mizubayashi TATSUMI, Kasuga SHIGEMITSU, Matsumoto TAKASHI, Yano MASAHIRO (2009) Development of genome-wide simple sequence repeat markers using whole-genome shotgun sequences of sorghum (*Sorghum bicolor* (L.) Moench). DNA Res 16(3):187–193

Zhang Y, Fan X, Liang D, Guo Q, Zhang X, Nie M, Li C, Meng S, Zhang X, Xu P, Guo W (2023) The identification of a yield-related gene controlling multiple traits using GWAS in Sorghum (Sorghum bicolor L.). Plants 12(7):1557

Zheng LY, Guo XS, He B, Sun LJ, Peng Y, Dong SS, Liu TF, Jiang S, Ramachandran S, Liu CM, Jing HC (2011) Genome-wide patterns of genetic variation in sweet and grain sorghum (*Sorghum bicolor*). Genome Biol 12(11):1–15

Zhu Y, Peng D, Lin C, Nie G, Xu W, Huang L, Luo F, Peng J, Zhang X (2018) Development of SSR markers based on transcriptome sequence and analysis of genetic diversity in *Sorghum sudanense*. Acta Pratacul Sin 27(5):178–189

Zou G, Zhai G, Feng Q, Yan S, Wang A, Zhao Q, Shao J, Zhang Z, Zou J, Han B, Tao Y (2012) Identification of QTLs for eight agronomically important traits using an ultra-high-density map based on SNPs generated from high-throughput sequencing in sorghum under contrasting photoperiods. J Exp Bot 63(15):5451–5462

Chapter 5
Advances in Pearl Millet Genomics for Nutritional Traits Improvement

Sandeep Nanjundappa, Mahalingam Govindaraj, Tara C. Satyavathi, and Nepolean Thirunavukkarasu

Abstract Pearl millet (*Pennisetum glaucum* (L.) R. Br.) is a highly climate-resilient cereal extensively cultivated in arid and semi-arid agroecological zones. It plays a vital role in the diets of nutritionally vulnerable populations, particularly those suffering from iron and zinc deficiencies, key contributors to hidden hunger and widespread malnutrition. Over the past two decades, significant progress in genomic research has revolutionized pearl millet improvement by providing deep insights into its genome structure, organization, and functional complexity. Initially limited to traditional molecular markers like RFLPs, RAPDs, SSRs, and AFLPs, pearl millet genomics has now advanced to include high-density SNP arrays, genotyping-by-sequencing (GBS), and the development of platinum-grade reference genomes and pangenome resources. These advancements have facilitated the construction of dense genetic linkage maps and enabled the identification of quantitative trait loci (QTLs) and marker-trait associations through genome-wide association studies (GWAS), significantly accelerating the dissection of complex traits. Such progress has led to the discovery of candidate genes controlling important agronomic and nutritional traits. For instance, genes like *PgZIP (zinc transporter), PgNAS3 (nicotianamine synthase), PgFER (ferritin)*, and lipid-degrading enzymes such as *PgTAG lipase, PgLOX*, and *PgPLA2* have been linked to iron/zinc accumulation, protein quality, and flour rancidity. Modern approaches such as genomic selection (GS), pangenome analysis, and integrated multi-omics are reshaping predictive breeding strategies. However, challenges persist due to the genome's complexity, including high repetitive DNA, structural variation, and heterogeneity. Future directions involving long-read sequencing, haplotype-resolved assemblies, CRISPR/Cas

S. Nanjundappa · T. C. Satyavathi · N. Thirunavukkarasu (✉)
Global Centre of Excellence on Millets (Shree Anna), ICAR-Indian Institute of Millets Research, Hyderabad, India
e-mail: nepolean@millets.res.in

M. Govindaraj
Alliance Bioversity International and International Center for Tropical Agriculture (CIAT), Cali, Colombia

M. Govindaraj (ed.), *Breeding Biofortified Millets for Nutritional Security*, Sustainability Sciences in Asia and Africa,
https://doi.org/10.1007/978-981-95-9827-4_5

genome editing, and AI-driven breeding will be crucial for unlocking pearl millet's full genetic potential to support climate resilience and nutritional security.

Keywords Pearl millet · Biofortification · Genomics · SNPs · Multi-omics · Accelerated Selection

5.1 Introduction

Global food and nutrition security in the twenty-first century is a challenging task due to the combined impacts of climate change and economic vulnerabilities (Mirzabaev et al. 2023). Pearl millet (*Pennisetum glaucum* (L.) R. Br.) is critical to tackling this issue, particularly in the severe dryland agro ecosystems of Sub-Saharan Africa and South Asia (Yadav et al. 2024a). Pearl millet, known for its amazing tolerance to heat, drought, and nutrient-poor soils, not only provides food but also sustains the lives of millions in some of the world's most marginalized and resource-poor locations (Yadav et al. 2024b). It is regarded as a "Powerhouse of Nutrition" due to its high iron and zinc content, rich protein profile, balanced amino acid composition, and enhanced amounts of vitamins such as niacin, folate, and riboflavin (Kanatti et al. 2014; Anitha et al. 2019).

Due to its excellent nutritional profile, pearl millet is gaining popularity worldwide, including in developed countries. Its low glycemic index, gluten-free nature, and abundance of bioactive compounds confer additional health benefits, positioning it as a functional food in the fight against "hidden hunger" and lifestyle-related disorders (Jacob et al. 2024). Despite its nutritional superiority, the commercial utilization of its flour is constrained by poor keeping quality and the development of off-odor during storage. Nutritional enhancement efforts in pearl millet now target a suite of interrelated traits such as Fe, Zn, protein quality, amino acid balance, vitamin density, and extended grain shelf life to simultaneously address malnutrition and improve market value (Govindaraj et al. 2019).

For pearl millet to fully realize its potential as a nutrient-rich crop, advances in grain nutrient content must be combined with an understanding of nutrient homeostasis, as both are required for long-term gains. Equally crucial is how efficiently the plant acquires nutrients from the soil, translocates them through its tissues, and finally deposits them in bioavailable forms that the human body can absorb (Sharma et al. 2025). Pearl millet thrives even in nutrient-poor soils and still captures essential minerals, owing to its robust root system and finely regulated transport processes, and soil fertility matters when it goes below critical limits (Govindaraj et al. 2019). Strengthening this natural balance of nutrient homeostasis offers an opportunity not only to raise nutrient levels in the grain but also to ensure that these nutrients are truly accessible in human diets (Krishnan and Meera 2018).

Genomics has emerged as a transformative approach to dissecting the genetic basis of these traits, enabling more precise and accelerated improvement strategies (Nepolean et al. 2018). A comprehensive understanding of the pearl millet genome is essential for unravelling the molecular architecture of nutritional traits such as iron, zinc, protein quality, and vitamins. These traits are highly complex and polygenic, influenced by intricate gene networks and their interactions with environmental factors. Without genomic insights, conventional breeding approaches often face limitations in achieving stable and substantial gains due to the low heritability and genotype-by-environment interactions associated with these traits (Dwivedi et al. 2024). Furthermore, insights into allelic diversity, structural variations, and gene expression dynamics allow breeders to design strategies that combine biofortification with essential agronomic traits, thereby ensuring nutritional security without compromising yield.

In the last two decades, pearl millet genomics has advanced considerably, transitioning from the use of simple molecular markers to high-resolution, genome-wide approaches. Early efforts relied on restriction fragment length polymorphism (RFLP), random amplified polymorphic DNA (RAPD), simple sequence repeat (SSR), and amplified fragment length polymorphism (AFLP) markers for diversity analysis and quantitative trait loci (QTL) mapping. However, the advent of next-generation sequencing (NGS) technologies marked a paradigm shift, enabling the generation of a high-quality reference genome and extensive marker resources. Today, advanced platforms such as Illumina, PacBio, and Oxford Nanopore, combined with Hi-C scaffolding and optical mapping, deliver near-complete assemblies with high contiguity and accuracy (Kasule et al. 2024). These genomic resources, complemented by transcriptomic, proteomic, and metabolomic datasets, have accelerated the application of genome-wide association studies (GWAS), genomic selection (GS), and haplotype-based breeding. Emerging approaches such as pangenome construction, structural variant mapping, and CRISPR-based functional genomics are further expanding the toolkit, allowing researchers to capture the full spectrum of genetic variation underlying agronomic and nutritional traits in pearl millet. These integrative, genomics-led biofortification strategies are poised to deliver nutrient-dense pearl millet cultivars that can serve as a sustainable solution for enhancing dietary diversity and resilience in nutritionally insecure and vulnerable populations.

5.2 Impact on Malnutrition and Relevance for Genetic Biofortification

Micronutrient deficiencies are widespread in regions relying heavily on cereal-based diets. These deficiencies, often termed "hidden hunger," disproportionately affect vulnerable groups such as children, women of reproductive age, and the elderly, leading to severe health and socioeconomic consequences (WHO 2020). Globally, half of preschool aged children and two-thirds of nonpregnant women of reproductive age suffer from at least one micronutrient deficiency, with iron and zinc being the most prevalent (Stevens et al. 2022) Currently, 40% of pregnant

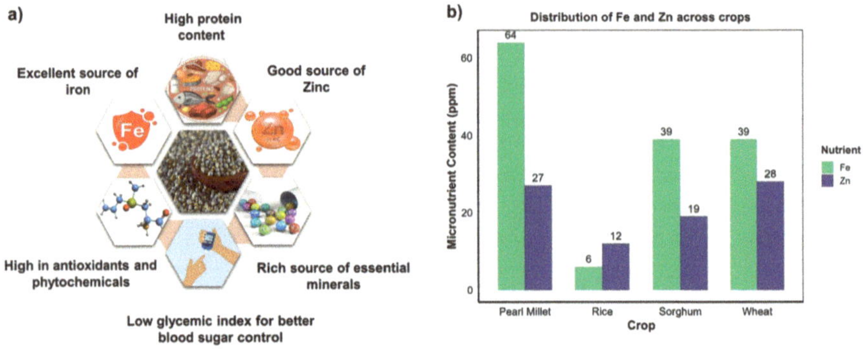

Fig. 5.1 (**a**) Nutritional benefits and (**b**) Comparative micronutrient (Fe and Zn) content of pearl millet with major cereals: Rice, Sorghum, and Wheat

women and 42% of children below 5 years of age are anemic and 17% of the global population is at risk of inadequate Zn intakes (WHO 2021). Nutrient deficiencies contribute to a wide range of health problems, including impaired growth and development, weakened immunity, reduced cognitive function, and lower work capacity, collectively diminishing individual well-being and national productivity (Black et al. 2013).

Pearl millet is inherently nutritious, encompassing essential macro and micronutrients, vitamins, and health-promoting phytochemicals. It contains higher levels of Fe (40–80 ppm) and Zn (30–60 ppm) compared to other cereals (Fig. 5.1). Additionally, it offers considerable protein (10–15%) and essential amino acids, as well as dietary fiber and B complex vitamins (Rai et al. 2013). In rural areas with limited dietary diversity and heavy dependence on staple cereals, such as pearl millet, genetic biofortification provides a cost-effective, sustainable, and culturally acceptable approach to increasing micronutrient intake without changing established food habits. Enhancing nutrient density through breeding and genomics offers a sustainable solution to combat micronutrient malnutrition. The introduction of nutrient-dense cultivars such as biofortified pearl millet can help reduce hidden hunger while also improving socio-economic resilience through enhanced labor productivity, reduced healthcare costs, and higher market value of nutrient-rich grains (Bouis and Saltzman 2017).

5.3 Genetic Nature, Variability, and Molecular Pathways Underlying Nutritional Traits in Pearl Millet

Enhancement of grain nutritional traits has become a recent breeding priority, driven by the global recognition of widespread micronutrient deficiencies, alongside the need to improve other key nutrients such as protein quality, amino acid balance, and essential vitamins (Bouis et al. 2024). However, the success of such efforts ultimately depends on the presence of adequate genetic variation within the germplasm to support substantial gains in trait enhancement. Substantial genotypic variation exists for these traits, with landraces, breeding lines, and wild accessions serving as valuable reservoirs of alleles for improvement. Extensive variability in Fe and Zn content across germplasm and breeding materials has been well documented. Assessment of 428 pearl millet accessions (core collection) revealed wide ranges in grain nutrient concentrations: Fe (34–90 mg/kg), Zn (30–74 mg/kg), Ca (85–249 mg/kg), Mn (10–17 mg/kg), Cu (4–7 mg/kg), Mo (0.6–2.2 mg/kg), Ni (0.7–1.9 mg/kg), Mg (1127–1837 mg/kg), Na (8–22 mg/kg), K (3167–5133 mg/kg), P (2900–4733 mg/kg), and S (1123–1703 mg/kg) (Govindaraj et al. 2020). High-throughput phenotyping tools, such as energy-dispersive X-ray fluorescence (XRF), have further boosted these breeding efforts by making it easy to rapidly screen large numbers of germplasm for mineral content (Govindaraj et al. 2024).

Nutritional traits in pearl millet are predominantly polygenic and subject to genotype-by-environment interactions (Kanatti et al. 2014; Govindaraj et al. 2016). Iron and zinc concentrations exhibit moderate-to-high heritability, indicating good potential for genetic enhancement (Velu et al. 2008). Environmental factors, including soil nutrient status and moisture availability, further influence their expression. Screening of over 120 Indian commercial hybrids revealed Fe concentrations of 46–56 ppm and Zn concentrations of 37–44 ppm (Rai et al. 2016). The predominance of additive genetic variance for these traits (Govindaraj et al. 2016; Kanatti et al. 2014) highlights the need to improve both parental lines in hybrid breeding programmes to enhance micronutrient content. ICTP 8203, an OPV with high grain Fe content, was improved through three cycles of progeny-based mass selection, producing Dhanashakti, released in 2014 as India's first biofortified crop cultivar (Rai et al. 2016). Similarly, Chakti was released in West Africa in 2018 (Gangashetty et al. 2021).

Improving flour storability and quality has also become a key focus. A recent study evaluated 93 diverse pearl millet germplasm accessions, revealing significant variability in hydrolytic and oxidative lipid degradation both in freshly milled flour and after 10 days of storage. Using parameters such as comprehensive acid value, peroxide value, and lipase and lipoxygenase activities, accessions were classified into low, medium, and high rancidity groups. This led to the development of a rancidity matrix (Goswami et al. 2020), enabling the identification of low-rancid germplasm for varietal improvement.

At the molecular level, nutritional traits are governed by highly complex metabolic pathways involving nutrient uptake, transport, storage, and homeostasis. Amino acid and protein content are regulated through interconnected pathways such as the *aspartate family pathway, shikimate pathway*, and *glutamine–glutamate cycle*, coordinated by biosynthetic enzymes, transporters, and regulatory transcription factors (e.g., *bZIP, MYB*) (Heidarzadeh 2025). Mineral accumulation involves multiple metal transport and homeostasis pathways, including *ZIP (Zinc-/Iron-regulated transporter-like proteins)*, *NRAMP (Natural Resistance-Associated Macrophage Protein)*, and *CAX (cation exchangers)*, coupled with chelation and compartmentalization in vacuoles or seed tissues (Stanton et al. 2022). Nutrient metabolism intersects with broader pathways, such as sulfur assimilation, nitrogen assimilation, folate biosynthesis, and lipid metabolism, which are integrated with antioxidant and stress response networks (e.g., *glutathione and ascorbate pathways*) (Wu et al. 2021). These pathways are highly interconnected, with multiple feedback loops, cross-talk, and environmental responsiveness, making them inherently complex (Mishra et al. 2025). Overall, the integration of genetic variation, comprehensive insights into molecular pathways, and high-throughput phenotyping offers a robust framework for dissecting the genetics and genomics underlying nutritional traits in pearl millet.

5.4 Genomic Resources in Pearl Millet

Over the past few decades, pearl millet genomics has evolved from the era of classical markers to cutting-edge genome-enabled breeding, significantly enhancing our understanding of complex traits. Early efforts relied on isozymes, RFLPs, RAPDs, and AFLP markers, which laid the foundation for assessing genetic diversity and constructing preliminary linkage maps (Singh and Nara 2023a, b). These approaches, although limited in resolution, enabled the identification of genomic regions associated with key agronomic and nutritional traits, setting the stage for marker-assisted selection. The development of SSR markers further enhanced the ability to dissect genetic variation, providing robust tools for diversity analysis and the selection of favorable alleles.

Early efforts to sequence the pearl millet genome involved shotgun and BAC-based approaches, resulting in the draft genome of the inbred line Tift $23D_2B_1$-P1-P5 (1.79 Gb, 38,579 annotated genes) (Varshney et al. 2017). Although highly fragmented, consisting of thousands of contigs and numerous gaps, this draft genome provided a foundational framework for identifying QTLs and candidate genes controlling grain nutrient composition, quality traits, and stress resilience, marking the beginning of genome-assisted breeding in pearl millet. Salson et al. (2023) reported a more complete and higher-quality assembly of Tift $23D_2B_1$-P1-P5, with a genome size of 1.8 Gb and 36,898 genes, of which 98% aligned with the previous version, covering 99% of the genome, using Oxford Nanopore Technologies (ONT). Building on this foundation, de novo assemblies of additional lines such as Tift, ICMR 06777, and 843 B, generated using PacBio HiFi long reads, bionano optical maps, and advanced scaffolding, have achieved exceptional contiguity (scaffold N50: 144–279 Mb), enabling accurate detection of structural variants and haplotypes (Ramu et al. 2023). To date, more than 20 pearl millet de novo assemblies have been reported, representing a substantial advancement in genomic resources for functional genomics and breeding (Table 5.1).

Concurrently, pangenome initiatives are capturing the broader genetic diversity across landraces, breeding lines, and biofortified varieties, revealing presence–absence variations and copy number variations that were absent in the Tift reference genome (Yan et al. 2023; Chen et al. 2023). These structural and sequence-level variations are critical for enhancing both the concentration and bioavailability of micronutrients while minimizing anti-nutritional factors.

Advances in genome assembly have gone hand in hand with the evolution of high-resolution molecular markers. Diversity arrays technology (DArT), single-nucleotide polymorphism (SNP) arrays, genotyping by sequencing (GBS) platforms, and haplotype-based markers have progressively enabled the fine-mapping of trait-associated loci, facilitating the pyramiding of high Fe/Zn, yield, and stress-resilience traits (Kumar et al. 2024; Kanfany et al. 2020; Supriya et al. 2011). Well-characterized germplasm resources, including PMiGAP panels, core and mini-core collections, bi-parental Recombinant inbred lines (RILs), chromosome segment substitution lines (CSSLs), and association panels, have been instrumental in conducting GWAS and QTL mapping to dissect grain nutrient traits (Sehgal et al. 2015; Srivastava et al. 2021). However, challenges remain in developing nutrient-focused core collections enriched for high-variation accessions in Fe, Zn, protein quality, and other biofortification traits, which would accelerate allele discovery and breeding.

Integrated platforms such as "*The Millet Genome Database (MilletDB)*," complemented by transcriptomic resources such as "*PMDTdb*," provide detailed insights into gene expression, genotype–phenotype associations, and pangenome analyses,

Table 5.1 Summary of publicly available pearl millet genome assemblies

Sl. No.	Sequencing strategy/ Technology	Accessions/Identifiers	Highlights	References
1	Illumina-only (short reads)	Tift 23D2B1-P1-P5	First draft genome; basic assembly	Varshney et al. (2017)
2	Hybrid short + long reads (Illumina PE + ONT)	Tift 23D2B1-P1-P5	Improved contiguity over short-read assembly	Salson et al. (2023)
3	PacBio HiFi + Bionano + Omni-C	Tift 23D2B1-P1-P5, ICMR 06777, 843B	High-quality, near-complete assemblies	Ramu et al. (2023)
4	PacBio HiFi + Bionano + Hi-C + Illumina	PI 186338, PI 250656/ K551, PI 343841, PI 521612, PI 526529, PI 527388, PI 537069, PI 583800, PI 587025, Tifleaf3	Chromosome-scale assemblies with high accuracy	Yan et al. (2023)
5	Graph-based pangenome + SV analysis	10 de novo assembled accessions + 1 existing assembly	424,085 structural variants, including PAVs, CNVs, INVs; RWP-RK TF family expansion linked to heat tolerance; ER-related and HSP genes respond to heat; SVs associated with agronomic traits, domestication, and adaptation	Yan et al. (2023)

thereby facilitating candidate gene discovery and trait-based breeding for agronomic traits and nutritional enhancement (Jaiswal et al. 2018; Sun et al. 2023). At present, the pearl millet research community is at a transformative stage, with reference genomes, pangenomes, high-density markers, and multi-omics platforms collectively enabling unprecedented precision in understanding and improving this climate-resilient crop. However, challenges persist, including the need for broader germplasm integration, functional validation of candidate genes, and linking genomic variation to complex traits across diverse environments. Future directions involve leveraging graph-based pangenomes, machine learning, and functional genomics, combined with trait-focused breeding populations, to accelerate the development and deployment of nutrient-rich, stress-resilient, and high-yielding pearl millet varieties (Fig. 5.2).

5.5 Tools and Platforms for Developing Comprehensive Genomic Resources

The development of genomic resources in pearl millet has been accelerated by the availability of diverse genotyping platforms designed for different stages of trait discovery and breeding. These tools have enabled functional genomics, candidate gene discovery, and the design of robust biofortification pipelines.

5.5.1 Advances in Molecular Marker Development

5.5.1.1 Early Low-Throughput Markers

Initial studies in pearl millet primarily used RAPD and RFLP markers. RAPD markers were used for assessing genetic diversity among landraces and elite lines, helping to fingerprint germplasm collections (Govindaraj et al. 2009). RFLP markers enabled the construction of early linkage maps and the identification of QTLs for agronomic traits such as flowering time and drought tolerance (Bidinger et al. 2007). Later, AFLP markers offered higher polymorphism detection and genome coverage, supporting diversity analysis and mapping of traits including grain yield and disease resistance (Mariac et al. 2010).

5.5.1.2 Microsatellite Markers

SSR markers, also known as microsatellites, have played a pivotal role in pearl millet genomics due to their desirable attributes such as high reproducibility, co-dominant inheritance, multi-allelic nature, and high level of polymorphism. Their abundance and genome-wide distribution have made them a preferred choice for

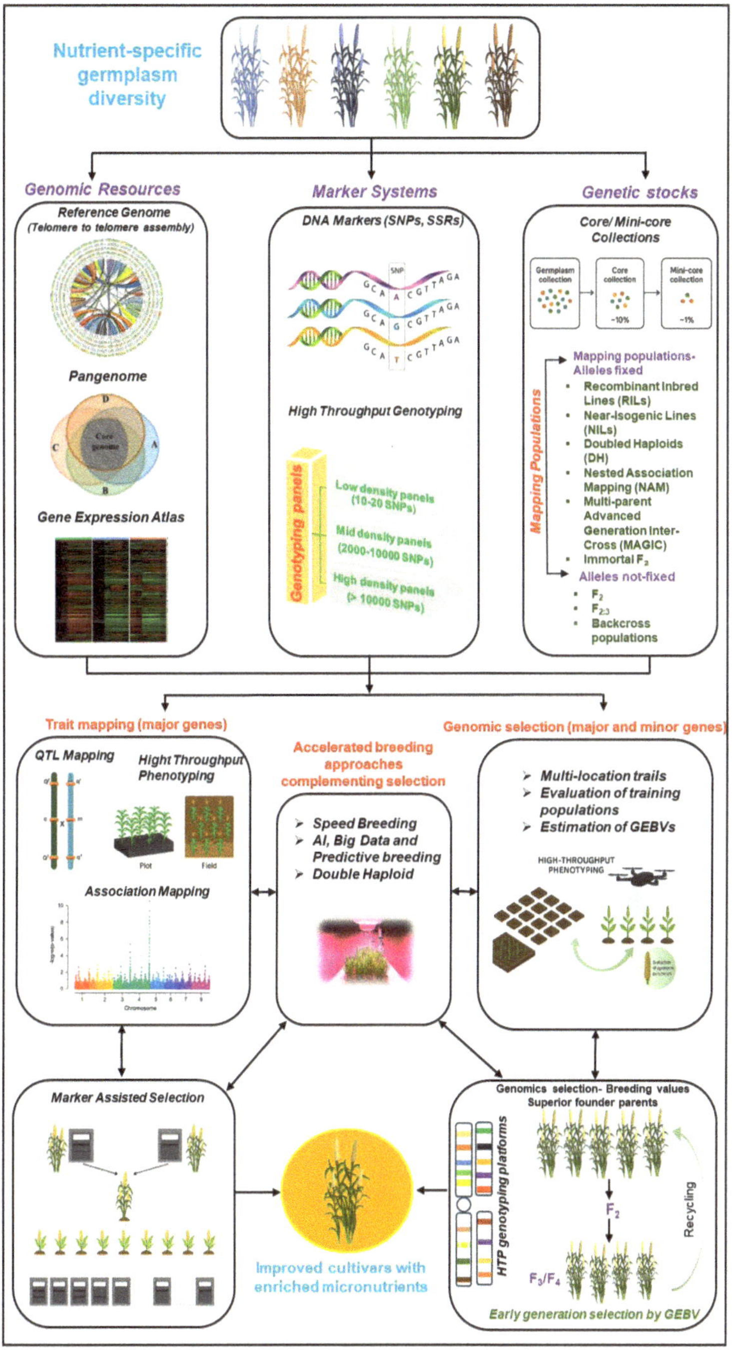

Fig. 5.2 Genomic resources and integrated breeding approaches for enhancing nutritional traits in pearl millet

genetic diversity analysis and molecular breeding prior to the availability of reference genomes. SSRs have been extensively utilized for the characterization of landraces and elite lines, assessment of genetic purity, and parental line selection for hybrid breeding programs. Moreover, SSR-based linkage maps have facilitated the identification and mapping of QTLs associated with key traits such as drought tolerance, iron and zinc accumulation, and grain quality parameters (Senthilvel et al. 2008; Mittal and Dubey 2009). Although high-throughput SNP genotyping platforms have gained prominence, SSR markers continue to be highly relevant in seed purity testing and conservation of genetic resources, thereby maintaining their significance in pearl millet improvement programs.

5.5.1.3 Diversity Arrays Technology (DArT)

DArT markers have provided a cost-effective solution for medium-density genotyping in pearl millet. As a hybridization-based marker system, DArT detects DNA polymorphisms through microarray platforms without requiring prior sequence information, making it particularly suitable for species with limited genomic resources. These markers have been widely used to screen large germplasm collections, assess genetic diversity, and establish population structure (Jaccoud et al. 2001). Furthermore, DArT-based genotyping has facilitated early-stage mapping of key agronomic and nutritional traits such as iron and zinc accumulation, flowering time, and drought tolerance (Ambawat et al. 2016; Allan et al. 2020). Despite being gradually replaced by high-density SNP genotyping platforms, DArT markers remain valuable for diversity studies, pre-breeding programs, and initial QTL discovery.

5.5.1.4 SNP-Based Genotyping

SNP markers have revolutionized genomic research by facilitating high-density, genome-wide genotyping, which significantly enhances the resolution and accuracy of genetic analyses. SNP arrays and customized panels have been extensively employed in GWAS to identify loci linked to critical agronomic and nutritional traits, including grain micronutrient content and yield components (Morgil et al. 2020). Beyond association mapping, these markers serve as the foundation for genomic prediction models and early-stage genomic selection programs, enabling breeders to predict genetic merit and select superior parental lines with greater precision and efficiency.

5.5.1.5 Genotyping-by-Sequencing

GBS-enabled high-density genome-wide SNP discovery without prior marker development. In pearl millet, GBS has been used to perform GWAS for nutritional traits, yield, and drought tolerance across diverse germplasm collections (Hu et al. 2015). Targeted sequencing strategies, including exome capture and amplicon-based panels, can be employed to interrogate candidate genes governing micronutrient homeostasis. These methodologies facilitate high-resolution detection of functional variants and enable precise fine mapping to accelerate trait improvement.

5.5.2 High-Throughput Sequencing Technologies

Advances in NGS have transformed allele discovery and functional genomics in pearl millet by enabling precise characterization of genetic variation, facilitating the identification of trait-associated genes, markers, and accelerating the development of nutrient-rich and stress-resilient varieties.

5.5.2.1 Short-Read Sequencing Technologies

Platforms such as Illumina, Ion Torrent, and BGISEQ provide high-throughput short-read data, ideal for whole-genome resequencing (WGRS). WGRS in pearl millet has been used to identify millions of SNPs across diverse accessions, supporting association studies for traits such as grain micronutrients and yield components. Exome capture complements short-read sequencing by targeting coding regions, enhancing the detection of functional variants associated with nutritional traits. RNA-seq profiling has been applied to study gene expression changes under nutrient stress, identifying genes involved in micronutrient transport, protein synthesis, and vitamin biosynthesis (Mundada et al. 2022).

5.5.2.2 Long-Read Sequencing Technologies

NGS platforms such as Oxford Nanopore and PacBio generate long-read sequencing data that markedly enhance genomic analyses. These platforms improve genome assembly quality by producing highly contiguous reference genomes, enabling precise detection of structural variants by revealing large-scale genomic rearrangements often overlooked by short-read approaches and facilitating the resolution of complex gene families, including tandem repeats and multi-copy genes associated with nutrient regulation (Logsdon et al. 2020). By combining long-read sequencing with advanced assembly algorithms, recent de novo pearl millet genome projects have delivered substantial improvements in contiguity, accuracy, and haplotype reconstruction (Ramu et al. 2023).

5.5.2.3 Hybrid Assemblies

The integration of short- and long-read sequencing data enriches genomic datasets and enables more comprehensive analyses. Hybrid genome assemblies combine the base-level accuracy of short reads with the continuity of long reads, while variant calling across scales captures genetic variation ranging from SNPs to large structural variants. Furthermore, allele-level haplotype resolution provides the precision needed to track parental contributions in breeding populations.

5.5.3 Genotyping Platforms

The accelerated development of genomic resources in pearl millet hinges on the strategic deployment of diverse genotyping platforms, each tailored to specific breeding and research stages. SSRs remain valuable for diversity assessment and parental line characterization, while SNP arrays and sequencing-based approaches enable high-density genome-wide detection of variants. DArT offers a cost-effective option for medium-density genotyping, supporting germplasm screening and trait mapping. Genotyping panels of varying densities, low-density genotyping panels (LDP), medium-density genotyping panels (MDP), and high-density genotyping panels (HDP) are strategically applied depending on the stage and objectives of breeding programs.

5.5.3.1 Low-Density Genotyping Panels

These streamlined marker sets, typically comprising fewer than 10 to 20 SNPs, serve as a highly affordable and efficient option for early-stage selection and rapid germplasm screening. Particularly valuable for pre-breeding and backcross programs, they enable breeders to quickly filter large populations and drive decision-making during initial trait introgression phases.

5.5.3.2 Mid-Density Genotyping Panels

Platforms such as the 4K AgriSeq panel developed by ICAR–IIMR, which generally range from approximately 2000 to 10,000 markers, offer a balanced solution between cost and genomic resolution (Semalaiyappan et al. 2023). Ideally suited for medium-scale studies, they allow breeders to capture major-effect loci with sufficient genome-wide coverage. This capacity enhances haplotype resolution and supports the implementation of marker-assisted selection and early-stage genomic selection.

5.5.3.3 High-Density Genotyping Panels

Comprising more than 10,000 markers, typically in the form of dense SNP arrays or GBS datasets, high-density platforms provide unparalleled precision in mapping complex traits. They facilitate meticulous fine mapping and GWAS, enabling breeders to elucidate intricate relationships among traits such as Fe/Zn accumulation, protein composition, grain storability, and yield performance (Negro et al. 2019). Together, these sequencing platforms form the backbone of contemporary molecular breeding, enabling strategic allele selection, pyramiding of nutrient-enhancing traits, and the design of robust biofortification pipelines.

5.6 Trait Mapping Approaches for Nutritional Improvement in Pearl Millet

The genetic architecture of nutritional traits in pearl millet, such as grain iron and zinc concentration, protein content, and anti-nutritional factors, is often complex, governed by multiple loci with small to moderate additive effects (Govindaraj et al. 2022). Identifying the genomic regions underlying these traits is essential for designing effective marker-assisted and genomic selection strategies. Trait mapping methodologies primarily involve QTL mapping and association studies, which use molecular markers to establish a relationship between genotype and phenotype. QTL mapping, based on biparental populations, offers high power to detect loci with relatively large effects but provides limited allelic diversity and coarse mapping resolution (Collard et al. 2005). In contrast, GWAS leverage natural populations and historical recombination to identify marker–trait associations (MTAs) with higher resolution, but require careful control for population structure and relatedness (Myles et al. 2009). Together, these complementary approaches have provided a robust framework for dissecting the genetic basis of nutritional quality in pearl millet.

5.6.1 Identification and Mapping of QTL Regions

QTL mapping uses molecular markers to link genotype and phenotype, providing insights into the number, location, and effect size of underlying loci. Traditional biparental populations such as F_2, $F_{2:3}$, backcross (BC), recombinant inbred lines (RILs), doubled haploids (DH), and near-isogenic lines (NILs) are widely employed for QTL mapping. These populations offer high power to detect loci with relatively large effects but are generally limited in allelic diversity and mapping resolution (Collard et al. 2005). To capture broader genetic variation and improve mapping resolution, advanced multi-parent populations such as multi-parent advanced generation intercrosses (MAGIC) and nested association mapping (NAM) populations have been developed. MAGIC populations combine alleles from multiple founders through successive intercrossing, increasing recombination and allelic diversity, while NAM populations use structured families derived from diverse parental lines to enhance both mapping precision and statistical power (Scott et al. 2020). These diverse mapping populations, when used alongside appropriate marker systems, provide a comprehensive framework for dissecting the genetic architecture of nutritional traits, supporting both marker-assisted and genomic selection strategies in pearl millet.

QTL mapping has helped in discovering genomic regions associated with nutritional traits, allowing marker-assisted selection and genomic breeding strategies. Multiple studies using RIL populations produced from diverse parental crosses have reported significant QTLs for grain Fe and Zn content. Kumar et al. (2016) utilized a linkage map comprising 305 markers, including SSRs and DArT, covering seven linkage groups to identify significant QTLs on linkage groups 3 and 5. These QTLs explained up to 36% and 19% of phenotypic variation for Zn and Fe, respectively. Kumar et al. (2018) mapped 19 QTLs for grain iron and zinc content in a pearl millet RIL population derived from an *Iniadi* germplasm. These QTLs explained up to 32% of phenotypic variance, with key loci co-localized on linkage groups 1 and 7. Similarly, Singhal et al. (2021) reported 22 QTLs across 3 years and multiple locations, with phenotypic variance explained ranging from 2.85% to nearly 26%. Importantly, co-localization of QTLs for Fe and Zn on linkage groups 2 and 3 suggests shared genetic control or pleiotropy, which is advantageous for simultaneous improvement of both micronutrients (Kumar et al. 2016; Singhal et al. 2021) (Fig. 5.3).

Despite these advances, the QTLs identified explain only a fraction of the total phenotypic variance, reflecting the complex polygenic inheritance of micronutrient accumulation. To address this, meta-QTL analyses integrating data from multiple studies have refined the genetic intervals and identified consensus genomic regions. Gupta et al. (2024) conducted such an analysis using 191 QTLs reported over two decades, revealing 34 Meta-QTL regions with substantially narrowed confidence intervals. Meta-QTL3.1 has been shown to possess ABC transporter-like activity coupled with ATPase activity, playing a crucial role in iron and zinc uptake in leaves and root tissues (Gupta et al. 2024). These Meta-QTLs harbor candidate genes with functions related to nutrient transport, disease resistance, and seed development, underscoring their potential utility in breeding programs (Gupta et al. 2024).

Candidate genes identified within QTL regions include *ferritin, ABC transporters*, and various ion transporters responsible for Fe, Zn, K, and Mg uptake and translocation (Kumar et al. 2016; Gupta et al. 2024; Borrill et al. 2014). These discoveries provide a robust foundation for genomics-assisted breeding aimed at improving pearl millet nutritional quality to combat micronutrient malnutrition in vulnerable populations. Enhancing the precision and applicability of findings requires integrating QTL mapping with high-resolution genomic tools and multi-environment validation. To achieve this, QTL mapping is increasingly

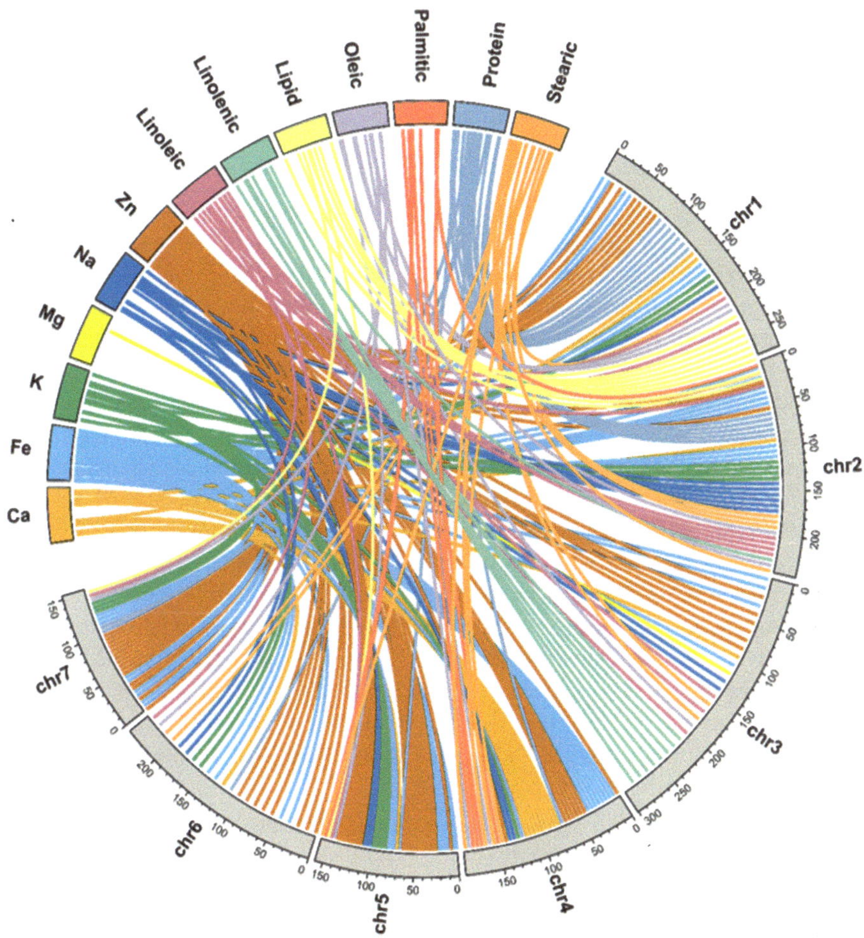

Fig. 5.3 Genomic associations of nutritional traits in pearl millet. The outermost track represents the seven pearl millet chromosomes (Chr1–Chr7). The second track displays major nutritional traits, including mineral nutrients (Fe, Zn, Ca, K, Mg, and Na), as well as protein- and lipid-related traits (linoleic, linolenic, oleic, palmitic, and stearic acids, and total lipid content). Colored links within the circle connect trait-associated loci to their corresponding chromosomal positions, as determined from various QTL and GWAS studies

complemented by approaches such as fine mapping and candidate gene identification to pinpoint causative variants for nutritional improvement.

5.6.2 Genome-Wide Association Studies

GWAS are a powerful approach for identifying MTAs in natural or breeding populations, particularly for complex traits such as grain iron and zinc concentration, protein content, and anti-nutritional factors in pearl millet. While QTL mapping in biparental populations is limited by low recombination and allelic diversity, GWAS leverages historical recombination and natural allelic variation to achieve high-resolution mapping across the genome (Myles et al. 2009). By analysing large panels of genetically diverse accessions, GWAS can detect both common and rare alleles contributing to trait variation, providing a more comprehensive understanding of the genetic architecture underlying nutritional traits.

GWAS typically employs diverse germplasm panels, including landraces, breeding lines, cultivars, and natural populations. Structured populations, such as diversity panels or multi-parent populations such as MAGIC and NAM, can also be used to enhance statistical power and reduce confounding due to population structure. Proper accounting for relatedness and population stratification is critical to avoid false-positive associations. When combined with high-density marker platforms such as SNP arrays or GBS, GWAS provides a robust framework for discovering genomic regions associated with key nutritional traits and supporting precision breeding in pearl millet (Liang et al. 2025).

GWAS in pearl millet have identified numerous significant MTAs linked to essential mineral content, including iron, zinc, calcium, magnesium, potassium, and sodium. Anuradha et al. (2017) identified favorable alleles through association mapping of 130 pearl millet lines across diverse Indian agro-climatic zones. Markers such as *Xpsmp 2261* and *Xipes 0180* showed robust association with grain iron and zinc content, enabling the identification of promising lines with multiple favorable alleles for breeding programs. Similarly, Pujar et al. (2020) uncovered 78 MTAs for iron, zinc, and protein content, including co-segregated SNPs on chromosomes 4, 5, and 7, with candidate genes coding for late embryogenesis abundant proteins, MYB domains, and iron ion-binding proteins, reinforcing the potential for genomics-assisted biofortification. The study discovered significant broad-sense heritability for Fe (93%), Zn (90%), and PC (96%), as well as strong positive correlations across the three traits, indicating the possibility of simultaneous improvement (Pujar et al. 2020). Key potential genes near key markers included those that encode *MYB transcription factors, LEA proteins*, and *PPR proteins*, all of which are likely involved in micronutrient accumulation and transport. Eleven high-performing lines were discovered, outperforming global nutrition targets, and the study emphasizes the use of these SNPs in marker-assisted breeding procedures for both hybrids and open-pollinated variants to improve pearl millet nutrition. Singh et al. (2024) reported 74 MTAs using over 456K SNPs across the pearl millet genome, with 59 candidate genes located within 50 Kb of significant SNPs. Key candidate genes include those encoding proteins with Brix domains associated with calcium and serrate RNA effector molecule-like proteins linked to iron accumulation, glycohydrolase domains relevant to potassium, and C_2H_2-type zinc finger proteins implicated in sodium regulation.

Furthermore, GWAS analyses of other nutritional traits such as lipids, fatty acids, proteins, amino acids, and starch composition have expanded the scope of biofortification in pearl millet. Mittal et al. (2025) identified 64 significant MTAs and 52 candidate genes associated with lipid content and fatty acid profiles, including genes influencing the linoleic and oleic acid concentrations. Singh et al. (2024) reported 544 significant MTAs linked to total protein and 18 amino acids, with 286 candidate genes underlying these traits. Yadav et al. (2022) dissected starch-related traits and found 902 SNPs associated with starch biosynthesis pathways, identifying candidate genes encoding starch synthases and amylases.

Recent advances have further refined mapping resolution through haplotype-based GWAS, which considers local linkage disequilibrium (LD) structure to group correlated SNPs into haplotypes, thereby capturing more genetic variation than single-SNP analyses (Schaid et al. 2018). Haplotype-based analyses not only pinpoint superior allelic combinations but also provide breeders with defined genetic blocks for introgression (Joshi et al. 2023), expediting the development of nutritionally superior pearl millet varieties. Integration of QTL mapping, GWAS, and haplotype-based approaches coupled with high-throughput genotyping platforms offers a comprehensive toolkit for accelerating the genetic improvement of nutritional traits in pearl millet. Beyond micronutrients, MTAs for seed protein content and reduced anti-nutritional factors, such as phytic acid, have also been reported, underscoring the multifaceted nature of nutritional improvement.

5.7 Genomic Approaches for Accelerated Genetic Gain in Micronutrient-Enriched Pearl Millet

5.7.1 Marker-Assisted Selection (MAS) for Nutritional Traits

MAS has been widely employed in crop breeding for the targeted improvement of nutritional traits, particularly when these traits are controlled by a few major-effect loci. Polymerase chain reaction (PCR) based SSR markers have been extensively used in various genetic and mapping studies of pearl millet due to their abundance in the genome, high polymorphism, and ease of assay (Hasan et al. 2021). With the reduction in genotyping costs, SNP-based allele-specific screening methods, such as Kompetitive Allele Specific Primers (KASP) assays (He et al. 2014), are increasingly being adopted.

MAS has been successfully applied to introgress favorable alleles for grain iron and zinc content from donor parents into elite backgrounds. For example, the validated QTL on linkage group 3 (LG3) for higher grain Fe and Zn contents (Kumar et al. 2016) has been a key target in marker-assisted breeding (MAB) programs. Similar successes have been reported in other crops, such as the use of *OsNAS* in rice to enhance biofortification and the introgression of *crtRB*1 and *lcyE* alleles in maize to increase provitamin A content (Babu et al. 2013; Naveen et al. 2014). While MAS is highly effective for traits governed by large-effect QTLs, its efficiency diminishes for complex, polygenic traits, for which genome-wide approaches such as genomic selection are more suitable for achieving genetic gain.

5.7.2 Precision Breeding Through Breeding Value-Based Selection

GS enables the estimation of genomic estimated breeding values (GEBVs) using genome-wide marker information, making it especially suitable for complex traits such as micronutrient content, which are influenced by multiple small-effect loci (Meuwissen et al. 2001). The concept of breeding value, the additive genetic merit of an individual, is central to GS-driven breeding programs. Estimated breeding values are routinely used to rank parental lines before crossing to ensure the accumulation of favorable alleles in progeny (Xu et al. 2017). Maize breeding programs extensively utilise GEBVs to select inbred parents for yield and quality traits, maximizing genetic gain (Rice and Lipka 2021). Sparse testing designs where subsets of genotypes are phenotyped in different environments and their genomic predictions combined have proven cost-effective while maintaining high prediction accuracy (Atanda et al. 2022).

In pearl millet, predictions using ridge regression best linear unbiased prediction (RR-BLUP) have achieved median prediction accuracies of 0.73–0.74 for 1000-grain weight, 0.87–0.89 for days to flowering, 0.48–0.51 for grain yield, and 0.72–0.73 for plant height (Liang et al. 2018). Parallel research in wheat, maize, oat, barley, and soybean demonstrates the broader applicability of GS to enhance nutritional traits across cereals. In wheat, Bayesian regression models have demonstrated medium to high prediction accuracies for grain Fe, Zn, and β-carotenoid concentrations (Meher et al. 2023; Joukhadar et al. 2021). Maize genomic prediction models have achieved high accuracy for carotenoid traits, including provitamin A, while oat genomic prediction has effectively targeted seed fatty acid composition (Brzozowski et al. 2023). These examples highlight the potential to apply similar genomic and breeding strategies to accelerate nutritional improvement in pearl millet.

Line development programs leveraging GS integrate phenotypic and genotypic data from both inbred and hybrid lines to improve the accuracy of GEBVs and accelerate selection cycles (Gaynor et al. 2017). GS enables breeders to predict the performance of hybrids based on parental genotypes, facilitating the selection of superior parental combinations for high iron and zinc content. By incorporating GS at early generation stages, such as F_3 or F_4, breeders can effectively discard inferior lines and prioritize promising candidates before resource-intensive evaluations, thereby reducing costs and time. This strategy can significantly shorten the breeding cycle, aiming to develop nutrient-rich lines within 4–5 years compared to 8–10 years using traditional methods. Early-generation genomic selection combined with rapid cycling enhances genetic gain for complex nutritional traits such as micronutrient content (Beyene et al. 2021).

5.8 Accelerated Breeding for Rapid Line Development and Nutritional Improvement

5.8.1 Speed Breeding for Nutritional Enhancement

Speed breeding, which involves manipulating photoperiod, temperature, and light intensity to achieve multiple generations per year, can be integrated with MAS and GS to rapidly advance pearl millet lines (Sharma et al. 2024). Controlled-environment facilities have demonstrated the feasibility of achieving up to 5–6 generations annually in crops such as wheat, barley (Watson et al. 2018), and millets such as finger millet (Sajja et al. 2025), with similar protocols now being optimized for pearl millet. Speed breeding coupled with genomic predictions for nutritional traits will enable the rapid fixation of favorable alleles for nutritional traits, thereby facilitating the development of biofortified varieties within significantly reduced breeding cycles. The integration of genomic tools with accelerated growth conditions thus represents a transformative strategy for enhancing nutritional traits in millets and other cereals.

5.8.2 Doubled Haploid Technology to Accelerate Breeding and Enhance Nutritional Quality

Doubled haploid (DH) technology offers an efficient strategy for producing completely homozygous lines within a single generation, thereby significantly accelerating the breeding cycle. While its application in pearl millet is still at an early stage, advancements in tissue culture and androgenization protocols are paving the way for routine use. Incorporating DH lines into breeding programs targeting nutritional traits can substantially improve selection efficiency by rapidly fixing favorable alleles and reducing genetic variation in early generations. DH technology has been successfully implemented at a large scale in cereals such as maize, barley, and wheat, demonstrating its potential to accelerate genetic gain in crop improvement programs (Ren et al. 2017; Chaikam et al. 2019). When combined with MAS or GS, DH technology enables breeders to develop biofortified varieties with enhanced iron, zinc, and other micronutrient concentrations in record time. Further research is required to standardize DH protocols in pearl millet, but its successful implementation in cereals such as maize and wheat underscores its transformative potential in accelerating genetic gain for nutritional improvement.

5.9 Integrated Breeding Approaches for Developing Biofortified Varieties

The quest to develop nutritionally superior pearl millet begins in the field, where diverse genotypes are carefully evaluated for traits such as grain iron, zinc, and protein content across a range of environments. These trials capture the natural variability within the crop and reveal promising lines that perform well under both favorable and stress conditions. Parallel to this, high-throughput genotyping platforms such as GBS and SNP arrays are used to explore the genetic landscape, uncovering loci and candidate genes linked to nutrient accumulation. Comparative genomics with crops such as rice, sorghum, and foxtail millet expands this knowledge base, allowing breeders to draw on conserved genes as well as novel discoveries.

From this foundation, the breeding process blends precision and prediction. MAS ensures the targeted introgression of major-effect QTLs, while GS captures the complexity of polygenic traits. Modern predictive models, powered by machine learning and artificial intelligence, integrate genotype × environment interactions to forecast performance with remarkable accuracy. Cross-crop datasets enrich these predictions, while speed breeding protocols and strategic trial designs compress breeding cycles, making it possible to deliver nutrient-rich varieties faster than ever before without sacrificing yield or resilience. Genomics plays a pivotal role by enabling the identification of genes and molecular pathways associated with nutritional traits, facilitating marker development, and allowing genome-wide selection strategies for complex traits.

5.10 Multi-omics and Epigenetic Landscapes: Driving Advances in Pearl Millet Genomics

5.10.1 Functional Genomics Strategies for Enhancing Nutritional Traits in Pearl Millet

Functional genomics is a powerful discipline that unravels the molecular basis of complex traits by linking genetic variation to phenotypic expression. Unlike structural genomics, which focuses on the sequencing and organisation of genomes, functional genomics emphasises how genes and their regulatory networks contribute to biological processes (Lappalainen 2015). Among the available tools, RNA sequencing (RNA–Seq) has emerged as a central strategy for candidate gene discovery in nutritional genomics. Transcriptome profiling across contrasting genotypes enables the identification of transcription factors, transporters, and metabolic pathway genes involved in nutrient uptake and storage. By integrating functional genomics with well-characterized germplasm, researchers can accelerate the identification of key regulatory networks that influence nutrient partitioning and bioavailability. Govindaraj et al. (2020) developed a core collection of 39 diverse pearl millet accessions with multiple nutrient-rich profiles, providing a robust resource for transcriptomic studies.

The pathway of mineral homeostasis in plants is illustrated in Fig. 5.4, showing the regulated processes of nutrient uptake, transport, storage, and remobilization. Key transporters such as *IRT1, HKT1,* and *NRT1.1* mediate root uptake, while *HMA4* and *YSL1/3* enable long-distance transport. Intracellular regulation involves vacuolar sequestration (*VIT1*), remobilization (*NRAMP3/4*), and seed loading via *NAS* genes, which collectively shape the nutritional profile of pearl millet grains. Functional genomics provides powerful tools to dissect and validate these pathways (Chen et al. 2021; Fang et al. 2020). Transcriptome profiling, gene expression analysis, and comparative genomics can reveal key regulators and functional variants within transporter and storage gene families. For instance, differential expression of *IRT1* or *NAS* genes between high and low-Fe/Zn genotypes helps explain variation in nutrient accumulation. By revealing expression dynamics, allelic diversity, and regulatory networks, functional genomics connects these genes and pathways to practical breeding applications.

Stage-specific transcriptomic analyses in pearl millet have identified critical gene networks regulating grain iron and zinc content, revealing dynamic expression

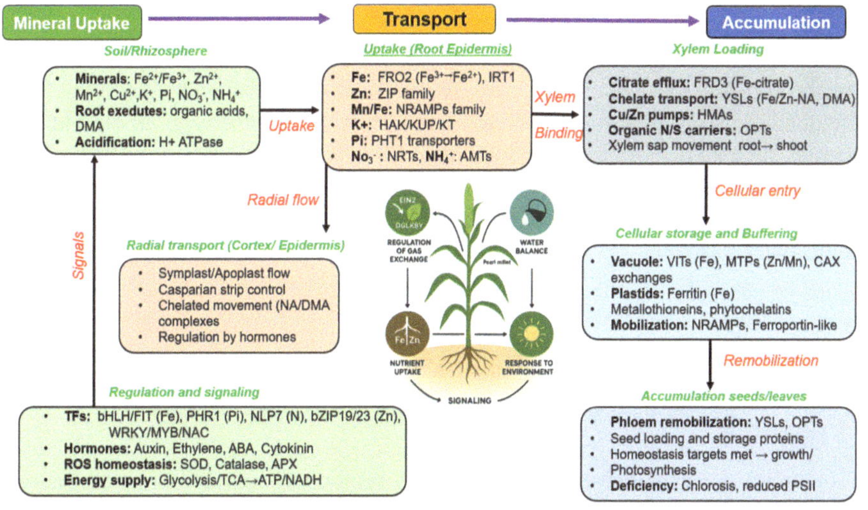

Fig. 5.4 Soil-to-seed pathway: an integrated overview of mineral uptake, transport, storage, and homeostasis in plants

of metal transporter families, chelator biosynthesis enzymes, and associated metabolic pathways during panicle development stages such as initiation, flowering, and milking (Satyavathi et al. 2022). High Fe and Zn genotypes showed up-regulation of key genes including *nicotianamine synthase (NAS), SAM synthases, mugineic acid* pathway genes (*DMA, IDS3, IDS2*) and transporters such as *FER-like transcription factors, H+-ATPases, heavy metal ATPases (HMA), ZIF1, and Zn transporters (ZIP3, ZIP5)*, which are involved in efficient micronutrient uptake, transport and accumulation. Functional annotation highlighted enriched biological processes such as oxidation–reduction, metabolic pathways, and ion binding, suggesting a complex network that underpins mineral homeostasis (Satyavathi et al. 2022). These findings provide molecular targets for biofortification breeding to enhance nutrient density in pearl millet grains.

Complementing these insights, Goud et al. (2022) identified 37,093 differentially expressed genes in response to Fe and Zn stress, including those involved in the mugineic acid pathway and transport mechanisms, such as metal transporter families and enzymes related to chlorophyll biosynthesis and carbohydrate metabolism. Genes such as *NAS, IDS2,* and transporters involved in Fe and Zn homeostasis were confirmed, emphasising the cross-talk between Fe and Zn uptake pathways. Additionally, a genome-wide transcriptome study by Moin et al. (2024) identified a family of 44 phospholipase genes in pearl millet, many of which exhibit grain-specific expression patterns. Notably, genes such as *PgPLD-alpha1-1, PgPLD-alpha1-5, PgPLD-delta1-7a, PgPLA1-II-1a,* and *PgPLD-delta1-2a* showed predominant expression in grains and harbored promoter variations influencing transcription factor binding, suggesting their transcriptional regulation is critical for grain quality traits. These phospholipases are implicated in seed lipid metabolism and are believed to play a pivotal role in managing lipid degradation processes linked to flour rancidity, thereby impacting the shelf life and nutritional stability of milled pearl millet flour. Understanding the regulation of these enzymes provides promising targets for breeding interventions aimed at enhancing grain storage quality and consumer acceptability. Extensive research has been conducted to elucidate the distinct pathways through which micronutrients reach the grain in diploid crop species such as rice, maize, and barley (Sperotto et al. 2012; Dhaliwal et al. 2019; Che et al. 2023). The conservation of these pathways across species enables predictions about Fe and Zn transport in pearl millet, where considerably less information is currently available.

5.10.2 Proteomics and Metabolomics in Nutritional Genomics of Pearl Millet: Insights from Rancidity Mechanisms

Proteomics and metabolomics serve as complementary approaches to unravel the molecular basis of nutrient accumulation, stability, and deterioration in pearl millet, linking genetic variation to biochemical phenotypes. Proteomic studies, employing methods such as two-dimensional gel electrophoresis (2D–GE) and liquid chromatography-tandem mass spectrometry (LC–MS/MS), have identified nutrient-responsive proteins involved in the transport, storage, and regulation of micronutrients such as iron and zinc (Roychowdhury et al. 2023). These proteins also include stress-responsive and seed storage classes that influence nutrient stability and bioavailability, presenting promising targets for functional validation and breeding improvement (Sachdev et al. 2023).

Metabolomics extends this understanding by profiling the biochemical networks governing nutrient accumulation and degradation processes in pearl millet grains. Advanced techniques such as liquid chromatography–high resolution mass spectrometry (LC–HRMS) have been applied to characterize diverse metabolites, including primary and secondary metabolites, flavonoids, alkaloids, and anti-nutritional factors such as phytates. Importantly, metabolite profiling has uncovered metabolic signatures associated with key grain quality traits, such as rancidity, a critical factor limiting the shelf life and consumer acceptance of pearl millet flour (Yogendra et al. 2024).

Recent comparative metabolomic analyses of pearl millet genotypes exhibiting varying rancidity levels revealed significant differential accumulation of fatty acid-derived metabolites, including fatty aldehydes, fatty alcohols, fatty esters, fatty acyls, and fatty amides, which are markedly elevated in high-rancidity lines. Four key fatty acids, stearic, palmitic, linoleic, and linolenic acids, were quantitatively confirmed to play pivotal roles in rancidity development. These findings elucidate how lipid metabolism, particularly the breakdown of triacylglycerols (TAGs) and phospholipids via lipases and phospholipases, contributes to rancidity by releasing free fatty acids and volatile organic compounds that degrade flour quality (Yogendra et al. 2024).

The biochemical degradation is further linked to enzymatic activities affecting phospholipid membranes, such as those encoded by phospholipase gene families (e.g., *PgPLD-alpha1-1, PgPLD-alpha1-5, PgPLD-delta1-7a*), recently identified in pearl millet and associated with rancidity control. Natural variation in lipase genes (*PgTAGLip1* and *PgTAGLip2*) also modulates free fatty acid accumulation post-milling, offering molecular targets for reducing rancidity and improving grain shelf life (Yogendra et al. 2024). This integrated omics strategy advances precision breeding efforts aimed not only at enhancing micronutrient content and bioavailability but also at improving grain quality traits such as rancidity resistance, ultimately supporting the development of nutritionally superior and climate-resilient pearl millet varieties.

5.10.3 Epigenetics and Regulatory Mechanisms

Epigenetics refers to heritable changes in gene expression that occur without altering the underlying DNA sequence, primarily through mechanisms such as DNA methylation, histone modifications, and non-coding RNAs. These chemical and structural modifications influence chromatin architecture, thereby regulating the accessibility of genes to the transcriptional machinery. By adding an additional layer of regulation beyond the genetic code, epigenetics significantly increases genome complexity, enabling dynamic responses to developmental cues and environmental stimuli (Eichten et al. 2014). This complexity allows the same genetic sequence to produce diverse phenotypes and adaptive traits, which is crucial for processes such as stress tolerance, nutrient regulation, and seed development in plants.

In rice, DNA methylation patterns are dynamically altered in response to iron deficiency, with widespread hypermethylation observed particularly in the CHH context that regulates genes involved in iron homeostasis (Sun et al. 2021). Such epigenetic changes coordinate with small RNAs and transposable element activity to fine-tune nutrient acquisition and distribution. Histone modifications also play a significant role in nutrient homeostasis. In *Arabidopsis*, the symmetric demethylation of histone *H4 arginine 3* (*H4R3sme2*), catalysed by *Shk1 binding protein 1 (SKB1)*, negatively regulates iron uptake by repressing key transcription factors under iron-sufficient conditions (Xing et al. 2015). Similarly, histone acetylation by the General Control Non-repressed Protein 5 (GCN5) influences iron translocation by modulating genes involved in citrate transport, essential for iron mobility within the plant. Mutations in these epigenetic regulators result in altered iron distribution and accumulation phenotypes, emphasizing the critical role of chromatin-based mechanisms in maintaining nutrient balance (Xing et al. 2015).

Small RNAs like microRNAs (miRNAs), siRNAs, and long non-coding RNAs (lncRNAs) regulate nutrient traits by modulating gene expression through mRNA degradation, translation inhibition, or chromatin remodeling (Valdés-López et al. 2010; Wang et al. 2017). While these regulatory RNAs are well-studied in model plants, data for pearl millet are still lacking. Insights from model crops provide a valuable framework for understanding epigenetic regulation in pearl millet. Exploring its epigenomic landscape will be crucial to identify new regulatory elements and enhance nutrient use efficiency and biofortification through breeding.

5.11 Predictive Breeding and Artificial Intelligence (AI)

Predictive breeding powered by AI is rapidly transforming crop improvement programs by integrating large-scale genomics and phenomics datasets to model trait performance with high precision (Farooq et al. 2024). In pearl millet, the large genome size (~1.8 Gb) necessitates the generation and management of substantial high-density genomic data, including SNP arrays, whole-genome resequencing, and structural variant information. These datasets, when combined with multi-environment phenotyping data, provide the inputs for training AI models to predict key nutritional traits such as grain iron, zinc, and protein content. Machine learning algorithms, including random forests, gradient boosting, and deep learning neural networks, can capture complex, nonlinear genotype-by-environment interactions

(Danilevicz et al. 2022). Moreover, cross-species nutritional genomics data from related cereals such as sorghum, foxtail millet, and maize can supplement these models, leveraging conserved genomic regions and orthologous gene functions to improve prediction accuracy.

Beyond predicting trait values, AI can facilitate the functional annotation and prediction of candidate genes involved in nutrient metabolism. For example, known Fe/Zn transporter genes from rice (OsZIP family) (Zhao et al. 2022) and maize (Li et al. 2013) can serve as references to identify orthologs in pearl millet, with predictions informed by sequence conservation, promoter architecture, and gene interaction networks (Yang et al. 2025). Integrating these genomic insights with detailed phenotypic datasets, including traits measured across multiple environments, developmental stages, and stress conditions, enables the development of predictive models that forecast both gene function and plant performance. Crop growth simulators linked with AI-driven genomic predictions can thus guide selection decisions even before field trials are conducted (Zhu et al. 2025). By combining large-scale genomic data, high-resolution phenotyping, and AI analytics, predictive breeding can streamline selection cycles, target nutritional ideotypes, and significantly enhance the efficiency and precision of pearl millet improvement pipelines.

5.12 Challenges and Future Perspectives

Developing biofortified pearl millet cultivars presents numerous obstacles, owing mostly to the complexity of the genome and the trade-offs between nutritional enhancement, yield stability, and stress tolerance. Pearl millet's big genome (~1.8 Gb) and high repetitive content make genome assembly and gene function characterization challenging. While superior nutritional quality lines have been produced, our understanding of the genetic basis for these qualities is limited due to a lack of knowledge of the genome, structural variations, and regulatory factors. Environmental factors such as soil type, temperature, and postharvest storage all influence phenotypic expression, making it difficult to reliably relate genotype to nutritional phenotype. Furthermore, grain quality characteristics, including size, color, and flavor, which are important for farmer and customer acceptance, complicate biofortification operations. Future progress will require the generation of high-quality reference genomes, long-read sequencing, and pangenome approaches to capture structural diversity and missing regions in existing assemblies. Integrating multi-omics data, including transcriptomics, proteomics, metabolomics, and high-throughput phenomics, will enable a more complete understanding of the molecular networks controlling nutritional traits. Despite present genomic gaps, AI-driven predictive analytics and cross-species genomic knowledge can help to identify candidate genes and enhance breeding tactics. Collaborative networks of breeders, molecular biologists, nutritionists, and computer scientists will be required to convert partial genomic knowledge into effective, farmer-preferred, and nutritionally enhanced cultivars. Future breeding initiatives can increase pearl millet nutrition, production, and resilience more precisely and consistently by tackling the intrinsic constraints of genomic complexity and sequencing limits.

5.13 Conclusion

Despite present genomic gaps, AI-driven predictive analytics and cross-species genomic knowledge can help to identify candidate genes and enhance breeding tactics. Although major cereals have been thoroughly researched for features such as grain iron and zinc content, protein quality, and amino acid composition, there have been few comparable genomic and functional assessments of pearl millet. Recent developments in genomics, transcriptomics, proteomics, metabolomics, and high-throughput phenomics, along with predictive breeding and AI-driven analytics, are providing unparalleled insights into the organism's genetic architecture. Integrated breeding procedures that combine conventional selection, MAS, GS, and speed breeding offer very efficient pathways for producing biofortified cultivars with high nutrient density, yield stability, and resistance to environmental challenges. These discoveries have the potential to expedite genetic progress and address the critical demand for climate-smart, nutrition-sensitive crop development.

Elite lines produced using these combined methodologies must eventually demonstrate their performance in real-world scenarios. When tested on farmers' fields, they are evaluated not only for yield and stress tolerance, but also for grain quality, postharvest stability, and consumer acceptance. Translating scientific discoveries into scalable breeding pipelines necessitates strong germplasm resources, multidisciplinary collaboration among breeders, molecular geneticists, nutritionists, and bioinformaticians, as well as informed policymaking. Pearl millet can transform from a subsistence staple to a strategic crop for combating hidden hunger and malnutrition, increasing dietary diversity, and strengthening resilience to climate change, contributing to global food and nutrition security.

References

Allan V, Vetriventhan M, Senthil R, Geetha S, Deshpande S, Rathore A, Kumar V, Singh P, Reddymalla S, Azevedo VCR (2020) Genome-wide DArTSeq genotyping and phenotypic based assessment of within and among accessions diversity and effective sample size in the diverse sorghum, pearl millet, and pigeonpea landraces. Front Plant Sci 11:587426. https://doi.org/10.3389/fpls.2020.587426

Ambawat S, Senapathy S, Hash C, Nepolean T, Vengaldas R, Eshwar K, Sharma R, Thakur RP, Rao VP, Yadav RC, Srivastava R (2016) QTL mapping of pearl millet rust resistance using an integrated DArT- and SSR-based linkage map. Euphytica 209:461–476. https://doi.org/10.1007/s10681-016-1671-9

Anitha S, Govindaraj M, Kane-Potaka J (2019) Balanced amino acid and higher micronutrients in millets complements legumes for improved human dietary nutrition. Crop Clim Health. https://doi.org/10.1002/cche.10227

Anuradha N, Satyavathi CT, Bharadwaj C, Nepolean T, Sankar SM, Singh SP et al (2017) Deciphering genomic regions for high grain iron and zinc content using association mapping in pearl millet. Front Plant Sci 8:412. https://doi.org/10.3389/fpls.2017.00412
Atanda SA, Govindan V, Singh R, Robbins KR, Crossa J, Bentley AR (2022) Sparse testing using genomic prediction improves selection for breeding targets in elite spring wheat. Theor Appl Genet 135(6):1939–1950. https://doi.org/10.1007/s00122-022-04085-0
Babu R, Rojas NP, Gao S, Yan J, Pixley K (2013) Validation of the effects of molecular marker polymorphisms in LcyE and CrtRB1 on provitamin A concentrations for 26 tropical maize populations. Theor Appl Genet 126(2):389–399. https://doi.org/10.1007/s00122-012-1987-3
Beyene Y, Gowda M, Pérez-Rodríguez P, Olsen M, Robbins KR, Burgueño J, Prasanna BM, Crossa J (2021) Application of genomic selection at the early stage of breeding pipeline in tropical maize. Front Plant Sci 12:685488. https://doi.org/10.3389/fpls.2021.685488
Bidinger FR, Nepolean T, Hash CT, Yadav RS, Howarth CJ (2007) Quantitative trait loci for grain yield in pearl millet under variable postflowering moisture conditions. Crop Sci 47:1–9. https://doi.org/10.2135/cropsci2006.07.0465
Black RE et al (2013) Maternal and child undernutrition and overweight in low-income and middle-income countries. Lancet 382(9890):427–451. https://doi.org/10.1016/S0140-6736(13)60937-X
Borrill P, Connorton JM, Balk J, Miller AJ, Sanders D, Uauy C (2014) Biofortification of wheat grain with iron and zinc: integrating novel genomic resources and knowledge from model crops. Front Plant Sci 5:53. https://doi.org/10.3389/fpls.2014.00053
Bouis HE, Saltzman A (2017) Improving nutrition through biofortification: a review of evidence from HarvestPlus, 2003 through 2016. Glob Food Sec 12:49–58. https://doi.org/10.1016/j.gfs.2017.01.009
Bouis H, Foley J, Lividini K, Jumrani J, Reinke R, Van Der Straeten D, Zagado R, Boy E, Brown LR, Mudyahoto B, Alioma R, Hussain M, Pfeiffer WH (2024) Biofortification: future challenges for a newly emerging technology to improve nutrition security sustainably. Curr Dev Nutr 8(12):104478. Springer. https://doi.org/10.1016/j.cdnut.2024.104478
Brzozowski LJ, Campbell MT, Hu H, Yao L, Caffe M, Gutiérrez L, Smith KP, Sorrells ME, Gore MA, Jannink J-L (2023) Genomic prediction of seed nutritional traits in biparental families of oat (Avena sativa). Plant Genome. https://doi.org/10.1002/tpg2.20370
Chaikam V, Molenaar W, Melchinger AE, Boddupalli PM (2019) Doubled haploid technology for line development in maize: technical advances and prospects. Theor Appl Genet 132(12):3227–3243. https://doi.org/10.1007/s00122-019-03433-x
Che J, Ricachenevsky FK, Deng F (2023) Editorial: micronutrients movement from soil to the grains: role of plant membrane transporters. Front Plant Sci 14:1179674. https://doi.org/10.3389/fpls.2023.1179674
Chen SY, Gu TY, Qi ZA, Yan J, Fang ZJ, Lu YT, Li H, Gong JM (2021) Two NPF transporters mediate iron long-distance transport and homeostasis in Arabidopsis. Plant Commun 2(6):100244. https://doi.org/10.1016/j.xplc.2021.100244
Chen J, Liu Y, Liu M, Guo W, Wang Y, He Q et al (2023) Pangenome analysis reveals genomic variations associated with domestication traits in broomcorn millet. Nat Genet 55(12):2243–2254. https://doi.org/10.1038/s41588-023-01571-z
Collard BCY, Jahufer MZZ, Brouwer JB, Pang ECK (2005) An introduction to markers, quantitative trait loci (QTL) mapping and marker-assisted selection for crop improvement: the basic concepts. Euphytica 142(1):169–196. https://doi.org/10.1007/s10681-005-1681-5
Danilevicz MF, Gill M, Anderson R, Batley J, Bennamoun M, Bayer PE, Edwards D (2022) Plant genotype to phenotype prediction using machine learning. Front Genet 13:822173. https://doi.org/10.3389/fgene.2022.822173
Dhaliwal SS, Naresh RK, Mandal A, Singh R, Dhaliwal MK (2019) Dynamics and transformations of micronutrients in agricultural soils as influenced by organic matter build-up: a review. Environ Sustain Indic 1–2:100007. https://doi.org/10.1016/j.indic.2019.100007

Dwivedi SL, Heslop-Harrison P, Amas J, Ortiz R, Edwards D (2024) Epistasis and pleiotropy-induced variation for plant breeding. Plant Biotechnol J 22:2788–2807. Springer. https://doi.org/10.1111/pbi.14405

Eichten SR, Schmitz RJ, Springer NM (2014) Epigenetics: beyond chromatin modifications and complex genetic regulation. Plant Physiol 165(3):933–947. https://doi.org/10.1104/pp.113.234211

Fang XZ, Liu XX, Zhu YX, Ye JY, Jin CW (2020) The K^+ and NO_3^- interaction mediated by NITRATE TRANSPORTER1.1 ensures better plant growth under K^+-limiting conditions. Plant Physiol 184(4):1900–1916. https://doi.org/10.1104/pp.20.01229

Farooq MA, Gao S, Hassan MA, Huang Z, Rasheed A, Hearne S, Prasanna B, Li X, Li H (2024) Artificial intelligence in plant breeding. Trends Genet 40(10):891–908. https://doi.org/10.1016/j.tig.2024.07.001

Gangashetty PI, Riyazaddin M, Sanogo MD, Inousa D, Issoufou KA, Asungre PA, Sy O, Govindaraj M, Ignatius AI (2021) Identification of high-yielding iron-biofortified open-pollinated varieties of pearl millet in West Africa. Front Plant Sci 12:688937. https://doi.org/10.3389/fpls.2021.688937

Gaynor RC, Gorjanc G, Bentley AR, Ober ES, Howell P, Jackson R, Mackay IJ, Hickey JM (2017) A two-part strategy for using genomic selection to develop inbred lines. Crop Sci. https://doi.org/10.2135/cropsci2016.09.0742

Goswami S, Asrani P, Ali TPA et al (2020) Rancidity matrix: development of biochemical indicators for analysing the keeping quality of pearl millet flour. Food Anal Methods 13(10):2147–2164. https://doi.org/10.1007/s12161-020-01831-2

Goud CA, Satturu V, Malipatil R, Viswanath A, Semalaiyappan J, Kudapa H, Rathod S, Rathore A, Govindaraj M, Thirunavukkarasu N (2022) Identification of iron and zinc responsive genes in pearl millet using genome-wide RNA-sequencing approach. Front Nutr 9:884381. https://doi.org/10.3389/fnut.2022.884381

Govindaraj M, Selvi B, Arun Prabhu D, Rajarathinam S (2009) Genetic diversity analysis of pearl millet (Pennisetum glaucum [L.] R. Br.) accessions using molecular markers. Afr J Biotechnol 8(22):6046–6052

Govindaraj M, Rai KN, Shanmugasundaram P (2016) Intra-population genetic variance for grain iron and zinc contents and agronomic traits in pearl millet. Crop J 4(1):48–54. https://doi.org/10.1016/j.cj.2015.11.003

Govindaraj M, Rai KN, Cherian B, Pfeiffer WH, Kanatti A, Shivade H (2019) Breeding biofortified pearl millet varieties and hybrids to enhance millet markets for human nutrition. Agriculture 9:106. https://doi.org/10.3390/agriculture9050106

Govindaraj M, Rai KN, Kanatti A, Upadhyaya HD, Shivade H, Rao AS (2020) Exploring the genetic variability and diversity of pearl millet core collection germplasm for grain nutritional traits improvement. Sci Rep 10:21177. https://doi.org/10.1038/s41598-020-77818-0

Govindaraj M, Kanatti A, Rai KN, Pfeiffer WH, Shivade H (2022) Association of grain iron and zinc content with other nutrients in pearl millet germplasm, breeding lines, and hybrids. Front Nutr 8:746625. https://doi.org/10.3389/fnut.2021.746625

Govindaraj M, Pujar M, Pravalika V (2024) High throughput phenotyping for grain zinc: sampling and analytical overview. In: Govindaraj M et al (eds) Breeding zinc crops for better human health. Springer, Cham. https://doi.org/10.1007/978-3-031-84342-6_1

Gupta S, Rangari SK, Sahu A, Naik YD, Satyavathi CT, Punnuri S, Thudi M (2024) Meta-QTL analysis reveals important genomic regions for biotic stresses, nutritional quality, and yield-related traits in pearl millet. CABI Agric Biosci 5:36. https://doi.org/10.1186/s43170-024-00136-7

Hasan N, Choudhary S, Naaz N, Sharma N, Laskar RA (2021) Recent advancements in molecular marker-assisted selection and applications in plant breeding programmes. J Genet Eng Biotechnol 19(1):128. https://doi.org/10.1186/s43141-021-00231-1

He C, Holme J, Anthony J (2014) SNP genotyping: the KASP assay. Methods Mol Biol 1145:75–86. https://doi.org/10.1007/978-1-4939-0446-4_7

Heidarzadeh A (2025) Role of amino acids in plant growth, development, and stress responses: a comprehensive review. Discov Plants 2:237. https://doi.org/10.1007/s44372-025-00322-0

Hu Z, Mbacké B, Perumal R, Guèye MC, Sy O, Bouchet S, Prasad PVV, Morris GP (2015) Population genomics of pearl millet (Pennisetum glaucum (L.) R. Br.): comparative analysis of global accessions and Senegalese landraces. BMC Genomics 16:1048. https://doi.org/10.1186/s12864-015-2255-0

Jaccoud D, Peng K, Feinstein D, Kilian A (2001) Diversity arrays: a solid state technology for sequence information independent genotyping. Nucleic Acids Res 29(4):E25. https://doi.org/10.1093/nar/29.4.e25

Jacob J, Krishnan V, Antony C, Bhavyasri M, Aruna C, Mishra K, Nepolean T, Satyavathi CT, Visarada KBRS (2024) The nutrition and therapeutic potential of millets: an updated narrative review. Front Nutr 11:1346869. https://doi.org/10.3389/fnut.2024.1346869

Jaiswal S, Antala TJ, Mandavia MK, Chopra M, Jasrotia RS, Tomar RS et al (2018) Transcriptomic signature of drought response in pearl millet (Pennisetum glaucum (L.)) and development of web-genomic resources. Sci Rep 8:3382. https://doi.org/10.1038/s41598-018-21732-7

Joshi G, Soe YP, Palanog A, Hore TK, Nha CT, Calayugan MI, Swamy BM (2023) Meta-QTLs and haplotypes for efficient zinc biofortification of rice. Plant Genome:e20315. https://doi.org/10.1002/tpg2.20315

Joukhadar R, Thistlethwaite R, Trethowan RM, Hayden MJ, Stangoulis J, Cu S, Daetwyler HD (2021) Genomic selection can accelerate the biofortification of spring wheat. Theor Appl Genet 134(11):3339–3350. https://doi.org/10.1007/s00122-021-03900-4

Kanatti A, Rai KN, Radhika K, Govindaraj M, Sahrawat KL et al (2014) Grain iron and zinc density in pearl millet: combining ability, heterosis and association with grain yield and grain size. Springerplus 3:763. https://doi.org/10.1186/2193-1801-3-763

Kanfany G, Serba DD, Rhodes D, St. Amand P, Bernardo A, Gangashetty PI et al (2020) Genomic diversity in pearl millet inbred lines derived from landraces and improved varieties. BMC Genomics 21:469. https://doi.org/10.1186/s12864-020-06857-5

Kasule F, Diack O, Mbaye M, Kakeeto R, Fallon Econopouly B (2024) Genomic resources, opportunities, and prospects for accelerated improvement of millets. Theor Appl Genet 137:273. Springer. https://doi.org/10.1007/s00122-024-04777-9

Krishnan R, Meera MS (2018) Pearl millet minerals: effect of processing on bioaccessibility. J Food Sci Technol 55(9):3362–3372. https://doi.org/10.1007/s13197-018-3305-9

Kumar S, Hash CT, Thirunavukkarasu N, Singh G, Rajaram V, Rathore A et al (2016) Mapping quantitative trait loci controlling high iron and zinc content in self and open pollinated grains of pearl millet (Pennisetum glaucum (L.) R. Br.). Front Plant Sci 7:1636. https://doi.org/10.3389/fpls.2016.01636

Kumar S, Hash CT, Thirunavukkarasu N, Mahendrakar MD, Satyavathi CT, Singh G, Rathore A, Yadav RS, Gupta R, Srivastava RK (2018) Mapping grain iron and zinc content quantitative trait loci in an Iniadi-derived immortal population of pearl millet. Front Plant Sci 9:596. https://doi.org/10.3389/fpls.2018.00596

Kumar S, Singh S, Kumar R, Gupta D (2024) The Genomic SSR Millets Database (GSMDB): enhancing genetic resources for sustainable agriculture. Database 2024:baae114. https://doi.org/10.1093/database/baae114

Lappalainen T (2015) Functional genomics bridges the gap between quantitative genetics and molecular biology. Genome Res 25(10):1427–1431. https://doi.org/10.1101/gr.190983.115

Li S, Zhou X, Huang Y, Zhu L, Zhang S, Zhao Y, Guo J, Chen J, Chen R (2013) Identification and characterization of the zinc-regulated transporters, iron-regulated transporter-like protein (ZIP) gene family in maize. BMC Plant Biol 13:114. https://doi.org/10.1186/1471-2229-13-114

Liang Z, Gupta SK, Yeh C-T, Zhang Y, Ngu DW, Kumar R, Patil HT, Mungra KD, Yadav DV, Rathore A et al (2018) Phenotypic data from inbred parents can improve genomic prediction in pearl millet hybrids. G3 (Bethesda) 8(7):2513–2522. https://doi.org/10.1534/g3.118.200242

Liang YL, Yan PS, Li WX, Ding WN, Li GX et al (2025) Bridging genomic insights to agricultural practice: GWAS applications in mining nutrient utilization efficiency genes for crop improvement. Plant Cell Environ. https://doi.org/10.1111/pce.70027

Logsdon GA, Vollger MR, Eichler EE (2020) Long-read human genome sequencing and its applications. Nat Rev Genet 21(10):597–614. https://doi.org/10.1038/s41576-020-0236-x

Mariac C, Jehin L, Saïdou A-A, Thuillet A-C, Couderc M, Sire P, Jugdé H, Adam H, Bezançon G, Pham J-L, Vigouroux Y (2010) Genetic basis of pearl millet adaptation along an environmental gradient investigated by a combination of genome scan and association mapping. Mol Ecol 19:3720–3735. https://doi.org/10.1111/j.1365-294X.2010.04893.x

Meher PK, Gupta A, Rustgi S, Mir RR, Kumar A, Kumar J, Balyan HS, Gupta PK (2023) Evaluation of eight Bayesian genomic prediction models for three micronutrient traits in bread wheat (Triticum aestivum L.). Plant Genome 16(4):e20332. https://doi.org/10.1002/tpg2.20332

Meuwissen THE, Hayes BJ, Goddard ME (2001) Prediction of total genetic value using genome-wide dense marker maps. Genetics 157(4):1819–1829. https://doi.org/10.1093/genetics/157.4.1819

Mirzabaev A, Bezner Kerr R, Hasegawa T, Pradhan P, Wreford A, Tirado von der Pahlen MC, Gurney-Smith H (2023) Severe climate change risks to food security and nutrition. Clim Risk Manag 39:100473. https://doi.org/10.1016/j.crm.2022.100473

Mishra A, Bhat A, Kumari S, Sharma R, Braynen J, Tadesse D, El Alaoui S, Seaver S, Grosjean N, Ware D, Xie M, Paape T (2025) Time-series multi-omics analysis of micronutrient stress in Sorghum bicolor reveals iron and zinc crosstalk and regulatory network conservation. Plant Biol. https://doi.org/10.1111/plb.70038

Mittal N, Dubey AK (2009) Microsatellite markers-a new practice of DNA based markers in molecular genetics. Phcog Rev 3(6):235–246

Mittal GK, Singh S, Yadav D, Beckmann M, Mur LAJ, Yadav RS (2025) Genome-wide association study for dissecting lipid and fatty acid variation in a global collection of pearl millet germplasm. J Food Compos Anal 144:107679. https://doi.org/10.1016/j.jfca.2025.107679

Moin M, Bommineni PR, Tyagi W (2024) Exploration of the pearl millet phospholipase gene family to identify potential candidates for grain quality traits. BMC Genomics 25(1):581. https://doi.org/10.1186/s12864-024-10504-x

Morgil H, Can Gercek Y, Tulum I (2020) Single nucleotide polymorphisms (SNPs) in plant genetics and breeding. In: Abdurakhmonov IY (ed) Plant breeding—current and future views. IntechOpen, London. https://doi.org/10.5772/intechopen.91886

Mundada PS, Kadam SB, Pable AA, Barvkar VT (2022) Recent advances and applicability of GBS, GWAS and GS in millet crops. In: Genotyping by sequencing for crop improvement. Wiley. https://doi.org/10.1002/9781119745686.ch12; Chapter 12.

Myles S, Peiffer J, Brown PJ, Ersoz ES, Zhang Z, Costich DE, Buckler ES (2009) Association mapping: critical considerations shift from genotyping to experimental design. Plant Cell 21(8):2194–2202. https://doi.org/10.1105/tpc.109.068437

Naveen KG, Pavan JK, Rakhi S, Rajeswari A, Ashwathanarayana R, Berhanu DB et al (2014) Identification of putative candidate gene markers for grain zinc content using recombinant inbred lines (RIL) population of IRRI38 × Jeerigesanna. Afr J Biotechnol 13(6):657–663. https://doi.org/10.5897/ajb2013.13334

Negro SS, Millet EJ, Madur D, Bauland C, Combes V et al (2019) Genotyping-by-sequencing and SNP-arrays are complementary for detecting quantitative trait loci by tagging different haplotypes in association studies. BMC Plant Biol 19(1):318. https://doi.org/10.1186/s12870-019-1926-4

Nepolean T, Kaul J, Mukri G, Mittal S (2018) Genomics-enabled next-generation breeding approaches for developing system-specific drought tolerant hybrids in maize. Front Plant Sci 9:361. https://doi.org/10.3389/fpls.2018.00361

Pujar M, Gangaprasad S, Govindaraj M, Gangurde SS, Kanatti A, Kudapa H (2020) Genome-wide association study uncovers genomic regions associated with grain iron, zinc and protein content in pearl millet. Sci Rep 10(1):19473. https://doi.org/10.1038/s41598-020-76674-w

Rai KN, Yadav OP, Rajpurohit BS, Patil HT, Govindaraj M, Khairwal IS, Rao AS (2013) Breeding pearl millet cultivars for high iron density with zinc density as an associated trait. J SAT Agric Res 11:1–7

Rai KN, Yadav OP, Govindaraj M, Pfeiffer WH, Yadav HP, Rajpurohit BS et al (2016) Grain iron and zinc densities in released and commercial cultivars of pearl millet (Pennisetum glaucum). Indian J Agric Sci 86(3):11–16

Ramu P, Srivastava RK, Sanyal A et al (2023) Improved pearl millet genomes representing the global heterotic pool offer a framework for molecular breeding applications. Commun Biol 6:902. https://doi.org/10.1038/s42003-023-05258-3

Ren J, Wu P, Trampe B, Tian X, Lübberstedt T, Chen S (2017) Novel technologies in doubled haploid line development. Plant Biotechnol J 15(11):1361–1370. https://doi.org/10.1111/pbi.12805

Rice BR, Lipka AE (2021) Diversifying maize genomic selection models. Mol Breed 41(5):33. https://doi.org/10.1007/s11032-021-01221-4

Roychowdhury R, Das SP, Gupta A, Parihar P, Chandrasekhar K, Sarker U, Kumar A, Ramrao DP, Sudhakar C (2023) Multi-omics pipeline and omics-integration approach to decipher plant's abiotic stress tolerance responses. Genes 14(6):1281. https://doi.org/10.3390/genes14061281

Sachdev N, Goomer S, Singh LR, Pathak VM, Aggarwal D, Chowhan RK (2023) Current status of millet seed proteins and its applications: a comprehensive review. Appl Food Res 3(1):100288. https://doi.org/10.1016/j.afres.2023.100288

Sajja S, Pranati J, Shyamala S, Vinutha KS, Reddy R, Joshi P, Pannem S, Rakshit P, Jadhav Y, Kumar CVS, Mayes S (2025) Rapid Ragi: a speed breeding protocol for finger millet. Plant Methods 21:84. https://doi.org/10.1186/s13007-025-01273-8

Salson M, Orjuela J, Mariac C, Zekraouï L, Couderc M, Arribat S, Rodde N, Faye A, Kane NA, Tranchant-Dubreuil C, Vigouroux Y, Berthouly-Salazar C (2023) An improved assembly of the pearl millet reference genome using Oxford Nanopore long reads and optical mapping. G3: Genes|Genomes|Genetics 13(5):jkad051. Springer. https://doi.org/10.1093/g3journal/jkad051

Satyavathi CT, Tomar RS, Ambawat S, Kheni J, Padhiyar SM, Desai H, Bhatt SB, Shitap MS, Meena RC, Singhal T, Sankar SM, Singh SP, Khandelwal V (2022) Stage specific comparative transcriptomic analysis to reveal gene networks regulating iron and zinc content in pearl millet [Pennisetum glaucum (L.) R. Br.]. Sci Rep 12:276. https://doi.org/10.1038/s41598-021-04388-0

Schaid DJ, Chen W, Larson NB (2018) From genome-wide associations to candidate causal variants by statistical fine-mapping. Nat Rev Genet 19(8):491–504. https://doi.org/10.1038/s41576-018-0016-z

Scott MF, Ladejobi O, Amer S, Bentley AR, Biernaskie J et al (2020) Multi-parent populations in crops: a toolbox integrating genomics and genetic mapping with breeding. Heredity 125:396–416. https://doi.org/10.1038/s41437-020-0336-6

Sehgal D, Skot L, Singh R, Srivastava RK, Das SP, Taunk J et al (2015) Exploring potential of pearl millet germplasm association panel for association mapping of drought tolerance traits. PLoS One 10(5):e0122165. https://doi.org/10.1371/journal.pone.0122165

Semalaiyappan J, Selvanayagam S, Rathore A, Gupta SK, Chakraborty A, Gujjula KR et al (2023) Development of a new AgriSeq 4K mid-density SNP genotyping panel and its utility in pearl millet breeding. Front Plant Sci 13:1068883. https://doi.org/10.3389/fpls.2022.1068883

Senthilvel S, Jayashree B, Mahalakshmi V, Kumar PS, Nakka S, Nepolean T, Hash C (2008) Development and mapping of simple sequence repeat markers for pearl millet from data mining of expressed sequence tags. BMC Plant Biol 8:119. https://doi.org/10.1186/1471-2229-8-119

Sharma A, Pandey H, Misra V, Devadas VS, Kesavan AK, Heisnam P, Vashishth A, Wangsu M, Singh D, Mall AK, Kumar R (2024) Harnessing speed breeding in controlled environment ecosystem for millets sustainability. Plant Breed. https://doi.org/10.1111/pbr.13258

Sharma N, Chaudhary A, Sharma N, Shukla S, Joshi NC, Vishwakarma K, Singh PK, Mishra A (2025) Plant nutrient requirements and nutrient homeostasis in plants. In: Soil health and nutrition management. CABI, pp 1–20. https://doi.org/10.1079/9781800624597.0001

Singh M, Nara U (2023a) Genetic insights in pearl millet breeding in the genomic era: challenges and prospects. Plant Biotechnol Rep 17:15–37. Springer. https://doi.org/10.1007/s11816-022-00767-9

Singh M, Nara U (2023b) Genetic insights in pearl millet breeding in the genomic era: challenges and prospects. Plant Biotechnol Rep 17(1):15–37. https://doi.org/10.1007/s11816-022-00767-9

Singh S, Manwaring HR, Naveen A, Hegarty M, Yadav RS (2024) Genetic dissection of minerals and phytate content in pearl millet germplasm panel using genome-wide association study. Food Energy Secur 13(4):e565. https://doi.org/10.1002/fes3.565

Singhal T, Satyavathi CT, Singh SP, Kumar A, Sankar SM, Bhardwaj C, Mallik M, Bhat J, Anuradha N, Singh N (2021) Multi-environment quantitative trait loci mapping for grain iron and zinc content using bi-parental recombinant inbred line mapping population in pearl millet. Front Plant Sci 12:659789. https://doi.org/10.3389/fpls.2021.659789

Sperotto RA, Ricachenevsky FK, Waldow VD, Fett JP (2012) Iron biofortification in rice: it's a long way to the top. Plant Sci 190:24–39. https://doi.org/10.1016/j.plantsci.2012.03.004

Srivastava RK, Satyavathi CT, Mahendrakar MD, Singh RB, Kumar S, Govindaraj M, Ghazi IA (2021) Addressing iron and zinc micronutrient malnutrition through nutrigenomics in pearl millet: advances and prospects. Front Genet 12:723472. https://doi.org/10.3389/fgene.2021.723472

Stanton C, Sanders D, Krämer U, Podar D (2022) Zinc in plants: integrating homeostasis and biofortification. Mol Plant 15(1):65–85. https://doi.org/10.1016/j.molp.2021.12.008

Stevens GA, Beal T, Mbuya MNN, Luo H et al (2022) Micronutrient deficiencies among preschool-aged children and women of reproductive age worldwide: a pooled analysis of individual-level data from population-representative surveys. Lancet Glob Health 10(11):e1590–e1599. https://doi.org/10.1016/S2214-109X(22)00367-9

Sun S, Zhu J, Guo R, Whelan J, Shou H (2021) DNA methylation is involved in acclimation to iron-deficiency in rice (Oryza sativa). Plant J. https://doi.org/10.1111/tpj.15318

Sun M, Yan H, Zhang A, Jin Y, Lin C et al (2023) Milletdb: a multi-omics database to accelerate the research of functional genomics and molecular breeding of millets. Plant Biotechnol J 21(11):2348–2357. https://doi.org/10.1111/pbi.14136

Supriya A, Senthilvel S, Nepolean T, Eshwar K, Rajaram V, Shaw R et al (2011) Development of a molecular linkage map of pearl millet integrating DArT and SSR markers. Theor Appl Genet 123(2):239–250. https://doi.org/10.1007/s00122-011-1580-1

Valdés-López O, Yang SS, Aparicio-Fabre R, Graham PH, Reyes JL, Vance CP, Hernández G (2010) MicroRNA expression profile in common bean (Phaseolus vulgaris) under nutrient deficiency stresses and manganese toxicity. New Phytol 187(3):805–818. https://doi.org/10.1111/j.1469-8137.2010.03320.x

Varshney RK, Shi C, Thudi M, Mariac C, Wallace J et al (2017) Pearl millet genome sequence provides a resource to improve agronomic traits in arid environments. Nat Biotechnol 35(10):969–976. https://doi.org/10.1038/nbt.3943

Velu G, Rai KN, Sahrawat KL, Sumalini K (2008) Variability for grain iron and zinc contents in pearl millet hybrids. J SAT Agric Res 6:4

Wang T, Zhao M, Zhang X, Liu M, Yang C, Chen Y, Chen R, Wen J, Mysore KS, Zhang WH (2017) Novel phosphate deficiency-responsive long non-coding RNAs in the legume model plant Medicago truncatula. J Exp Bot 68(21–22):5937–5948. https://doi.org/10.1093/jxb/erx384

Watson A, Ghosh S, Williams MJ, Cuddy WS, Simmonds J, Rey M-D, Md Hatta MA, Hinchliffe A, Steed A, Reynolds D et al (2018) Speed breeding is a powerful tool to accelerate crop research and breeding. Nat Plants 4(1):23–29. https://doi.org/10.1038/s41477-017-0083-8

World Health Organization (2020) Micronutrients. Available at: https://www.who.int/health--topics/micronutrients. Accessed 9 Aug 2025

World Health Organization (2021) The state of food security and nutrition in the world 2018: building climate resilience for food security and nutrition. http://www.fao.org/3/i9553en/i9553en.pdf. Accessed 6 Mar 2025

Wu D, Saleem M, He T, He G (2021) The mechanism of metal homeostasis in plants: a new view on the synergistic regulation pathway of membrane proteins, lipids and metal ions. Membranes 11:984. https://doi.org/10.3390/membranes11120984

Xing J, Wang T, Ni Z (2015) Epigenetic regulation of iron homeostasis in Arabidopsis. Plant Signal Behav 10(12):e1064574. https://doi.org/10.1080/15592324.2015.1064574

Xu Y, Li P, Zou C, Lu Y, Xie C, Zhang X, Prasanna BM, Olsen MS (2017) Enhancing genetic gain in the era of molecular breeding. J Exp Bot 68(11):2641–2666. https://doi.org/10.1093/jxb/erx135

Yadav CB, Srivastava RK, Beynon S, Englyst K, Gangashetty PI, Yadav RS (2022) Genetic variability and genome-wide marker association studies for starch traits contributing to low glycaemic index in pearl millet. Food Energy Secur 11(1):e341. https://doi.org/10.1002/fes3.341

Yadav OP, Gupta SK, Govindaraj M, Singh DV, Verma A, Sharma R, Mahala RS, Srivastava SK, Birthal PS (2024a) Strategies for enhancing productivity, resilience, nutritional quality, and consumption of pearl millet (*Pennisetum glaucum* (L.) R. Br.) for food and nutritional security in India. Crop Sci 64:1–19. https://doi.org/10.1002/csc2.21346

Yadav OP, Singh DV, Kumari V, Prasad M, Seni S, Singh RK, Sood S, Kant L, Rao BD, Madhusudhana R, Bhat BV, Gupta SK et al (2024b) Production and cultivation dynamics of millets in India. Crop Sci 64(5):2459–2484. https://doi.org/10.1002/csc2.21207

Yan H, Sun M, Zhang Z et al (2023) Pangenomic analysis identifies structural variation associated with heat tolerance in pearl millet. Nat Genet 55:507–518. https://doi.org/10.1038/s41588-023-01302-4

Yang L, Wang H, Zou M, Chai H, Xia Z (2025) Artificial intelligence-driven plant bio-genomics research: a new era. Trop Plants 4:e015. https://doi.org/10.48130/tplants.2025.e015

Yogendra K, Sanivarapu H, Avuthu T, Gupta SK, Durgalla P, Banerjee R, Raman A, Tyagi W (2024) Comparative metabolomics to unravel the biochemical mechanism associated with rancidity in pearl millet (Pennisetum glaucum L.). Int J Mol Sci 25(21):11583. https://doi.org/10.3390/ijms252111583

Zhao YN, Li C, Li H, Liu XS, Yang ZM (2022) OsZIP11 is a trans-Golgi-residing transporter required for rice iron accumulation and development. Gene 836:146678. https://doi.org/10.1016/j.gene.2022.146678

Zhu W, Li W, Zhang H, Li L (2025) Big data and artificial intelligence-aided crop breeding: progress and prospects. J Integr Plant Biol 67(3):722–739. https://doi.org/10.1111/jipb.13791

Chapter 6
Pearl Millet Biofortification Research for Food and Nutritional Security in West Africa

Riyazaddin Mohammed, Bassirou Sani Boubacar Gaoh, Epiphane Lamine, Zakari Hassane, Prakash I. Gangashetty, and Mahalingam Govindaraj

Abstract Micronutrient malnutrition, or hidden hunger, is a preventable food-system-based health problem persisting in developing countries. Conventional interventions such as dietary diversification, food fortification, and supplementation have shown limited success in low- and middle-income countries (LMICs). The dietary deficiency of vitamins and minerals contributes to lower human productivity and becomes a major impediment to socioeconomic development. Biofortification is a sustainable and cost-effective strategy to increase micronutrient concentration in food grains through plant breeding. In West Africa, pearl millet biofortification research focuses on enhancing the nutritional quality of this widely consumed staple crop to improve food and nutritional security. Biofortified varieties aim to improve the nutritional profile of pearl millet without compromising its yield or adaptability. To sum up, ICRISAT's pearl millet biofortification research in West Africa seeks to combat micronutrient deficiencies, enhance food security, and improve the nutritional status of populations through the development and promotion of nutrient-dense millet varieties that are well-suited to the region's farming conditions. This requires a national variety release standard (for iron and zinc), like India and Nigeria, to mainstream nutrition in high-yielding backgrounds.

Keywords Biofortification · Pearl millet · Micronutrient malnutrition · Nutritional insecurity · Grain Fe and Zn

R. Mohammed (✉) · B. S. B. Gaoh · E. Lamine · Z. Hassane
International Crops Research Institute for the Semi-Arid Tropics, Niamey, Niger
e-mail: riyazaddin.mohammed@icrisat.org

P. I. Gangashetty
International Crops Research Institute for the Semi-Arid Tropics, Hyderabad, Telangana, India

M. Govindaraj
Alliance of Bioversity International and the International Center for Tropical Agriculture (CIAT), Cali, Colombia

M. Govindaraj (ed.), *Breeding Biofortified Millets for Nutritional Security*, Sustainability Sciences in Asia and Africa,
https://doi.org/10.1007/978-981-95-9827-4_6

6.1 Introduction

Food and nutritional insecurity are significant threats to the world's growing population. Current global food production systems struggle to deliver nutritious, healthy diets to a growing world population, with millions facing food insecurity and micronutrient deficiencies despite overall increases in food availability (Hussain et al. 2025). Micronutrient malnutrition is a major health issue affecting two billion people globally. Deficiencies of minerals such as iron (Fe), zinc (Zn), iodine (I), folic acid, and vitamin A have been reported as a widespread food-related health problem, affecting most people worldwide, particularly children, adolescents, and pregnant women (WHO 2002). This issue is more significant when most of the population has limited access to diversified and nutrient-rich foods such as meat, fruits, and vegetables, which are generally rich sources of micronutrients (Velu et al. 2007). Resource-poor households, lacking access to food diversification, are more vulnerable to micronutrient malnutrition. Even after many scientific breakthroughs and policy advances, a large proportion of people still cannot access or afford diets of adequate quality, contributing to persistent malnutrition and undernutrition (FAO, WFP, and UNICEF 2025).

Micronutrient deficiencies can have serious health and economic consequences; they increase the risk for infectious diseases, compromise child growth and development, thus limiting human potential worldwide. Micronutrient deficiencies are most prevalent in South Asia and Sub-Saharan Africa, but there is surprisingly high prevalence even in high-income countries (Lowe 2025). In West and Central Africa, the situation remains challenging in most countries, considering the effects of climate change, economic slowdown, inequities, unaffordability, political instability, and the increase in insecurity. Recurrent and multiple shocks lead to increased vulnerabilities and a worrying malnutrition situation for women and children. Malnutrition and food insecurity remain alarmingly high in West and Central Africa in 2025, with more than two-thirds of households unable to afford healthy diets and around 8 out of 10 children aged 6–23 months not consuming the minimum number of food groups needed for optimal growth and development (WFP and UNICEF 2025).

The "hidden hunger" seriously affects millions of women and children in Africa. Almost half of African children under five suffer from vitamin A deficiency; 60% suffer from anaemia, often caused by iron deficiency; and 25% are deficient in zinc. Iron deficiency affects 2 billion people (over 30% of the world's population, WHO 2016). In Niger, 55.5% of children under 5 and 46.1% of women aged 15–49 in Niger suffer from Fe deficiency-induced anemia (WHO 2025). Iron deficiency anemia can lead to higher maternal mortality, low-birth weight, lower resistance to infections, less psychomotor and cognitive development, lower school performance, and less physical work capacity. Zinc deficiency, on the other hand, can lead to weak immunity, mental depression, impotence, lack of appetite, dyslexia, growth retardation, and poor wound healing.

6.2 Pearl Millet Production and Nutrition

The world population is projected to reach 9.7 billion by 2050 and based on the World Population Prospects 2024 medium-fertility projections, West Africa's population is expected to grow from roughly 467 million in 2025 to about 735 million by 2050, an increase of about 57% over 25 years (UN-DESA 2024). Such population growth will increase the food requirement in production and quality to feed the ever-growing population. Historically, plant breeding has been focused on improving grain and fodder yields rather than improving the nutritional quality of crops (Singh et al. 2016). However, contemporary there is growing recognition that enhancing micronutrient density in staple crops through plant breeding is essential to reduce micronutrient malnutrition, particularly among women of reproductive age and young children (Paul et al. 2024).

Among cereals consumed worldwide, pearl millet is known as a rich source of grain Fe and Zn. West Africa is the largest producer of pearl millet in Africa, where it is grown in countries like Niger, Mali, Burkina Faso, Chad, Nigeria, Ghana, and Senegal. In total, about 20 million hectares of land in Africa are dedicated to millet cultivation. This number fluctuates depending on climate conditions, agricultural practices, and market demand, but West Africa is the largest region for millet production, contributing around 80% of the continent's total production. Niger is the leading producer of pearl millet in Africa, with a production of over 3.3 million tons in the year 2023 (FAOSTATS 2024). Other major producers include Mali (1.94 million tons), Nigeria (1.55 million tons), Senegal (1.35 million tons), and Burkina Faso (0.86 million tons). Overall, Africa produces roughly 10–15 million metric tons of millet annually, with a significant portion being pearl millet. While exact statistics on biofortified iron-rich millet are not easily separated, biofortified varieties have been integrated into local agricultural systems.

Traditionally, pearl millet is an important dietary source of iron in regions where diets are heavily reliant on millet as a staple food. Depending on genotype and growing environment, the iron concentration of conventional pearl millet ranges from approximately 1.6 to 9.6 mg per 100 g of grain, contributing meaningfully to daily iron intake. Iron-biofortified pearl millet varieties have been developed to contain higher iron concentrations, typically around 6–8 mg per 100 g, representing a roughly two-fold increase compared with commonly consumed varieties. Evidence from human absorption studies indicates that total iron absorption from iron-biofortified pearl millet meals is approximately double that from regular millet meals, although lower than that from post-harvest iron-fortified millet products (Cercamondi et al. 2013). Given that iron-deficiency anemia remains highly prevalent in Africa, particularly among women and young children, biofortification of pearl millet offers a sustainable, food-based strategy to improve iron intake and nutritional status through everyday diets.

6.3 Importance of Biofortification: Nutritional Improvement

As staple crops are the main source of nutrients, biofortification is a sustainable, cost-effective approach to reach the target nutrients needed by people in their natural form (Bouis et al. 2011). During the last few decades, significant progress has been made in the international crop breeding community to boost the nutrient concentration of staple crops through a biofortification approach (Pfeiffer and McClafferty 2007). Initial efforts have focused on the development of Zn-dense wheat (*Triticum aestivum* L.), rice (*Oryza sativa* L.), and cowpea (*Vigna unguiculata* L.) (Andersson et al. 2017), and Fe-dense pearl millet (*Pennisetum glaucum* L.) and common bean (*Phaseolus vulgaris* L.) (Saltzman et al. 2017).

Among the crops, millets such as pearl millet (*Pennisetum glaucum*), finger millet (*Eleusine coracana*), foxtail millet (*Setaria italica*), proso millet (*Panicum miliaceum*), kodo millet (*Paspalum scrobiculatum*), little millet (*Panicum sumatrense*), and barnyard millet (*Echinochloa frumentacea*) are traditionally important food crops for millions of people living in the dry lands across the globe. They are recognized as high-energy, nutritious cereals that can help reduce malnutrition, which is emerging as a major health problem due to the consumption of food with unbalanced nutrient quality. The grains of millet contain higher iron, zinc, protein, fibre, calcium, and mineral content, and so are often hailed as "noble" or "miracle" grains.

Millets are also well adapted to harsh agro-ecological conditions, thriving under high temperatures, low rainfall, and poor soils where other cereals fail. Their low water requirement and relatively small carbon footprint make them climate-resilient crops well suited to climate-smart agriculture in dryland regions (ICRISAT 2004; Gowda et al. 2022). Approximately 95% of global millet cultivation occurs in developing countries, with major production zones spanning the Sahelian belt of West Africa, the Horn of Africa, and South Asia (Dayakar Rao et al. 2011). Enhancing both grain yield and nutritional quality of millets, therefore, offers a direct pathway to improving food and nutritional security among resource-poor households that rely heavily on these crops.

HarvestPlus initiative of the Consultative Group on International Agricultural Research (CGIAR) embarked some years ago on addressing the micronutrient malnutrition issue through the breeding of improved crop cultivars with elevated levels of these micronutrients. Breeding millet for higher micronutrient concentration, an approach termed biofortification, has become an active goal of plant breeding globally (Bouis 2003; Welch 2002). Thus, this cost-effective approach is a sustainable way to address micronutrient malnutrition in low-income countries of West Africa (Stein et al. 2007; Bouis et al. 2011). In addition, increasing the quantity of micronutrients stored in seeds can result in improved seedling vigor and viability when these seeds are sown in micronutrient-poor soils, thus enhancing the performance of seedlings during crop establishment (Welch 1999).

Almost all identified iron millet sources are based on *iniadi* germplasm (early-maturing, large-seeded landrace materials from a geographic area adjoining Togo, Ghana, Burkina Faso, and Benin) or have a large proportion of *iniadi* germplasm in their parentage. This germplasm in an earlier study was found to be the most promising source of both high Fe and Zn contents (Velu et al. 2007). The major focus of the breeding programs based in WCA was to develop stable, high-yielding, high-iron and zinc open-pollinated varieties (OPVs), and hybrids for different agroecological zones using African germplasm. During the HarvestPlus Phase I (Discovery, 2003–2008), ICRISAT initiated screening of germplasm accessions in pearl millet, and found ranges of 30–76 mg/kg iron and 25–65 mg/kg zinc. High-iron genotypes were selected to initiate crosses in India and West Africa to improve both iron and zinc in pearl millet grains (Velu et al. 2007).

Screening efforts during the HarvestPlus discovery phase indicated the existence of large variability for grain Fe and Zn concentrations in pearl millet. Of 72 pearl millet accessions from West and Central Africa (WCA) assessed in Niger, few showed moderate ranges in mineral density (24.2 to 48.7 mg kg^{-1} for Fe and 19.8 to 43.4 mg kg^{-1} for Zn) (Pucher et al. 2014). A study focusing on the grain mineral density of 225 Sudanese pearl millet accessions evaluated in Sudan also found Fe and Zn ranging from 19.7 to 86.4 mg kg^{-1} for Fe and 13.5 to 82.4 mg kg^{-1} for Zn (Bashir et al. 2014).

These efforts culminated in the release of Africa's first biofortified pearl millet variety, "Chakti," in 2018, developed through collaboration between ICRISAT and NARS partners in Niger. Chakti contains approximately 65 mg kg^{-1} iron and 45 mg kg^{-1} zinc, nearly double the mineral density of commonly grown local varieties, without yield penalties (Gangashetty et al. 2021). Since then, additional biofortified pearl millet varieties have been released or promoted in Niger, Nigeria, Senegal, and Burkina Faso, and adoption is gradually increasing by conducting demonstrations, through seed system strengthening and nutrition-sensitive extension programs (Govindaraj et al. 2019; Saltzman et al. 2023). Although current adoption levels remain modest, ongoing research and policy support aim to expand the reach and impact of biofortified millets in West Africa.

6.4 Bioavailability of Iron and Zinc

The total concentration of micronutrients in staple crops does not directly reflect the amount absorbed and utilized by the human body; rather, bioavailability is influenced by interactions among dietary components that either inhibit or enhance mineral absorption (Murgia et al. 2023). Plant foods commonly contain antinutrients, such as phytic acid, polyphenols including tannins, and certain fibers, which form complexes with Fe and Zn, thereby reducing their intestinal absorption (Reddy et al. 2021). Conversely, dietary factors, including ascorbic acid (vitamin C), certain organic acids, and amino acids, can enhance mineral solubility and uptake in the gut (Troesch et al. 2018).

The major antinutrient factors in pearl millet are phytic acids, tannins, polyphenols, oxalates, and trypsin inhibitors. Recent research continues to demonstrate that phytic acid (inositol hexakisphosphate, IP6) is a major antinutrient in plant-based diets, especially where cereals and legumes are dominant, because it forms insoluble complexes with divalent minerals such as iron and zinc, reducing their bioavailability in the human gut (Singh et al. 2025; Murphy et al. 2026). Phytic acid's strong chelating properties are consistently identified as a key factor limiting the absorption of Fe, Zn, Ca, Mg, and other minerals, particularly in unrefined grain-based diets common in low-income regions. Although much of the older literature focused on its negative impacts, contemporary reviews also emphasize that processing methods (e.g., fermentation, germination) and enzyme treatments (phytase) can substantially reduce phytate content, thereby improving Fe and Zn bioavailability.

Tannins and other polyphenolic compounds likewise interact with minerals and proteins in the gastrointestinal tract, forming complexes that can further inhibit mineral solubility and absorption; these interactions reduce the bioavailability of Fe and Zn, as well as digestive enzyme activity, which in turn lowers overall nutrient uptake (Zayed et al. 2025). Despite their antinutrient effects, phenolic compounds, including some tannins and phytates, also exhibit antioxidant properties that may contribute to physiological benefits, highlighting the complex role these compounds play in human nutrition (Singh et al. 2025; Zayed et al. 2025).

6.5 Inheritance of Nutritional Traits

Many studies reported substantial variability in Fe and Zn densities across diverse germplasm (Velu et al. 2007, 2008; Gupta et al. 2009; Govindaraj et al. 2013). Building on this variation, Upadhyaya (2016) developed a high-iron pearl millet line (IP 17521 of Gnali) in India, using the Iniadi landraces from Togo, which achieved an Fe concentration of 106.9 mg/kg through conventional plant breeding. One of the challenges in conventional plant breeding is the potential trade-offs between grain quality and yield. However, in pearl millet, enhancing micronutrient content has been shown to improve seedling vigor and viability, leading to better crop establishment even under low-micronutrient soils (Upadhyaya 2016; Velu et al. 2007).

Studies on the heritability of grain Fe and Zn traits indicate a predominance of additive genetic effects, suggesting these traits can be effectively selected in breeding programs. High broad-sense heritability for Fe (65–80%) and Zn (65–80%) has been reported in multiple studies (Gupta et al. 2009; Velu 2006; Bashir et al. 2013; Govindaraj et al. 2011, 2013). Recent evaluations confirm these findings, reporting moderate-to-high heritability (55–75%) for grain Fe and Zn across diverse environments and breeding lines, which facilitates selection for biofortified cultivars without compromising agronomic performance (Singh et al. 2023; Hiremath et al. 2021).

The development of biofortified pearl millet cultivars has demonstrated that nutritional enhancement can be achieved alongside stable yield and acceptable agronomic traits, reinforcing the potential of genetic improvement to alleviate micronutrient deficiencies in millet-dependent populations (Govindaraj et al. 2024; Patel et al. 2025).

6.6 Molecular Breeding in Improving Micronutrient Content

Molecular breeding, including marker-assisted selection (MAS) and quantitative trait locus (QTL) mapping, integrates phenotypic and genotypic data to accelerate the development of improved crop varieties with desirable traits such as enhanced micronutrient density (Xu and Crouch 2025). In pearl millet, molecular approaches have evolved from early linkage maps to high-resolution SNP-based maps that underpin the identification of genomic regions controlling grain iron (Fe) and zinc (Zn) concentrations (Gupta et al. 2018; Varshney et al. 2024).

Early molecular efforts in pearl millet centered on low-density linkage maps constructed with RFLP markers, which provided initial frameworks for genetic analysis (Liu et al. 1994). Subsequent incorporation of SSRs and SNPs expanded linkage maps to over 1000 cM, enhancing genome coverage and facilitating more precise QTL discovery (Sehgal et al. 2012; Yadav et al. 2021). These advances have enabled the detection of stable QTLs for Fe and Zn accumulation, which can be targeted in breeding programs to improve micronutrient content without compromising agronomic performance (Sushil Kumar et al. 2016; Varshney et al. 2024).

QTL mapping studies using recombinant inbred line (RIL) populations identified co-localized QTLs for grain Fe and Zn on multiple linkage groups, indicating pleiotropic or closely linked genetic control of these traits (Sushil Kumar et al. 2016; Sharma et al. 2022). Such loci are promising candidates for MAS, offering opportunities to select for both micronutrients simultaneously. Importantly, recent genome-wide association studies (GWAS) have further refined marker-trait associations, providing breeders with robust SNP markers linked to micronutrient levels (Hiremath et al. 2023).

Although MAS enhances selection efficiency, it still relies on accurate phenotyping of micronutrient traits across environments, underscoring the need for integrated genotype–phenotype pipelines in biofortification programs (Gupta et al. 2018; Hiremath et al. 2023). In West Africa, a center of pearl millet diversity, application of MAS and genomic selection approaches holds significant promise for rapidly introgressing high-Fe and high-Zn alleles into elite open-pollinated varieties and hybrid parents, thereby strengthening regional nutritional security (Varshney et al. 2024; Singh et al. 2024).

6.7 Micronutrient Enrichment Using Transgenic Strategies

Genetic engineering, also known as transgenic crop technology, is achieved by inserting specific, functionally characterized genes into plant genomes to modify metabolic pathways associated with micronutrient accumulation and bioavailability. While conventional breeding remains foundational, transgenic and gene-editing approaches extend the scope of nutritionally relevant trait enhancement, particularly when natural genetic variation is limited for specific micronutrient traits (Borrell et al. 2023). For example, transgenic expression of phytase genes has been used to reduce phytic acid, a major inhibitor of iron and zinc absorption, thereby improving micronutrient bioavailability in cereals. In rice, constitutive expression of a microbial phytase resulted in significant phytate reduction and enhanced iron bioavailability in grain (Seth et al. 2021). Similarly, transgenic maize lines overexpressing *Aspergillus niger* phytase showed up to a 35% decrease in seed phytic acid, with corresponding increases in mineral availability (Huang et al. 2022).

In addition to antinutrient reduction, enhancing micronutrient uptake and partitioning pathways has been demonstrated in cereals. Overexpression of iron transporter genes such as OsIRT1 and OsYSL15 in rice increased iron (Fe) and zinc (Zn) concentrations in grain, validating transporter manipulation as a viable strategy for micronutrient enrichment (Zhang and Li 2020). More recently, transgenic deployment of vacuolar iron transporter genes and nicotianamine synthase genes has shown promise in elevating grain Fe and Zn concentrations without adverse effects on agronomic performance (Mahajan and Singh 2024; Wei et al. 2025). These advances underscore the potential of transgenic and gene-editing strategies to complement conventional breeding and accelerate the development of nutritionally enhanced staples.

6.8 Modern Omics Tools to Enrich Micronutrient Content in Grain

Advances in next-generation sequencing (NGS) and high-throughput genotyping platforms have transformed the discovery and deployment of genomic loci associated with nutrient traits in crops. Methods such as genotyping-by-sequencing (GBS) and high-density SNP arrays enable comprehensive characterization of genetic variation, facilitating the identification of markers and candidate genes for micronutrient enhancement (Varshney et al. 2021).

Genome-wide association studies (GWASs) leverage diverse germplasm panels to uncover genomic regions associated with trait variation without the need for biparental mapping populations. In cereals, GWAS has successfully detected loci linked to grain Fe and Zn concentrations, informing marker-assisted selection and genomic selection strategies. For example, GWAS in barley identified Zn-associated loci on chromosome 6HL, providing targets for biofortification breeding (Mamo et al. 2014). Similarly, GWAS in rice has elucidated the genetic architecture of grain micronutrients using high-density SNP data across hundreds of accessions (Zhao et al. 2011).

More recently, multi-omics integration combining genomics with transcriptomics, metabolomics, and ionomics has refined the functional dissection of micronutrient pathways. Integrated analyses in maize and wheat have identified gene networks and regulatory elements controlling Zn and Fe homeostasis, improving the resolution of candidate targets for breeding (Wang et al. 2022; Singh et al. 2023).

Additionally, genomic prediction models incorporating omics data have increased the accuracy of selecting micronutrient-dense genotypes across environments, accelerating biofortification pipelines (Crossa et al. 2024).

These modern omics tools, when coupled with robust phenotyping, enable the discovery and deployment of genomic regions and biomarkers that enhance micronutrient content, facilitating precision breeding in pearl millet and other staple cereals.

6.9 Status of Millet Biofortification in WCA

To address micronutrient malnutrition in West and Central Africa, pearl millet breeders have released varieties with moderate grain iron (Fe) and zinc (Zn) concentrations, such as GB8735 and ICRI–Tabi. Since 2015, HarvestPlus has provided substantial support to intensify breeding efforts aimed at developing high grain Fe and Zn pearl millet varieties adapted to diverse agro-climatic zones of the WCA region. Current breeding programs target micronutrient levels above 45 mg/kg, resulting in several high-yielding, nutrient-rich open-pollinated varieties (Bashir et al. 2022; Varshney et al. 2024).

The first Fe-biofortified pearl millet variety released in Africa, "Chakti," was introduced in 2018 (Fig. 6.1). It matures in 65 days, contains 65 mg/kg of grain Fe and 45 mg/kg of grain Zn, and has a grain yield potential of up to 1.5 t/ha (HarvestPlus 2018). Its early-maturing trait and enhanced nutritional profile facilitated rapid adoption in the Sahel, allowing some farmers with irrigation facilities to harvest two crops per year (Reca Niger 2023).

Since its release, Chakti has been scaled up in collaboration with ICRISAT, national agricultural research systems (NARS), farmer organizations, and private seed companies. By 2024, production reached 716,457 kg of certified seed, and according to the 2025 national directory of improved seed varieties, Chakti is now the second most widely produced millet variety in Niger (Government of Niger

Fig. 6.1 Iron (Fe) biofortified pearl millet "Chakti"

Table 6.1 Superior biofortified pearl millet varieties identified for the Sahelian Zone

Genotype	Downy mildew percentage	Days to 50% flowering	Panicle length (cm)	Plant height (cm)	Grain yield (t/ha)	Grain Fe (mg/kg)	Grain Zn (mg/kg)
ICMP 167001	8.0	41.0	20.0	139.0	1.4	50.0	38.0
ICMP 177002	4.0	40.0	21.0	135.0	1.5	80.0	45.0
ICMP 187092	4.0	45.0	38.0	156.0	1.6	58.0	41.0
ICMP 187093	3.0	47.0	41.0	155.0	1.7	55.0	39.0
ICMP 207555	0.0	40.0	24.0	150.0	1.5	95.0	61.0
ICMP 207666	0.0	41.0	25.0	145.0	1.5	105.0	69.0
ICMP 207777	0.0	42.0	25.0	139.0	1.6	115.0	78.0
Chakti	0.0	40.0	18.0	135.0	1.5	65.0	45.0
ICMV 167006	0.0	46.0	15.0	150.0	1.4	45.0	35.0

2025). Additionally, the Lake Chad Research Institute released Chakti in Nigeria in 2023, where it is being upscaled through HarvestPlus programs. The variety is also registered in the ECOWAS catalog, allowing neighboring countries to adopt it without a formal national release. In Mali, it is locally known as "Sambalanyon," meaning "a variety sown late yet harvested at the same time with the local variety," reflecting its adaptation to local cropping calendars (Hama et al. 2022).

Breeding efforts in the WCA have produced several superior biofortified pearl millet varieties suitable for the Sahelian zone, combining high grain Fe and Zn content with disease resistance and acceptable agronomic traits (Table 6.1).

6.10 Challenges

A major constraint for millet biofortification in West and Central Africa (WCA) is insufficient or inconsistent funding for biofortification breeding programs, which can slow the development and deployment of climate-resilient, nutrient-dense varieties, and hybrids (Varshney et al. 2024). Despite recent releases of improved pearl millet varieties, adoption rates remain low, and varietal replacement rates (VRRs) are poor at the farmer level (Bashir et al. 2022; Hama et al. 2022). This limited uptake due farm-saved seeds of OPVs reduces VRR and the potential yield gains, compromises production stability, and exacerbates the vulnerability of smallholder farmers to climate change.

To address these challenges, efforts have focused on strengthening formal and informal seed systems through coordinated seed production, capacity development, delivery, and policy advocacy. Platforms such as the Hybrid Parents Research Consortium (HPRC) can facilitate the sharing of biofortified germplasm/parental material, breeding knowledge, and pre-command seed systems, allowing seed producers to forecast demand and ensure timely production of quality seed. These initiatives are critical to scaling biofortified pearl millet in WCA, ensuring that improved varieties reach farmers widely and efficiently. Finally, improved breeding lines screening for nutrition using XRF methods, as well as national and regional variety release policy, will amplify the mainstreaming of biofortification in WCA.

6.11 Conclusion

Biofortification of pearl millet has produced varieties containing up to three times more grain iron (Fe) and zinc (Zn) than traditional (non-biofortfied) cultivars, directly addressing micronutrient deficiencies prevalent in the region. Enhanced Fe and Zn intake through biofortified millet contributes to improved health outcomes, particularly among women, children, and other vulnerable populations.

Research and development efforts in pearl millet biofortification have also influenced policy discussions, encouraged investment in biofortification strategies, and strengthened collaborations among governments, NGOs, and international organizations. Successful adoption has been supported by training programs, extension services, and community engagement, enabling farmers to understand the benefits and agronomic management of biofortified varieties.

In Niger, government programs such as FESTIMIL have promoted Fe-rich millet among poor households, generating awareness of its nutritional and health benefits. Biofortified millet is also cultivated in Mali, Burkina Faso, Nigeria, and Senegal, with fast-track seed production in Nigeria supported by HarvestPlus programs. Scaling similar approaches across WCA has the potential to substantially improve food security and public health in the region.

References

Andersson MS, Saltzman A, Virk PS, Pfeiffer WH (2017) Biofortification issue chapter 5. Afr J Food Agric Nutr Dev 17(2):11905–11935

Bashir EM, Govindaraj M, Velu G (2013) Genetic control of grain iron and zinc concentration in pearl millet. J Plant Nutr 36(14):2210–2220. https://doi.org/10.1080/01904167.2013.828291

Bashir EM, Abdelbagi MA, Adam M, Mohamed A, Ismail I, Parzies HK, Haussmann BIG (2014) Patterns of pearl millet genotype-by-environment interaction for yield performance and grain iron (Fe) and zinc (Zn) concentrations in Sudan. Field Crop Res 166:245

Bashir EM, Govindaraj M, Velu G (2022) Progress in biofortification of pearl millet for iron and zinc in West Africa. Front Plant Sci 13:905432. https://doi.org/10.3389/fpls.2022.905432

Borrell AK, Hammer GL, Habyarimana E (2023) Biotechnological approaches to enhance nutrient density in staple crops. Annu Rev Plant Biol 74:159–182. https://doi.org/10.1146/annurev-arplant-070522-122343

Bouis HE (2003) Micronutrient fortification of plants through plant breeding: can it improve nutrition in man at low cost? Proc Nutr Soc 62:403–411. https://doi.org/10.1079/PNS2003262

Bouis HE, Hotz C, McClafferty B, Meenakshi JV, Pfeiffer WH (2011) Biofortification: a new tool to reduce micronutrient malnutrition. Food Nutr Bull 32(1_suppl1):31S–40S

Cercamondi CI, Egli IM, Mitchikpe E, Tossou F, Zeder C, Hounhouigan JD, Hurrell RF (2013) Total iron absorption by young women from iron-biofortified pearl millet composite meals is double that from regular millet meals but less than that from post-harvest iron-fortified millet meals. J Nutr 143(9):1376–1382. https://doi.org/10.3945/jn.113.176826

Crossa J, Pérez-Rodríguez P, Jarquin D (2024) Multi-omic genomic prediction models for micronutrient density in cereals. Theor Appl Genet 137:821–839. https://doi.org/10.1007/s00122-023-04412-7

Dayakar Rao B, Bhaskarachary K, Arlene Christina GDS, Sudha Devi G, Vilas AT (2011) Nutritional and health benefits of millets. Indian Institute of Millets Research (IIMR)

FAOSTAT (2024) Faostat: food and agriculture data. Retrieved 13 Dec 2025, from https://www.fao.org/faostat/en/#data/QCL

Food and Agriculture Organization (FAO), World Food Programme (WFP), & United Nations Children's Fund (UNICEF) (2025) The state of food security and nutrition in the world 2025: safeguarding against economic slowdowns and downturns. FAO. https://data.unicef.org/wp-content/uploads/2025/08/SOFI-Report-2025.pdf

Gangashetty PI, Govindaraj M, Kanatti A, Rai KN, Cherian B, Rao AS (2021) Development and dissemination of iron- and zinc-biofortified pearl millet cultivars in India and Africa. Front Sustain Food Syst 5:678324. https://doi.org/10.3389/fsufs.2021.678324

Government of Niger (2025) Annuaire national de disponibilité en semences de variétés améliorées 2025. Ministère de l'Agriculture

Govindaraj M, Bashir E et al (2011) Inheritance of micronutrient traits in pearl millet. J Genet 90:109–116. https://doi.org/10.1007/s12041-011-0020-2

Govindaraj M, Bashir E et al (2013) Genetic variability and heritability of grain Fe and Zn in pearl millet. Plant Breed 132:163–170. https://doi.org/10.1111/pbr.12003

Govindaraj M, Rai KN, Cherian B, Pfeiffer WH, Senthilvel S, Rao AS (2019) Breeding biofortified pearl millet varieties and hybrids to enhance grain iron and zinc density in India and Africa. Field Crop Res 233:44–55. https://doi.org/10.1016/j.fcr.2018.12.006

Govindaraj M, Rai KN, Kanatti A (2024) Biofortification of pearl millet: a sustainable strategy to alleviate micronutrient deficiencies and improve food security. Front Plant Sci 15:124567. https://doi.org/10.3389/fpls.2024.124567

Gowda CLL, Rai KN, Ramesh S, Govindaraj M (2022) Millets for climate-resilient agriculture and nutritional security. Front Plant Sci 13:918778. https://doi.org/10.3389/fpls.2022.918778

Gupta SK, Velu G, Rathore A et al (2009) Heritability of grain iron and zinc concentration in pearl millet. Field Crop Res 111:148–152. https://doi.org/10.1016/j.fcr.2009.01.006

Gupta PK, Singh S, Sharma R (2018) Genomic tools for enhancing micronutrient concentration in cereals and legumes. Front Genet 9:763. https://doi.org/10.3389/fgene.2018.00763

Hama A, Traoré S, Tchatchoua C (2022) Scaling up biofortified pearl millet in the Sahel: adoption, seed systems, and nutrition impacts. Int J Agric Res

HarvestPlus (2018) Africa's first iron-biofortified pearl millet variety aims to combat anemia. https://www.harvestplus.org/africas-first-iron-biofortified-pearl-millet-variety-aims-to-combat-anemia/

Hiremath SR, Rathore A, Yadav OP, Reddy BVS (2021) Genetic variability and trait associations for iron and zinc concentration in pearl millet. Front Plant Sci 12:756321. https://doi.org/10.3389/fpls.2021.756321

Hiremath SR, Yadav OP, Reddy BVS (2023) Genome-wide association studies reveal loci associated with grain iron and zinc concentration in pearl millet. Theor Appl Genet 136:1141–1154. https://doi.org/10.1007/s00122-023-04112-4

Huang X, Thompson JC, Lee DH (2022) Reduction of phytic acid in maize seed through transgenic overexpression of Aspergillus Niger phytase. Plant Biotechnol J 20(3):654–665. https://doi.org/10.1111/pbi.13609

Hussain MA, Li L, Kalu A, Wu X, Naumovski N (2025) Sustainable food security and nutritional challenges. Sustainability 17(3):874. https://doi.org/10.3390/su17030874

ICRISAT (2004) Sorghum, a crop of substance. International Crops Research Institute for the Semi-Arid Tropics.

Liu CJ, Witcombe JR, Pittaway TS, Nash M, Hash CT, Busso CS, Gale MD (1994) An RFLP-based genetic map of pearl millet (Pennisetum glaucum). Theor Appl Genet 89:481–487

Lowe NM (2025) Fortification or biofortification: complimentary strategies or duplication of effort? Proc Nutr Soc 84(1):65–74. https://doi.org/10.1017/S0029665124000041

Mahajan R, Singh S (2024) Genetic engineering of vacuolar iron transporters and nicotianamine pathways to enhance micronutrient content in cereals. Front Plant Sci 15:1170098. https://doi.org/10.3389/fpls.2024.1170098

Mamo BE, Barber B, Steffenson BJ (2014) Genome–wide association mapping of zinc and iron concentration in barley landraces from Ethiopia and Eritrea. J Cereal Sci 60:497–506

Murgia M et al (2023) Micronutrient bioavailability and human health: a review of physiology, influence of dietary factors, and food processing approaches. Crit Rev Food Sci Nutr. https://doi.org/10.1080/10408398.2023.2192910

Murphy R, O'Sullivan F, Keenan J (2026) The antagonistic influence of phytic acid on zinc absorption: an in vitro comparison of trace mineral sources. Nutrients 18(1):46. https://doi.org/10.3390/nu18010046

Patel J, Singh U, Chauhan G (2025) Biofortified pearl millet cultivars: agronomic performance and nutrient stability across target environments. Crop J 13(2):214–226. https://doi.org/10.1016/j.cj.2025.01.006

Paul M, Thriveni V, Niharika M, Upadhyay DK, Kiro D, Verma A, Patil S (2024) A review on biofortification of crops: a nutritional strategy for combating malnutrition. Eur J Nutr Food Saf 16(11):63–77. https://doi.org/10.9734/ejnfs/2024/v16i111577

Pfeiffer WH, McClafferty B (2007) Harvest plus: breeding crops for better nutrition. Crop Sci 47:88–105

Pucher A, Høgh-Jensen H, Jadah G, Hash CT, Haussmann BIG (2014) Micronutrient density and stability in West African pearl millet—potential for biofortification. Crop Sci 54:1709–1720

Reca Niger (2023) Biofortified millet varieties adoption and production in Niger. https://reca-niger.org/spip.php?article1488

Reddy NR, Sathe SK, Salunkhe DK (2021) Food antinutrients and their reduction strategies: an overview. J Food Sci Technol 58(6):1983–2000. https://doi.org/10.1007/s13197-020-04760-8

Saltzman A, Birol E, Bouis HE, Boy E, De Moura FF, Islam Y, Pfeiffer WH (2017) Biofortification issue chapter 2: biofortification: progress toward a more nourishing future. Afr J Food Agric Nutr Dev 17(2):11685–11708. https://doi.org/10.18697/ajfand.78.HarvestPlus02

Saltzman A, Birol E, Bouis HE, Boy E, De Moura FF, Islam Y, Pfeiffer WH (2023) Biofortification: progress toward a more nourishing future. Glob Food Sec 36:100689. https://doi.org/10.1016/j.gfs.2022.100689

Sehgal D, Rajaram V, Armstead IP, Vadez V, Yadav YP, Hash CT, Yadav RS (2012) Integration of gene-based markers in pearl millet genetic map for identification of candidate gene underlying drought tolerance quantitative trait loci. BMC Plant Biol 12(1):1–13

Seth P, Sharma R, Kapoor S (2021) Transgenic expression of microbial phytase in rice for improved iron bioavailability. J Agric Food Chem 69(12):3567–3577. https://doi.org/10.1021/acs.jafc.1c00734

Sharma V et al (2022) Co-localized QTLs controlling Fe and Zn in pearl millet. Front Plant Sci 13:879365. https://doi.org/10.3389/fpls.2022.879365

Singh U, Praharaj CS, Singh SS, Singh NP (2016) Biofortification of food crops. Springer. https://doi.org/10.1007/978-81-322-2716-8
Singh P, Chaturvedi S, Sharma R (2023) Heritability and genetic gains for grain micronutrients in pearl millet across environments. Field Crop Res 289:108800. https://doi.org/10.1016/j.fcr.2023.108800
Singh A et al (2024) Application of genomics-assisted breeding for nutritional enhancement in pearl millet. Crop J 12(1):96–108. https://doi.org/10.1016/j.cj.2023.11.003
Singh G et al (2025) Anti-nutritional factors: nutrient interactions, processing interventions, and health aspects. Food Chem 380:135830. https://doi.org/10.1016/j.foodchem.2025.135830
Stein AJ, Nestel P, Meenakshi JV, Qaim M, Sachdev HPS, Bhutta ZA (2007) Plant breeding to control zinc deficiency in India: how cost-effective is biofortification? Public Health Nutr 10:492–501. https://doi.org/10.1017/S1368980007223857
Sushil Kumar N, Yadav Y, Reddy BVS (2016) QTLs for grain iron and zinc in pearl millet identified using recombinant inbred line populations. Plant Breed 135:110–119. https://doi.org/10.1111/pbr.12340
Troesch B et al (2018) Role of nutrient interactions on iron and zinc bioavailability and implications for food fortification. Nutrients 10(6):678. https://doi.org/10.3390/nu10060678
United Nations (2024) World population prospects 2024: summary of results. Department of Economic and Social Affairs. https://population.un.org/wpp
Upadhyaya HD (2016) Development of high-iron pearl millet using Iniadi germplasm from Togo. Plant Breed 135:80–86. https://doi.org/10.1111/pbr.12345
Varshney RK, Roorkiwal M, Sinha P (2021) Genomics and breeding for micronutrient enrichment in cereals. Plant Biotechnol J 19(9):1683–1700. https://doi.org/10.1111/pbi.13584
Varshney RK et al (2024) Genomic selection and marker-assisted breeding for biofortification in pearl millet. Plant Biotechnol J 22(5):402–420. https://doi.org/10.1111/pbi.14012
Velu G (2006) Heritability of micronutrient traits in pearl millet. J SAT Agric Res 4:1–5
Velu G, Ortiz R et al (2007) Genetic variability and inheritance of grain iron and zinc in pearl millet. Crop Sci 47:2233–2241. https://doi.org/10.2135/cropsci2006.09.0567
Velu G, Ortiz R et al (2008) Assessment of iron and zinc concentration in pearl millet germplasm. Plant Genet Res 6:121–127. https://doi.org/10.1017/S1479262108002470
Wang H, Zhang C, Chen S (2022) Transcriptome-assisted GWAS highlights candidate genes for iron and zinc homeostasis in maize kernels. BMC Plant Biol 22:483. https://doi.org/10.1186/s12870-022-03821-y
Wei L, Chen Y, Xu Z (2025) Engineering nicotianamine synthases to enhance grain iron and zinc in cereal crops. Plant Mol Biol 109(2–3):193–207. https://doi.org/10.1007/s11103-024-01489-x
Welch RM (1999) Importance of seed mineral nutrient reserves in crop growth and development. In: Rengel Z (ed) Mineral nutrition of crops: fundamental mechanisms and implications. Food Products Press, pp 205–226
Welch RM (2002) Breeding strategies for biofortified staple plant foods to reduce micronutrient malnutrition globally. J Nutr 132:495S–499S
World Food Programme (WFP) & UNICEF (2025) Worsening hunger grips West and Central Africa amid persistent conflict and economic turmoil. https://www.wfp.org/news/worsening-hunger-grips-west-and-central-africa-amid-persistent-conflict-and-economic-turmoil
World Health Organization (WHO) (2002) The World Health Report: reducing risks, promoting healthy life. WHO
World Health Organization (WHO) (2016) Micronutrient deficiencies: iron deficiency anaemia. http://www.who.int/nutrition/topics/ida/en/
World Health Organization (WHO) (2025) Niger: WHO Africa country profile – nutrition. https://www.afro.who.int/fr/countries/niger
Xu Y, Crouch JH (2025) Marker-assisted and genomic selection in crop breeding: progress and prospects. Trends Plant Sci 30(2):145–158. https://doi.org/10.1016/j.tplants.2024.10.004
Yadav P et al (2021) Advances in SNP marker development and their application in pearl millet breeding. Theor Appl Genet 134:2325–2340. https://doi.org/10.1007/s00122-021-03808-x

Zayed A et al (2025) Tannin in foods: classification, dietary sources, and processing strategies to minimize anti-nutrient effects. Food Bioprocess Technol 18:9221–9249. https://doi.org/10.1007/s11947-025-04020-3

Zhang H, Li Q (2020) Overexpression of iron transport genes increases grain iron and zinc concentrations in transgenic rice. Plant Sci 290:110274. https://doi.org/10.1016/j.plantsci.2019.110274

Zhao K, Tung CW, Eizenga GC et al (2011) Genome-wide association mapping reveals a rich genetic architecture of complex traits in Oryza sativa. Nat Commun 2:467. https://doi.org/10.1038/ncomms1467

Chapter 7
Biofortified Pearl Millet Varieties Fast-Track Scaling and Impact in Nigeria

I. I. Angarawai, Z. G. S. Turaki, Y. Dollah, M. A. Dawud, P. Gangashetty, M. Riyazaddin, J. Jonah, and I. A. Hassan

Abstract Triple burden of malnutrition has been noted as the "new normal" (i.e., malnutrition or undernutrition). A well-diversified diet is nature's own way of meeting the micronutrient needs of a Sub-Saharan Africa population, where millets remain an important source for food security and livelihoods, due to their nutritional value and resilience to climate change. In recent years, there has been a renewed interest in millets (e.g., Pearl millet), with efforts being made to promote its biofortification, improve nutrition with productivity, and enhance market opportunities with potential for income generation. Pearl millet biofortified breeding lines introduced from ICRISAT-India to Nigeria through Niger, were evaluated jointly with NARES from 2018 to 2022 across the millet-growing states of Northern Nigeria. Data collected on grain yield, agronomic parameters, downy mildew infestation, and micronutrient content were statistically analyzed. Micronutrient content analysis indicated that Chakti with the highest iron content (74.2 ppm), along with the standard check ICMP 177002, was statistically superior relative to ICMP 187093 and ICMP1970115. Following their performance for grain yield (>2.5 t/ha) and low downy mildew incidences (<5%), they were recommended for registration and release for commercialization as biofortified iron pearl millet. Recognizing the urgency, HarvestPlus engaged the national variety release committee (NVRC) to

I. I. Angarawai (✉) · J. Jonah
International Crops Research Institute for the Semi-Arid Tropics, Kano, Nigeria
e-mail: Ignatius.Angarawai@icrisat.org

Z. G. S. Turaki · M. A. Dawud
Lake Chad Research Institute, Maiduguri, Nigeria

Y. Dollah · I. A. Hassan
HarvestPlus Program, International Food Policy Research Institute, c/o IITA Campus, Kano, Nigeria

P. Gangashetty
International Crops Research Institute for the Semi-Arid Tropics, Hyderabad, India

M. Riyazaddin
International Crops Research Institute for the Semi-Arid Tropics, Niamey, Niger

M. Govindaraj (ed.), *Breeding Biofortified Millets for Nutritional Security*, Sustainability Sciences in Asia and Africa,
https://doi.org/10.1007/978-981-95-9827-4_7

advance for a special release event, underscoring the role of biofortified crops in combating hidden hunger. This effort led to a groundbreaking decision: NVRC agreed to a HarvestPlus-funded special release event in January 2023, ensuring that iron pearl millet (IPM) varieties were made available to farmers without unnecessary delay.

Keywords Malnutrition · Nutrition · Biofortification · Pearl millet · Micronutrient

7.1 Introduction

The triple burden of malnutrition has been noted as the "new normal" (i.e., malnutrition, bad nutrition, or under-nutrition). World Health Organization (WHO) (2024), indicated globally in 2022 that, 149 million children under 5 years of age were estimated to be stunted (too short for their age), 45 million were estimated to be wasted (too thin for their height), and 37 million were overweight or living with obesity and nearly half of deaths among children under 5 years of age are linked to undernutrition resulting from diet-related noncommunicable diseases due to lack of important vitamins and minerals or micronutrients.

Several vitamin A-rich varieties of cassava, maize, zinc-rich pearl millet, and orange sweet potato have been released in Nigeria. These crops are crucial in addressing high rates of deficiencies in vitamin A, zinc, and iron, particularly among vulnerable populations like children and pregnant women. Biofortified crops represent a promising strategy for addressing micronutrient deficiencies in Nigeria through continued efforts in research, development, and dissemination of pearl millet biofortified varieties in collaboration with NARIs, especially Lake Chad Research Institute (LCRI), Maiduguri, for its genetic improvement, and HarvestPlus, which coordinates Nigeria's staple crop biofortification program. Since 2013, an estimated four million plus farmers have been reached with biofortified crop varieties (seeds and vines), where millets remain important for food security and livelihoods, due to their nutritional value and resilience to climate change, particularly in regions where other crops may struggle to grow (Onuegbu et al. 2017; Dollah 2024).

Pearl millet (*Pennisetum glaucum* L.) is a versatile and nutritious grain with significant potential for addressing food security and improving human nutrition. It is an excellent source of energy, dietary fiber, and essential nutrients such as iron, magnesium, and phosphorus. It contains complex carbohydrates with a low glycemic index, making it suitable for individuals with diabetes or those seeking to manage blood sugar levels (Kamalesh et al. 2024). Millets' nutritional value, as reviewed by Jacob et al. (2024), may help prevent cancer, reduce the risk of cardiovascular disease, stop the growth of tumors, lower blood pressure, lower the rate at which fat is absorbed, delay gastric emptying, and increase gastrointestinal bulk. Millets are also known to be gluten-free and easily digestible, making intolerance, stomach ulcers, or excessive cholesterol, good source of antioxidants and have anticarcinogenic qualities. Additionally, pearl millet is gluten-free, which makes it a viable alternative for individuals with celiac disease or gluten intolerance. Pearl millet also possesses desirable textural and water-holding capacities, making it suitable for various food applications such as baking, extrusion, and fortification. Improving the micronutrient concentration of locally adapted staple food crops like millet will be a sustainable solution to address this malnutrition.

7.2 Biofortification in Pearl Millet

Biofortification is becoming more of the breeding process by which the nutrient density is increased through conventional plant breeding and modern biotechnology without altering the preferred traits of the farmers (Nestel et al. 2006). This process of increasing the concentration of micronutrients in the staple crops to have a measurable impact on human nutrition after consumption is, therefore, important for combating micronutrient malnutrition by increasing the concentration and bioavailability of essential micronutrients in staple crops using conventional plant breeding and agronomic practices. The technique has the potential to become a sustainable, inexpensive, and effective solution to Fe deficiency at the population level. Previous work by the research group has demonstrated that the consumption of Fe-biofortified pearl millet was able to improve the Fe status and cognitive function of Indian children (Finkelstein et al. 2015; Scott et al. 2018). Other studies have shown that the positive impact of the consumption of biofortified crops, had similar results to iron fortification and supplementation strategies to fight against hidden hunger and that the quantities of both iron and zinc were absorbed when iron- and zinc-biofortified pearl millet was fed to children aged 2 years as the major staple food and was more than adequate to meet the physiological requirements for these micronutrients (Kodkany et al. 2013).

Biofortified cultivars are being developed through conventional breeding; thus, grains and foods produced from such cultivars do not face the possible challenges of food regulations and consumer acceptance (Govindaraj et al. 2019). Biofortified pearl millet can significantly improve the iron intake of those who consume it regularly, potentially providing up to 80% of the daily iron needs for women of reproductive age and children (HarvestPlus 2024).

Pearl millet is significantly grown and consumed in northern Nigeria; hence, this study was conducted to evaluate the pearl millet varieties with high grain Fe and Zn contents, identified at ICRISAT Niamey breeding Hub (Gangashetty et al. 2021), for their stability performance across the environment, and the influence of environmental factors on the inheritance of these traits. This is in consonance with the Nigeria Millet Varieties Standards (NGMVS) of 2024, "additional requirements for pearl millet variety registration and release."

7.3 Field Evaluation

The experimental material consisted of 20 entries (18 pearl millet lines introduced from ICRISAT and 2 checks), including a pearl millet variety called "Chakti" (derived from the intra-population variability within ICTP8203, at ICRISAT India), evaluated from 2020 to 2022 for grain yields and grain micronutrient content, adopting Farmer Participatory Variety selection (FPVS) across 7 locations in Northern Nigeria. The material was sown in a randomized complete block design in three replications during rainy seasons. A basal dose of Fertilizers was applied at the rate of 60 kg N and 30 kg each of P_2O_5 and K_2O per hectare. Half of the nitrogen was applied basally, while the remaining 30 kg/ha was supplied 4 weeks later. Plots were kept weed-free by three hoe-weeding at 3, 6, and 9 weeks after sowing (WAS), while thinning to one stand/hill was carried out at 2 WAS. The genotypes were sown in 4-row plots with a row length of 3.0 m and a row-to-row distance of 0.75 m and 0.20 m between plants. The thinning of the test entries to one plant per hill was carried out 15–21 days after seedling emergence (DAE). Hand weeding of the test plots was carried out where necessary. Microdosing with urea was carried out 30 DAE.

7.4 Data Collection Methods

Data in advanced multi environment testing were collected on flowering time (days) determined from the first day of planting to the time when 50% of the plants had flowered; plant height (cm) from the base of the plant to tip and panicle length (cm) from base to tip of the panicle in centimeter at maturity; downy mildew incidence (%), agronomic appreciation score (1–5 scale), iron (ppm), zinc (ppm) and grain yield (t/ha) which results were analyzed and presented in this study. Data for On-Farm testing were collected on days to physiological maturity, plant height, panicle length, grain yield, and mineral content as in the ICRISAT Descriptor Manual. Grain samples for Fe and Zn analysis were collected before harvesting on the standing crop, eliminating the plants that were grounded and contaminated with soil. Proper care was taken in cleaning the grain samples for analysis to avoid contamination from metals and dust. Micronutrient analysis, iron (ppm) and zinc (ppm) density was determined using Inductively Coupled Plasma Optical Emission Spectrometry (ICP–OES) method as described by Wheal et al. (2011), a suite of 18 mineral elements commonly found in plant tissues as either essential elements (B, Ca, Cu, Fe, K, Mg, Mn, Mo, Na, Ni, P, S, Se, Zn), potentially toxic metals (Al, Cd, Cr), or soil contaminants (Ti).

7.5 Statistical Methods Used

The replicated data from multi-environments from each location are subjected to perform analysis of variance (ANOVA) and Additive Main Effects and Multiplicative Interaction (AMMI) analysis using GenStat software version 21.0 (21). The significant differences between the genotypes were tested by using the *F*-test (Duncan's Multiple Range Test at 5% probability level), and the genotypic means were compared by the least significant difference at $p \leq 0.05$. AMMI analysis permits the estimation of the interaction effect of a genotype in each environment, and it helps to identify genotypes best suited for specific environmental conditions. The data from 2022 On-farm testing were subjected to ANOVA using the statistical software Statistic 8.0, means compared using LSD at 5% probability level, and cost-benefit analysis was employed to compare the profitability of the millet candidates compared to the farmers' local.

7.6 Results

Multilocation testing AMMI model ANOVA (Table 7.1) indicated that Genotype effects (G), Environments (E), and G × E interaction were all highly significant ($P < 0.001$); in which GGE biplot further depicts that PC 1 and PC 2 accounted for 74.16% and 9.24% of the total GE interaction (83.39%), respectively.

7.6.1 *Yield Superiority*

PC1 depicts genotype yield superiority (Fig. 7.1), where genotypes ICMP1970115 (2.79 t/ha), Chakti (2.71 t/ha), ICMP 187093 (2.67 t/ha), and Super SOSAT (2.61 t/ha), with PC1 scores >0, are featured as high-yielding as shown in Table 7.2.

Table 7.1 ANOVA table for AMMI model

Source	Df	SS	MS	VR	F PR
Treatments	167	129.49	0.775	11.32	<0.001
Genotypes	11	65.81	5.982	87.36	<0.001
Environments	13	17.54	1.349	10.50	<0.001
Block	28	3.60	0.128	1.88	0.0057
Interactions	143	46.14	0.323	4.71	<0.001
IPCA 1	23	19.37	0.842	12.30	<0.001
IPCA 2	21	9.54	0.454	6.64	<0.001
Residuals	99	17.23	0.174	2.54	<0.001
Error	308	21.09	0.068		
Total	503	154.18	0.307		

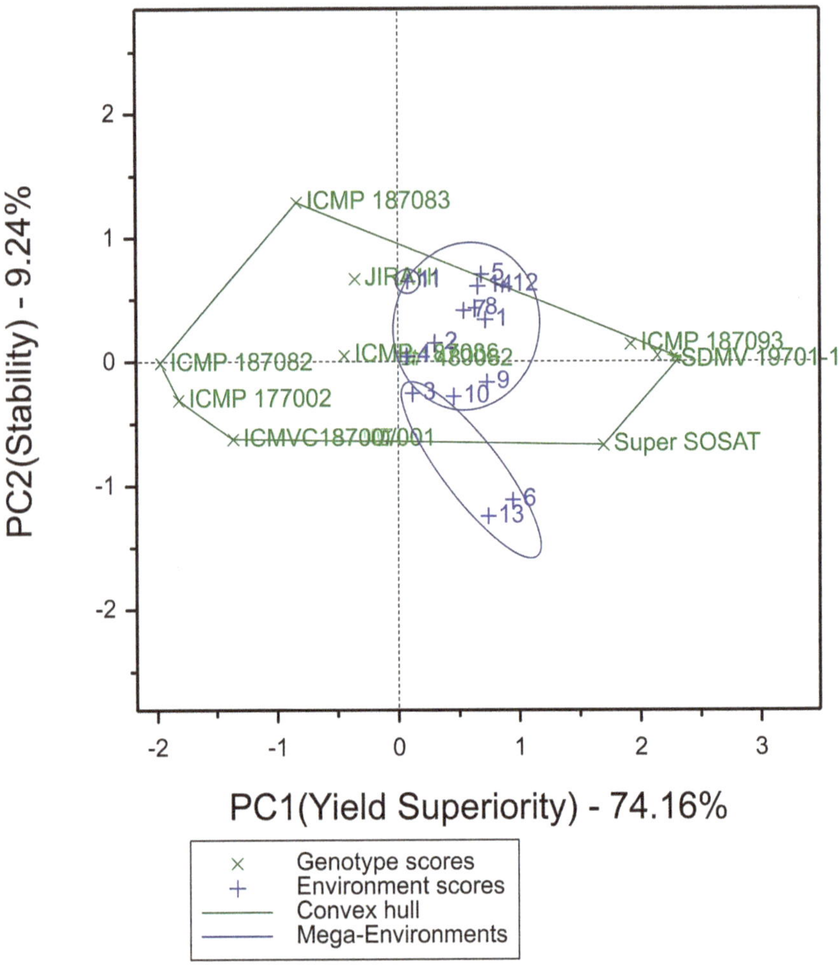

Fig. 7.1 GGE biplot of twelve genotypes in fourteen environments

Conversely, the genotypes JIRANI (2.15 t/ha), ICMP 167005 (2.13 t/ha), ICMP 187092 (2.13 t/ha), ICMP 187083 (2.04 t/ha), ICMV 167001 (1.94 t/ha), ICMV 167006 (1.88 t/ha), ICMP 187082 (1.80 t/ha), and ICMP 177002 (1.75 t/ha), with PC1 scores <0 were low yielding, in that yield rank order.

7.6.2 *Yield Stability*

PC2, as displayed in Fig. 7.1, indicates genotype yield stability; thus, the three top-most yielding entries (Table 7.2), ICMP1970115 had the highest mean yield of 2.79 t/ha, followed by Chakti (2.71 t/ha). Grains yield for Chakti, ICMP 187093

Table 7.2 Mean performances of the twelve millet entries for grain yield (t/ha) in multilocation trials for combined years and locations during the 2020 and 2021 cropping seasons

		Year			Location						
Entry	Genotype	2020	2021	Mean	Katsina	Damaturu	Gambawa	Kazaure	Maiduguri	Minjibir	Taura
1	ICMP 187082	1.75^{f-g}	1.86^{e-f}	1.80^{f-g}	1.45^{e}	1.98^{c}	2.53^{b-e}	2.23^{a-b}	1.48^{f}	1.49^{e-f}	1.46^{f}
2	ICMP 167005	2.19^{c}	2.08^{c-d}	2.13^{c}	2.30^{b}	1.92^{c-d}	2.42^{c-f}	$2.19^{a,b}$	$2.00^{c,d}$	2.20^{c}	1.95^{d-e}
3	**ICMP1970115**	**2.68^{a}**	**2.90^{a}**	**2.79^{a}**	**2.66^{a}**	**2.73^{a-b}**	**2.93^{a}**	**2.11^{b}**	**3.07^{a}**	**3.13^{a-b}**	**2.90^{a}**
4	ICMP 177002	1.69^{g}	1.80^{f}	1.75^{g}	1.31^{e}	1.73^{d-e}	2.42^{c-f}	1.78^{c}	1.64^{e-f}	1.70^{d-e}	1.64^{e-f}
5	**Chakti**	**2.61^{a-b}**	**2.81^{a-b}**	**2.71^{a-b}**	**2.67^{a}**	**2.81^{a}**	**2.72^{a-c}**	**2.24^{a-b}**	**3.19^{a}**	**3.04^{a-b}**	**2.33^{b-c}**
6	ICMP 187083	1.96^{e}	2.11^{c-d}	2.04^{c-d}	2.02^{c-d}	2.07^{c}	2.29^{d-f}	2.15^{b}	2.40^{b}	1.23^{f}	2.11^{c-d}
7	ICMP 187092	2.02^{d-e}	2.23^{c}	2.13^{c}	1.85^{d}	1.58^{e}	2.57^{b-d}	2.18^{ab}	$2.23^{b,c}$	2.29^{c}	2.19^{c-d}
8	ICMV 167006	1.89^{e-f}	1.87^{e-f}	1.88^{e-f}	1.40^{e}	1.95^{c-d}	$2.23^{e,f}$	2.20^{a-b}	1.74^{d-f}	2.18^{c}	1.44^{f}
9	JIRANI	$2.13^{c,d}$	2.17^{c}	2.15^{c}	2.24^{bc}	$1.94^{c,d}$	2.13^{f}	2.46^{a}	$2.27^{b,c}$	$2.01^{c,d}$	2.02^{c-d}
10	**ICMP 187093**	**2.49^{b}**	**2.85^{a-b}**	**2.67^{b}**	**2.60^{a}**	**2.57^{a-b}**	**2.91^{a}**	**2.20^{a-b}**	**2.93^{a}**	**2.87^{b}**	**2.61^{a-b}**
11	ICMV 167001	1.90^{e}	1.98^{d-e}	1.94^{d-e}	1.89^{d}	1.89^{c-d}	2.47^{b-e}	1.75^{c}	1.76^{d-e}	2.23^{c}	**1.59^{e-f}**
12	Super SOSAT	2.50^{b}	2.71^{b}	2.61^{b}	2.60^{a}	2.55^{b}	2.77^{a-b}	2.18^{a-b}	2.50^{b}	3.31^{a}	2.33^{b-c}
Mean		2.15	2.28	2.22	2.08	2.14	2.53	2.14	2.27	2.31	2.04
Se±		0.054	0.064	0.042	0.097	0.083	0.117	0.108	0.098	0.128	0.130
CV (%)		11.41	12.88	12.23	11.40	9.52	11.34	12.33	10.58	13.55	15.61

Means within a column bearing similar superscript letters are not significantly different at 5% probability of the Duncan's Multiple Range Test

(2.67 t/ha) and Super SOSAT (2.60 t/ha) were statistically at par, but significantly higher than the remaining entries. ICMP1970115, Chakti, and ICMP 187093 with PC2 scores above the zero abscissa were all high-yielding and showed good yield stability across environments. However, JIRANI, ICMP 167005, ICMP 187092, ICMP 187083, and ICMP 187082 were stable but low yielding; while ICMP 177002, ICMV 167006, and ICMV 167001 were both low-yielding and not stable.

7.7 End-Use Quality Testing

End-use quality testing was carried out to assess the nutritional values of the three promising millet candidates, preparatory to release, in terms of physical, proximate, and mineral qualities.

7.7.1 Micronutrient Content

The grain mineral contents, iron (Fe), zinc (Zn), calcium (Ca), potassium (K), phosphorus (P), and magnesium (Mg) of the six test samples are presented in Table 7.3. Iron content ranged from 24.6 to 74.2 ppm among genotypes, with all four test entries significantly exceeding the improved checks, JIRANI and Super SOSAT. Micronutrient grain content analysis revealed that Chakti had the highest iron content (~74.2 ppm), as well as the standard check for iron ICMP 177002, which were statistically superior to ICMP 187093 and ICMP1970115. However, Chakti with the highest iron content (74.2 ppm), along with the standard check for iron ICMP 177002 were also statistically superior relative to ICMP 187093 and ICMP1970115. This confirms earlier study of Gangashetty et al. (2021).

Table 7.3 Mineral composition of the three biofortified release candidates and standard checks

	Mineral composition					
Entry	Iron (ppm)	Zinc (ppm)	Calcium (ppm)	Potassium (ppm)	Phosphorus (ppm)	Magnesium (ppm)
ICMP1970115	**51.7**	**38.5**	**25.6**	**423.1**	**206.0**	**309.4**
ICMP 177002	73.2	47.2	24.8	445.2	216.8	279.4
Chakti	**74.2**	**44.7**	**39.4**	**413.6**	**201.3**	**303.0**
JIRANI	33.2	27.8	18.3	478.2	226.3	532.5
ICMP 187093	**52.7**	**45.6**	**31.5**	**464.6**	**232.8**	**311.4**
Super SOSAT	24.6	27.5	19.3	458.5	223.2	370.7
Mean	51.6	38.6	26.5	447.2	217.7	351.1
SE±	1.95	2.35	0.01	1.80	0.88	0.78
$LSD_{0.05}$	6.143	7.407	0.022	5.678	2.777	2.458
CV (%)	6.54	10.56	0.05	0.70	0.70	0.38

** = significant at 1% probability of the F-test, PPM = Parts per million

Thus, the sequence for iron content was: Chakti = ICMP 177002 > ICMP 187093 = ICMP1970115 > JIRANI > Super SOSAT. Zinc content differed significantly ($p < 0.05$) from 27.5 to 47.2 ppm among the entries, with all four test genotypes also significantly surpassing the two released checks. The highest zinc content was recorded in the standard check, ICMP 177002, though not statistically different from those of ICMP 187093 and Chakti, but higher than that of ICMP 1970115. Thus, there was no significant difference in zinc content among ICMP 187093, Chakti, and ICMP1970115 or between JIRANI and Super SOSAT. Results further indicated that in all the entries, calcium, potassium, and phosphorus ranged from 18.3 to 39.4 ppm, 413.6 to 478.2 ppm, and 201.3 to 232.8 ppm, respectively. These values were significantly higher in Chakti, ICMP 187093 and JIRANI. Similarly, JIRANI gave significantly higher magnesium, which ranged from 279.4 to 532.5 ppm among the genotypes. In summary, Chakti had the best iron (74.2 ppm) compared to the standard check, ICMP 177002 (73.2 ppm), and was equally the best in calcium content (39.4 ppm), and the third best in terms of zinc (44.7 ppm) after the standard check, ICMP 177002 (47.2 ppm) and ICMP 187093 (45.6 ppm). The latter recorded the highest phosphorus content (232.8 ppm). Following their performance in grain yield (>2.5 t/ha) and low downy mildew incidence (<5%), they were recommended for registration and release for commercialization as biofortified iron pearl millet (Dawud et al. 2023).

7.8 Scalable Seed System Development

Commercializing a new variety requires more than seed production but demands investment and market linkages. By integrating private sector partners into both seed and grain value chains, ICRISAT and HarvestPlus Nigeria set the stage for the rapid commercialization of Iron Pearl Millet (IPM) in 2024. Under the UK International Development Propcom+ program, HarvestPlus partnered with the International Crops Research Institute for the Semi-Arid Tropics (ICRISAT) and the Lake Chad Research Institute to introduce two varieties of IPM. These varieties not only address Nigeria's growing nutritional needs but also contribute to climate adaptation strategies for smallholder farmers. By integrating climate-smart production technologies and agronomic training, HarvestPlus is strengthening the resilience of millet farming communities in northern Nigeria.

Premier Seeds, a key partner, established an out-grower scheme of biofortified iron pearl millet variety (Chakti) on 64.5 ha with 11 communities from 10 local governments spread across 3 states (Kano, Jigawa, and Bauchi) of Nigeria valued at over NGN400 million (approximately USD 500,000 in 2023), providing farmers with essential inputs such as early generation seeds (EGS) and fertilizers on credit. The initiative proved highly successful, achieving an impressive 95% recovery rate and producing 300 metric tons of certified IPM seeds.

On the grain commercialization front, major millet processors like Dala and Halilco Foods were engaged to secure market opportunities for smallholder farmers. Through a targeted product development exercise, these processors adopted IPM as a raw material of choice, committing to off-take approximately 5000 metric tons of IPM grain in 2023. This assured demand provided critical incentives for farmers to adopt IPM at a scale to produce about 62 metric tons in Jigawa state, Nigeria.

7.9 Summary and Conclusion

The rapid adoption of biofortified pearl millet cultivars in Nigeria has great promise for enhancing food and nutritional security, particularly in areas prone to iron and zinc deficiencies. The effective development and release of nutrient-dense varieties, which contain up to 70–75 ppm iron and 35–45 ppm zinc, illustrates biofortification's ability to battle hidden hunger while sustaining resilient farming systems. Adoption is progressively expanding, notably in millet-growing states in northern Nigeria, because of multi-stakeholder engagement and increased understanding of the nutritional benefits. The critical gaps identified are

- Limited Access to Quality Seed: Despite progress, many smallholder farmers still lack timely access to certified biofortified seed due to weak seed distribution networks and low commercial seed production.
- Low Consumer Awareness: Awareness of the health benefits of biofortified millet remains low among end-users, limiting consumer demand and market pull.
- Weak Market Linkages: Limited value chain development and market incentives for biofortified grains discourage large-scale adoption and commercialization.
- Inadequate Policy Support: Biofortification is not yet fully mainstreamed into national nutrition, agriculture, and food procurement policies.
- Insufficient Extension Services: Many farmers lack adequate training and extension support to adopt and effectively manage biofortified varieties.

Possible Recommendations are

- Strengthen Seed Systems: Support public–private partnerships to scale up certified seed production and ensure last-mile delivery to farmers, particularly in remote areas.
- Enhance Awareness Campaigns: Implement nationwide behavior change communication (BCC) strategies to educate farmers, processors, and consumers about the nutritional and economic value of biofortified millet.
- Develop Inclusive Value Chains: Promote aggregation, processing, and branding of biofortified millet to create strong market incentives for farmers and SMEs.
- Integrate into National Programs: Institutionalize biofortification in national agricultural and nutrition policies, including school feeding, maternal health, and food aid programs.
- Invest in Capacity Building: Expand farmer training, extension services, and demonstration plots to accelerate adoption and good agronomic practices.

With targeted interventions to address these gaps, Nigeria can unlock the full potential of biofortified pearl millet to improve nutrition, strengthen livelihoods, and build climate-resilient food systems.

References

Dawud MA, Angarawai II, Govindaraj M, Gangashetty P, Riyazaddin M, Turaki ZGS, Ajeigbe HA, Mala KK, Kabir BGJ, Mohammed MI, Jonah J (2023) Proposal for registration and release of three millet varieties; Chakti (High grain FE and ZN content), ICMP1970115 (High grain yield) and ICMP187093 (Bristled panicle) AS LCIC MV5, LCIC MV6 and LCIC MV7 respectively. Technical reported submitted to National committee on Naming, Registration and release of Crops Varieties and Animals during their 31st Seating on 10th February 2023 at National Centre for Genetic Resources and Biotechnology (NACGRAB) Moor Plantation Ibadan, Nigeria. Dawud et al 2023 Iron pearl millet in Nigeria.pdf

Dollah Y (2024) Addressing malnutrition and improving public health in Nigeria. HarvestPlus presence in Nigeria. https://www.harvestplus.org/wp-content/uploads/2022/01/Nigeria.pdf

Finkelstein JL, Mehta S, Udipi SA et al (2015) A randomized trial of iron-biofortified pearl millet in school children in India. J Nutr 145:1576–1581. CrossRefGoogle ScholarPubMed

Gangashetty PI, Mohammed R, Moussa DS, Drabo I, Kassari, Peter AA, Ousmane S, Govindaraj M, Angarawai II (2021) Identification of high-yielding iron-biofortified open-pollinated varieties of pearl millet in West Africa. Front Plant Sci 22. https://doi.org/10.3389/fpls.2021.688937

Govindaraj M, Kedar NR, Cherian B, Wolfgang HP, Anand K, Harshad S (2019) Breeding biofortified pearl millet varieties and hybrids to enhance millet markets for human nutrition. Agriculture 9(5):106

HarvestPlus (2024) Commercializing iron pearl millet in one year: a success story of HarvestPlus in Nigeria. https://www.harvestplus.org/commercializing-iron-pearl-millet-in-one-year-a-success-story-of-harvestplus-in-nigeria/#:~:text=HarvestPlus%20Nigeria%20developed%20the%20ARC,Read%20more%20details%20here

Jacob J, Krishnan V, Antony C, Bhavyasri M, Aruna C, Mishra K, Nepolean T, Satyavathi CT, Visarada KBRS (2024) The nutrition and therapeutic potential of millets: an updated narrative review. Front Nutr 11:1346869. https://doi.org/10.3389/fnut.2024.1346869. PMID: 38746941; PMCID: PMC11091339

Kamalesh KM, Sunil M, Manvik J, Anant VD (2024) Nutritional and functional exploration of pearl millet and its processing and utilization: an overview. Food Hum. www.editorialmanager.com/foohum/journal_overview.ht

Kodkany BS, Bellad RM, Mahantshetti NS, Westcott JE, Krebs NF, Kemp JF et al (2013) Biofortification of pearl millet with iron and zinc in a randomized controlled trial increases absorption of these minerals above physiologic requirements in young children. J Nutr 143:1489–1493. https://doi.org/10.3945/jn.113.176677

Kramer CV, Allen S (2015) Malnutrition in developing countries. Paediatr Child Health 25:422–427

Nestel P, Bouis HE, Meenakshi JV, Pfeiffer W (2006) Biofortification of staple food crops. J Nutr 136:1064–1067

Onuegbu NC, Ihediohanma NC, Eze CF, Okafor DC, Ojukwu M (2017) Biofortification of local Staples in Nigeria: prospects and problems. J Food Biotechnol Res 1:5

Scott SP, Murray-Kolb LE, Wenger MJ et al (2018) Cognitive performance in Indian school-going adolescents is positively affected by consumption of iron-biofortified pearl millet: a 6-month randomized controlled efficacy trial. J Nutr 1:1462–1471

Wheal T, Matthew S, Fowles O, Lyndon TP (2011) A cost-effective acid digestion method using closed polypropylene tubes for inductively coupled plasma optical emission spectrometry (ICP-OES) analysis of plant essential elements. Anal Methods 3(12):2854–2863. https://doi.org/10.1039/c1ay05430a

World Health Organization (2024) Child malnutrition: 2022 facts and statistics. World Health Organization

Chapter 8
Finger Millet Biofortification Breeding Approaches and Progress: An Update

K. N. Ganapathy, Nithya Sree Koppula, Tara C. Satyavathi, Mahalingam Govindaraj, R. Venkateshwarlu, Karnam Venkatesh, C. Amasiddha, K. Hariprasanna, and C. M. Gopika

Abstract Finger millet (*Eleusine coracana*), is an important crop for its high nutritional value, and is an important climate-resilient crop with tremendous potential to alleviate micronutrient deficiencies in India and elsewhere. Biofortification breeding efforts performed by ICAR-Indian Institute of Millets Research (IIMR), Hyderabad, in conjunction with AICRP-Small Millets, have resulted in considerable increases in grain iron (Fe), zinc (Zn), and calcium (Ca). ICAR-IIMR has identified benchmark variety, such as GPU-28, with around 60 ppm Fe and 30 ppm Zn, to serve as control for national germplasm screening. Breeding product targets are set at 65 ppm Fe and 31 ppm Zn, with aspirational levels of 90 ppm Fe and 40 ppm Zn in breeding populations. Simultaneously, genetic stocks with critical features such as disease resistance and nutritional density have been conserved and defined. Among the biofortified cultivars introduced under the coordinated research program are CFMV1 (Indravati) with 428 mg/100 g Ca, 58 ppm Fe, and 44 ppm Zn, and CFMV2 with 454 mg/100 g Ca, 39 ppm Fe, and 25 ppm Zn. These results show that traditional breeding procedures are effective in producing nutrient-rich high-yielding finger millet cultivars. Moving forward, integrating genomic technologies, SNP-based selection, and speed breeding platforms will be critical to speeding the creation of commercially competitive biofortified varieties for long-term nutritional security. ICAR-IIMR, has spearheaded multiple initiatives to enhance nutrition, value addition, and farmer empowerment through biofortification and millet

K. N. Ganapathy (✉) · N. S. Koppula · T. C. Satyavathi · R. Venkateshwarlu · K. Venkatesh · C. Amasiddha · K. Hariprasanna
ICAR – Indian Institute of Millets Research, Rajendranagar, Hyderabad, India
e-mail: ganapathy@millets.res.in

M. Govindaraj
Alliance Bioversity International and International Centre for Tropical Agriculture (CIAT), Cali, Colombia

C. M. Gopika
College of Agriculture, University of Agricultural Sciences, Bengaluru, Karnataka, India

M. Govindaraj (ed.), *Breeding Biofortified Millets for Nutritional Security*, Sustainability Sciences in Asia and Africa,
https://doi.org/10.1007/978-981-95-9827-4_8

innovation in India. Notably, the Global Centre of Excellence on Millets is envisioned to support genomics-assisted breeding, speed breeding, gene editing, and advanced processing facilities for accelerated millets improvement for productivity and mainstreaming nutritional traits.

Keywords Biofortification · Calcium · Iron · Zinc and genomic markers

8.1 Introduction

8.1.1 Importance of Biofortification in Global Food Security

Biofortification is essential for achieving the global food security as it helps combat malnutrition and improve nutrient availability in regions affected by population growth and climate change. About 660 million people would experience severe hunger by 2030, according to updated projections from 2021, which showed concerning levels of the problem (Laborde et al. 2020). In developing countries such as Sub-Saharan Africa and Southern Asia, where diets are abundant in calories but deficient in essential vitamins and minerals, hidden hunger is a major issue (Wakeel et al. 2018; Beenish et al. 2018). The current food system's inability to provide fair access to nutrient-dense meals is highlighted by issues such as hidden hunger or malnutrition, which in turn have severe negative impacts on health (Gödecke et al. 2018).

An increase in population and resource shortages are major problems for developing nations to keep food production at the required speed (Zhu et al. 2007). The right amounts of nutrients in the food intake are essential for optimum human growth and development. Problems like stunted growth affect one-fifth of the children under 5 years of age (Chaudhary et al. 2020; Dwivedi et al. 2023). Micronutrient deficiencies affect more than two billion people worldwide (Díaz-Gómez et al. 2017; Mandial et al. 2023). All these alarming issues pave the way for a quick and mandatory resolution for addressing malnutrition globally.

Some strategies include developing crops that can withstand high levels of abiotic stress and changing one's diet, but biofortification has emerged as an efficient option (Wakeel and Labuschagne 2021; Bouis et al. 2019). Biofortification is a potential strategy for addressing widespread nutritional deficiencies, particularly in areas where there is no possibility for dietary diversification. Because staple crops are so important and accessible for major populations, an increase in the nutritional content of those crops would provide affordable solutions without any consumer compliance (Lowe 2021). Biofortification addresses a major worldwide health concern, reducing dependency on medical interventions, increasing workers' efficiency, and lowering healthcare expenditures (Kumar et al. 2019).

Although the second sustainable development goal of the United Nations is to eradicate extreme poverty and hunger while increasing food security (Olson et al.

2021; Saint Ville et al. 2019), various initiatives, such as the Harvest Plus Challenge Program, have been launched to tackle these problems (Bouis 2018). Although output increased after the success of the green revolution, it did not address malnutrition. So, there is a requirement for a second green revolution that prioritizes crop cultivars that are high in nutrients (Zhu et al. 2007).

8.1.2 Importance of Finger Millet in Global Nutrition

Finger millet is an important food staple of Africa and Southern Asia. Finger millet cropping area globally is estimated to be around 2.3 m ha with a production of 3.79 million tons and productivity of 1724 kg/ha (Bhat et al. 2023; Ganapathy et al. 2021a, b). The major finger millet growing countries in Africa are the sub-humid regions of Ethiopia, Malawi, Tanzania, Kenya, Uganda, Zambia, Zaire, and Zimbabwe. Similarly, in South Asia, the crop is mainly grown in India, which contributes nearly 50% to the global area, followed by Nepal, and to a certain extent, in Bhutan and Sri Lanka. Compared to other small millets, genetic gain in yield is much more pronounced in finger millet but not exploited to the extent observed in other major cereals. Often, the genetic potential is not fully expressed due to aberrant weather conditions, moisture scarcity, and poor management in many states. However, the diversity existing for grain yield, nutritional superiority, and its ability to tolerate a range of environmental conditions makes it a promising crop for the future. The grains of finger millet are known for its highest amount of calcium. Also, the content of iron, zinc, dietary fiber, and essential amino acids (Shobana et al. 2013). The grains are highly resistant to storage insect pests, and under minimum attention, the crop can stay in good condition up to 50 years (Iyengar et al. 1945). The stover after harvest of grains is a source of nutritive fodder to animals and is highly preferred due to the sweet-smelling stalks. Improved agronomic traits and yield under changing climate scenarios, high water use efficiency (Gupta et al. 2017), nutrient-dense grains (Puranik et al. 2017), and excellent grain storage quality (Gimode et al. 2016) are some of the traits that make finger millet a suitable food crop for semi-arid regions of the world (Vetriventhan et al. 2020). Finger millet is also described as a "crop for the poor," a "famine crop," or as a crop that achieves satisfactory yields even in low-input agriculture (National Research Council 1996; Gupta et al. 2017) (Fig. 8.1).

Finger millet is well known around the world for its nutritional benefits and ability to solve global health problems. Even though all kinds of millet are consumed worldwide, finger millet is the most popular, and plenty of dishes are made in different parts of the world. As a highly digestible, well-balanced grain, it is good to overcome obesity and malnutrition. It contains essential protein and carbohydrates, along with amino acids, which are inadequate in diets that consist merely of cereals (Kumar et al. 2016). In addition to nutritional qualities, this coarse grain can be used for

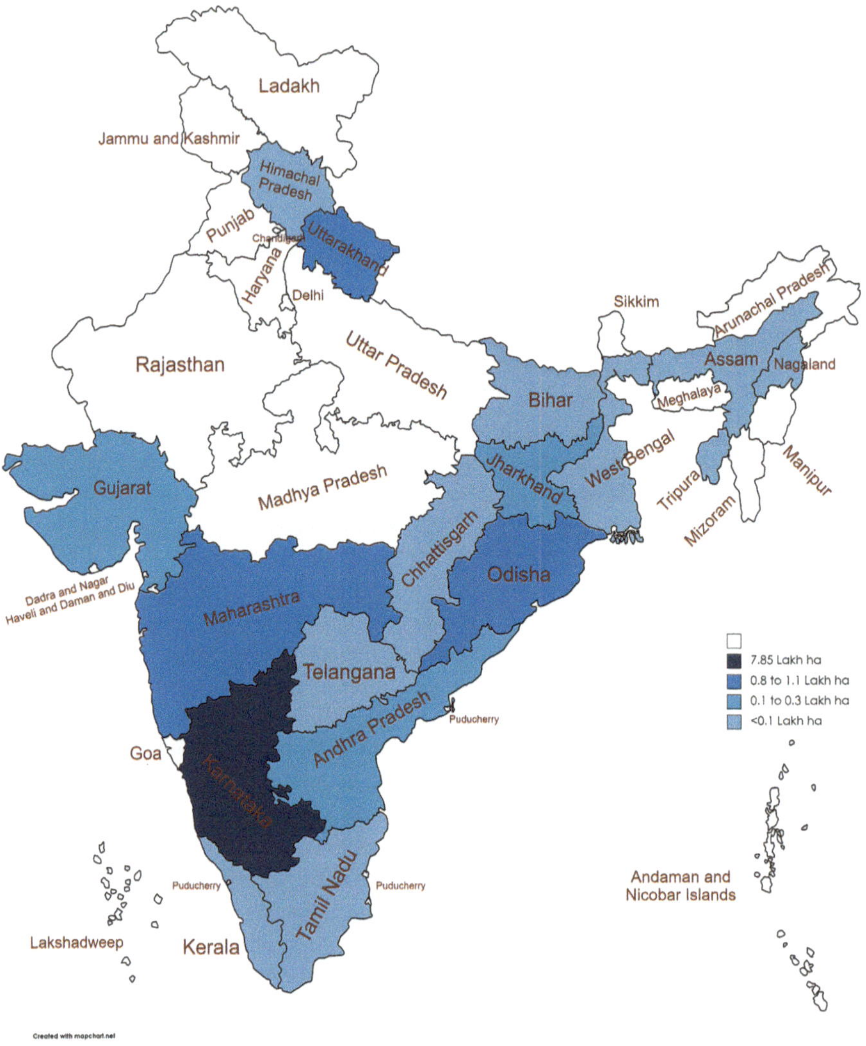

Fig. 8.1 Major finger millet growing states in India

baking, roasting, steaming, frying, boiling, fermentation, and making roti or chapattis and dumplings (Shrestha et al. 2024). These characteristics make finger millet a potential solution to the world's nutrition problems, particularly in underdeveloped nations where food security is a concern. It is very important for food and nutritional security in countries like Ethiopia, Kenya, China, India, and others since it is resilient and can thrive in marginal areas with few resources (Ignacimuthu and Ceasar 2012).

8.2 Finger Millet Nutritional Significance

8.2.1 Nutritional Composition of Finger Millet

Finger millet, also known as "future smart food" by the Food and Agriculture Organization (FAO 2018), is known for its superior nutritional profile, which surpasses the regularly consumed grains such as rice and wheat (Ceasar and Ignacimuthu 2011). It includes a substantial amount of important minerals, which are critical for fixing micronutrient deficiencies among populations with little to no access to diversified foods, thus making it a future smart food (Antony and Chandra 1998). The comprehensive nutritional composition of the finger millet is shown in Table 8.1, along with the related health advantages.

Table 8.1 Nutritional composition of finger millet

Component	% (or) mg/100 g	Health benefits	Reference
Calcium	344 mg	Supports bone health and helps prevent osteoporosis.	Kumar et al. (2016), Reddy et al. (2008)
Potassium	408 mg	Regulates blood pressure and reduces the risk of heart disease.	Kumar et al. (2016)
Phosphorus	283 mg	Promotes bone health, energy production, and cell repair	Ganesh et al. (2024)
Magnesium	53–217.8 mg	Regulates blood pressure and helps in muscle and nerve function	Fred et al. (2021)
Sodium	11 mg	Maintains fluid balance and nerve function	Ashik and Morya (2022)
Iron	3.9 mg	Red blood cell formation prevents anemia	Chaudhary (2024)
Zinc	2.3 mg	Supports the immune system, helps in protein synthesis	Mrudula et al. (2022)
Copper	0.47 mg	Helps in iron absorption and cardiovascular health	Chaudhary (2024)
Protein	5–8%	Contributes to muscle repair and growth	Udeh et al. (2017)
Carbohydrates	65–75%	Provides energy and helps maintain blood sugar levels	Gupta et al. (2017)
Dietary Fiber	15–20%	Improves digestion, reduces cholesterol, and helps regulate blood sugar	Ramashia et al. (2018), Udeh et al. (2017)
Fat	1.3% (mainly unsaturated)	Provides healthy fats for overall health	Udeh et al. (2017)
Bioactive Compounds	Ferulic acid, quercetin, feraxans	Offers anti-diabetic, anti-cancer, anti-inflammatory, and antioxidant effects	Udeh et al. (2017), Devi et al. (2014)

Table 8.2 Health benefits of finger millet

Health benefits	Explanation	Reference
Anti-diabetic	A low glycemic index helps manage blood sugar and reduce the risk of diabetes.	Saleh et al. (2013)
Anti-tumorogenic	Bioactive compounds help in cancer prevention.	Devi et al. (2014)
Anti--atherosclerogenic	Helps prevent heart disease by lowering cholesterol and improving blood circulation.	Shrestha (2024)
Antioxidant	Rich in antioxidants, protecting against oxidative stress and premature aging.	Devi et al. (2014)
Bone health	High calcium content aids in bone formation and maintenance, preventing diseases like osteoporosis.	Kumar et al. (2016)
Cholesterol-lowering	Fiber and bioactive compounds help lower bad cholesterol levels.	Shrestha (2024)
Improves digestive health	High dietary fiber helps with digestion and prevents constipation.	Ramashia et al. (2018)
Supports lactating mothers	Enhances milk production and improves the nutritional quality of milk.	Shrestha (2024)

8.2.2 Addressing Micronutrient Deficiencies and Its Health Benefits

Finger millet is an essential remedy for micronutrient deficiencies because of its high nutritional density, especially in areas with little variety in dietary sources. Weaning infants, those with lactose intolerance, and populations at risk for osteoporosis can all benefit from its high calcium content (Wijesinha-Bettoni and Burlingame 2013), which makes it a natural substitute for traditional calcium supplements (Arya and Bisht 2022). Finger millet is gluten-free, making it a great dietary choice for anyone with allergies to gluten or celiac disease (Taylor and Emmambux 2008). It contributes to the prevention of protein-energy malnutrition by supplying critical amino acids like lysine and methionine, which are frequently lacking in plant-based diets (Kumar et al. 2016). Understanding these health advantages is critical for promoting finger millet as a sustainable and nutritionally dense food source in biofortification programs and combating worldwide nutritional poverty. The health benefits associated with finger millet are highlighted in Table 8.2, which also includes extensive explanations and references.

8.3 Biofortification: Concept and Strategies

Biofortification is the process of increasing the nutritional content of crops by approaches such as traditional breeding, genetic engineering, or agronomic techniques to achieve high nutrient-rich yield (Kiran 2020). Finger millet, being climate-resilient, offering several health benefits, and capable of being prepared in diverse

Table 8.3 Variation for grain micronutrients (Ca, Fe, Zn) in released Indian finger millet varieties (based on one season data)

	Fe–XRF (ppm)	Fe–ICP (ppm)	Ca–XRF (ppm)	Ca–ICP (ppm)	Zn–XRF (ppm)	Zn–ICP (ppm)
Mean	34.05	24.96	3607.57	4083.32	19.81	18.43
Maximum	43.62 (OUAT 2)	37.24 (OUAT 2)	4718.45 (KMR 301)	4850.98 (KMR 204)	24.04 (KMR 204)	25.69 (VL 379)
Minimum	26.50 (MR 6)	13.52 (KMR 301)	2763.55 (Gujrat Nagli 2)	3210.74 (Gujrat Nagli 2)	15.76 (Indaf 9)	13.51 (KMR 301)
Top 5 varieties with higher levels	OUAT 2, Indira Ragi OEB 10, VR 847, VR 708	OUAT 2, ML 365, Indira Ragi, OEB 10, VR 708	KMR 301, Indira Ragi, OUAT 2, KMR 204, VR 936	KMR 204, KMR 301, Indira Ragi, VL 379, VR 936	KMR 204 Padmavathi, ML 365, OUAT 2, OEB 10	VL 379 Padmavathi, KMR 204, ML 365, VL 324

Unpublished data

culinary forms, is already a significant part of the diet in many regions where it is cultivated. These attributes make it highly suitable for biofortification, ensuring enhanced nutritional value in areas where it is a staple food. In order to improve the nutritional condition of people who rely on staple crops as their main source of nourishment, biofortification attempts to raise the micronutrient density of these crops. Particularly in areas where finger millet is already a staple grain, biofortification can increase its nutritional value and make it an even more effective instrument for promoting public health (Welch and Graham 2004) (Table 8.3).

8.4 Breeding Approaches and Their Progress for Finger Millet Biofortification

8.4.1 Breeding Approaches and Target Traits

Conventional breeding is vital for improving the nutritional profile of finger millet, notably by raising levels of key micronutrients, such as iron (Fe), zinc (Zn), and protein content. All major crop breeding programs are aimed at increasing grain yield levels with tolerance/resistance to insect pests and disease resistance. The development of improved cultivars targeting grain micronutrients was never prioritized previously. Moreover, combining high yield with improved grain micronutrients is a challenge. Selection for higher grain micronutrient content may negatively affect yield or other important agronomic and end-use characters. This happens if genes governing micronutrient content are linked with genes that have a deleterious effect on other desired yield or agronomic traits, or it could occur as a consequence of negative trait associations. Correlation between grain nutritional traits, especially between grain Fe and Zn, has largely been studied in pearl millet and other crops, and highly positive and significant correlations between Fe and Zn have been revealed (Gregorio et al. 2000; Ozkan et al. 2007; Velu et al. 2011). Development of cultivars with improved grain micronutrients sacrificing yield will not be accepted by farmers unless farmers do get a premium price for their produce. The genetic variation existing for the target trait, heritability, identification of promising donors, relationship between different grain micronutrients and agronomic traits, and the stability of the traits across environments would determine the scope for development of high-yield cultivars with improved grain micronutrients. Agronomic practices, soil composition, and climate all have a significant impact on nutrient absorption and accumulation. To ensure the durability of biofortified varieties, breeders must carefully choose genotypes that have both high nutritional content and consistent performance throughout a wide variety of environmental circumstances (Joshi et al. 2023). Genotype × environment (G × E) interaction has a significant impact on micronutrient content in different contexts, making biofortification a serious challenge (Table 8.4).

Finger millet is naturally rich in grain calcium; more emphasis should be given to the improvement of grain protein, iron, and zinc content. For grain protein, the levels in released cultivars are in the range of 7–8%, while substantial variation has been reported for protein content ranging from 11% to 12%. The grain iron is in the range of 25 to 28 ppm and grain zinc are in range of 18 to 20 ppm. The development of high calcium lines should be targeted for the development of niche products, while the normal range in cultivars should be maintained. Characterization of global germplasm is an essential first step in this process, where scientists evaluate a wide range of finger millet accessions to find notable variations in nutritional content (Backiyalakshmi et al. 2024). This diversity offers a solid basis for creating biofortified cultivars using selective breeding methods, such as mass and recurrent selections, that increase the frequency of advantageous alleles linked to nutrient content (Gebreyohannes et al. 2024). Despite its ability to yield improved genotypes, conventional breeding, especially in finger millet and small millets, poses challenges due to difficulties in anthesis behaviour and breeding processes. Conventional breeding demands more resources, land, labor, time, and infrastructure, and the same is true with small millets (Lenaerts et al. 2019). Conventional breeding can be further boosted by integrating molecular techniques that enhance the precision of trait selection in plant cultivars. By employing molecular markers and technologies

Table 8.4 Biofortified finger millet varieties

Variety	Grain yield (kg/ha)	Calcium (ppm)	Grain iron (ppm)	Zinc (ppm)	Area of adaptation
CFMV 1 (Indravathi)	3110	4280	58	ARS, Vizianagaram, India	AP, KA, TN, Puducherry, Odisha
CFMV 2 (Gira)	2950	4740	39	HRMS, Waghai, India	AP, Chhattisgarh, Gujarat, MH, Odisha

like genotyping-by-sequencing (GBS) and genome-wide association studies (GWASs), breeders can efficiently identify genomic regions associated with essential micronutrients, such as Fe and Zn (Chaudhary et al. 2025). While conventional breeding primarily relies on traditional techniques such as selection and crossbreeding using phenotypic traits, molecular breeding integrates advanced genetic tools and technologies to enhance the precision and efficiency of breeding programs.

8.4.2 Advanced Phenotyping Methods for Grain Micronutrients

Rapid screening of the germplasm for grain micronutrient screening is a prerequisite for successful biofortification breeding. A large number of instrumentation techniques are available and have been used for grain micronutrients assessment so far, plant breeders presently rely mostly on the use of Atomic Absorption Spectrometry (AAS). When a large number of samples are to be screened for a given micronutrient, a simple, rapid, and cost-efficient method can surely save time and resources, and can increase the breeding efficiency to enhance genetic gain for that trait. Advanced techniques, such as the use of X-ray Fluorescence (XRF), have been used for plant sample analysis. XRF is a non-destructive method for grain Fe and Zn contents estimation and is now being largely used in pearl millet (Paltridge et al. 2012) for rapid analysis of grain micronutrients. Setting up this table-top machine requires little recurring expenditure and provides non-destructive analysis of 300 samples per day at the cost of <USD 2.0 per sample compared to the ICP method (>18 USD/sample), which takes a month's time (Rai et al. 2012). The efficiency of XRF over ICP for high-throughput Fe and Zn estimation in pearl millet grain was demonstrated with large samples from several trials (Govindaraj et al. 2019). By adopting advanced strategies, future breeding programs can aim at developing improved genotypes for grain micronutrients by the identification of promising germplasm lines, incorporation of superior alleles into locally adapted varieties, and subsequent use in gene mapping (Sood et al. 2016).

8.4.3 Molecular Breeding and Genetic Approaches

Molecular breeding has revolutionized crop improvement by leveraging genetic insights to enhance desirable traits. This approach integrates advanced genotyping technologies and phenotypic data to identify genetic variations that contribute to nutritional enhancement. In finger millet, these methods are increasingly significant due to its potential as a nutrient-rich crop, especially in addressing calcium nutrient deficiencies. Association mapping studies in finger millet have uncovered critical quantitative trait loci (QTL) linked to important nutritional traits. In a breeding program, molecular markers like marker-assisted selection (MAS) offer an important edge over traditional markers for selecting and screening crop plants (Bailey-Serres et al. 2019). Additionally, markers related to calcium content have been identified, highlighting their potential use for biofortification after validation in target genotypes. These discoveries are paving the way for targeted breeding programs aimed at combining nutritional and agronomic priorities, such as blast resistance. Despite these advancements, limited germplasm diversity for grain micronutrients and inadequate phenotyping in earlier studies have hampered the development of robust markers. Addressing these challenges requires the use of diverse germplasm pools and advanced phenotyping techniques.

Protein content in finger millet varies significantly, ranging from 4.88% to 15.58%, with a noted negative correlation between calcium and protein levels (Singh and Srivastava 2006). This makes the simultaneous enhancement of both traits a challenging task. Early research using SSR markers identified genetic variations linked to tryptophan content, but faced limitations in studying calcium-related genetic markers. This led to a shift toward SNP markers, resulting in the identification of over 23,000 genetic markers associated with grain calcium and protein content. While this advancement holds promise for developing improved varieties, optimizing both traits simultaneously remains complex. Modern genotyping tools like genotyping-by-sequencing (GBS), diversity arrays technology (DArT), and genome-wide association studies (GWASs) have transformed genetic research by enabling precise analysis of quantitative traits. These techniques have been particularly effective in identifying genomic regions that regulate nutrient content, such as Fe and Zn, in major crops like maize, rice, and wheat (Pujar et al. 2020). GWAS and QTL mapping have proven instrumental in identifying chromosomal regions and specific alleles critical for biofortification (Connorton et al. 2017). For example, markers like OM5 and FM8 (linked to tryptophan) and FMO2EST1 (associated with protein content) have been identified in finger millet (Babu et al. 2014). Additionally, markers related to calcium content have been discovered, further aiding breeding efforts. Combining these techniques with marker-assisted selection (MAS) and genomic selection has enhanced the precision of breeding programs, enabling the integration of superior alleles into locally adapted varieties for improved yield and nutrition (Sood et al. 2016).

Genomic research in finger millet has lagged behind other major cereals due to limited resources. However, the release of its whole-genome sequence (WGS) for genotype ML-365 marks a turning point (Puranik et al. 2020). This resource has revealed high genetic similarity with foxtail millet and rice, enabling comparative genomic studies and accelerating breeding efforts (Hittalmani et al. 2017). The WGS has identified 96 orthologous genes associated with traits like nutrient signaling, disease resistance, and drought tolerance. These discoveries lay the groundwork for targeted breeding approaches to improve both the agronomic and nutritional qualities of finger millet. Simple sequence repeats (SSRs) and single-nucleotide polymorphisms (SNPs) have been pivotal in genetic studies of finger millet. Dida et al. pioneered the development of SSRs using HindIII, SalI, and PstI libraries (Dida et al. 2007). Additionally, tens of thousands of SNPs and SSRs were discovered in genotypes with varying calcium levels, demonstrating their potential for advancing genetic research. Advancements in sequencing technologies like Roche

454 and Illumina NGS have identified 10,327 SSRs and 23,285 SNPs in finger millet (Ingle et al. 2023). These markers are now being utilized to investigate traits such as calcium accumulation and overall nutritional improvement. The first genomic map of finger millet was developed using 31 SSR markers in conjunction with expressed sequence tags (ESTs), restriction fragment length polymorphism (RFLP), and amplified fragment length polymorphism (AFLP). Despite these advancements, the number of available ESTs remains low, with only 1934 identified. This underscores the need for further genomic research to enhance molecular mapping and breeding initiatives (Dida et al. 2007; Puranik et al. 2020).

In finger millet, transcriptome analyses have identified genes like EcCAX3, which is strongly correlated with calcium accumulation. Proteomic studies have further revealed calcium-binding proteins such as calcineurin-B and calreticulin, which play crucial roles in calcium retention within seeds (Singh et al. 2015). For iron and zinc, genes such as EcZIP1, EcIRT2, EcYSL2, EcFER1, and EcZTP29 have been identified as directly involved in their uptake and transport (Chandra et al. 2021). These findings highlight the genetic basis of nutrient homeostasis in finger millet and its potential for biofortification. However, finger millet contains antinutritional factors such as phytic acid (5.54–5.58 mg/g) and tannins (3.5 mg/g), which can reduce the bioavailability of minerals. Phytase-producing microbes have been employed to break down phytates, releasing phosphate and improving mineral uptake by plants and humans. This approach not only enhances nutrient availability but also boosts plant growth (Chaudhary et al. 2023; Sharma and Garg 2023).

Proteomic studies using techniques like 2D electrophoresis and MALDI–TOF have identified proteins involved in calcium accumulation, including calmodulin (CaM) and calreticulin. These discoveries shed light on the molecular mechanisms of calcium storage and transport in finger millet (Singh et al. 2015). Transgenic approaches have shown promise in enhancing finger millet's nutritional quality and stress tolerance. For example, *Agrobacterium*-mediated transformation has been used to introduce genes like mtlD (mannitol-1-phosphate dehydrogenase) for drought resistance (Hema et al. 2014). This shows the scope for Genome-editing tools like TALENs and CRISPR, which can be explored to alter the biosynthetic pathways of phytic acid, reducing antinutritional factors and improving nutrient accumulation, making finger millet a more sustainable and nutritious crop. Two biofortified finger millet varieties, CFMV 1 (Indravathi) and CFMV 2 (Gira), enriched in calcium, iron, and zinc (Hariprasanna 2023), are found to possess high nutritional content, showing a path for further development.

8.4.4 Genetic Variation for Grain Micronutrients in Improved Varieties

ICAR Indian Institute of Millets Research, in collaboration with the HarvestPlus Biofortification Challenge Program of the CGIAR, has undertaken an initiative to examine the genetic variation in improved cultivars and prospects of benchmarking for grain micronutrients in finger millet. Toward this direction, we have assembled the major cultivars developed and cultivated in diverse finger millet-growing regions. A set of 40 cultivars was analysed for grain calcium, grain iron, and grain zinc from the grain samples field tested at ICAR—Indian Institute of Millets Research, Rajendranagar, Hyderabad. The study revealed substantial levels of variation among the cultivars, as evidenced in Table 8.5. Two different methods were deployed for assessing the levels in the cultivated varieties. The grain iron ranged from 34.0 ppm to 43. 62 ppm using XRF, while using the ICP method, the levels were in the range of 24.96 to 37.24 ppm. For grain calcium, the levels ranged from 3607.57 to 4718.45 using XRF, while using the ICP method, the levels were in the range of 4083.3 to 4850.9 ppm. For grain zinc, the variation among varieties ranged from 19.8 ppm to 24 ppm, while using XCP, the variation levels were in the range of 18.43 to 25.69. Although the levels among both the methods showed variation, we observed a very good correlation between the two methods, providing a clue for using XRF as an effective non-destructive method, which can be best utilized for screening large breeding populations and developing varieties with improved grain yield and grain micronutrients (Table 8.5).

8.5 Future Aspects

The future of biofortification and its mainstreaming holds significant promise in addressing global malnutrition and hidden hunger. To enhance the nutritional value of traditional crops like finger millet, a coordinated multidisciplinary approach is essential. Considering the importance of finger millet in the context of a changing global climate and human nutrition, it would continue to be an important food crop for India and sub-Saharan Africa, and shows promise to expand the Indian and global food basket to meet the healthy food and nutritional demands of the growing population. Creating public awareness about the nutritional values of finger millet is required for further promotion of millets for consumption with improved post-harvest processing and product making. Cultivars with improved grain micronutrients have to be achieved without any trade-off with higher productivity. The biofortification program should aid in adding as essential to improve trait to the breeding program. Micronutrient traits screening is highly expensive when dealing with larger germplasm. With the availability of advanced phenotyping tools like XRF, the cost of biofortification breeding will be significantly reduced. To achieve it, a higher investment in breeding would be required on a long-term basis. Leveraging advanced genomic tools and advancements in next-generation sequencing to identify and utilize genetic diversity in nutrient content, particularly for calcium and other essential micronutrients. Geneticists and breeders must focus on harnessing unique alleles and improving processes for nutrient transport and storage in grains, aiming for higher bioavailability in the edible portions (seed/grains). HarvestPlus Understanding the interplay between various nutrients and factors such as antinutrients is crucial, necessitating pilot studies to confirm the bioavailability of these enhanced nutrients, especially among vulnerable population in near futures.

Table 8.5 Comparing the strategies for biofortifying finger millet, along with their respective advantages and challenges

Strategy	Description	Advantages	Challenges	Examples	References
Conventional breeding	Select and crossbreed high-nutrient germplasm/ varieties.	Natural genetic diversity. Public acceptance No GM involved.	Time consuming Linkage drag Absence of sufficient genetic variability in the primary gene pool	High--calcium finger millet. Calcium--rich germplasm.	Sharma et al. (2017), Collard and Mackill (2008)
Agronomic biofortification	Apply nutrient--enriched fertilizers/ amendments.	Quick nutrient impact. Applicable to existing crops. Uses current practices.	Pollutes soil over time due to an excessive buildup of wasted micronutrients Low micronutrient use efficiency Cost--effectiveness	Zinc/ iron/calcium fertilizers to soil or foliage. Nutri priming	Shrestha et al. (2024)
Biotechnological approaches	Genetic Engineering (Transgenic Approaches) Genome Editing (e.g., CRISPR–Cas9) Marker--Assisted Selection (MAS) RNA Interference (RNAi)	Precise nutrient enhancement. Faster than breeding. New traits from outside gene pool.	Low genomic information. Complexity of traits Cost of development Public opposition to GMOs	Transgenic millet with higher Nutrient content	Mane et al. (2024), Ceasar and Babu (2023)
Reduction of antinutrients	Reduce antinutrients (e.g., phytic acid, polyphenols, and tannins) via breeding/ genetic engineering.	Improves nutrient bioavailability.	Negative effects on crop yield off-targets in genetic engineering Stability over multiple generations	Low--phytate millet lines for better mineral absorption.	Vinoth and Ravindhran (2017)

8.6 Conclusion

Biofortification represents a sustainable cost-effective and impactful strategy to combat malnutrition and improve health outcomes globally. By focusing on the nutritional enhancement of staple crops, we can better address deficiencies and promote the well-being of both rural farming communities as farm subsistence and surplus can reach the urban communities consumers through special and consistent marketing strategies. Successful implementation of biofortification will require collaboration among government, policymakers, farmers, food scientists, and health professionals, ensuring that these improved crops meet both nutritional needs and market demands. HarvestPlus carried out calcium feasibility research in 2021 to investigate food-based strategies to boost calcium intake. The calcium concentrations attained in rice, maize, wheat, and finger millet flour and food products, as well as the contribution to the daily calcium requirements that these fortified foods may provide to vulnerable populations, were beneficial according to the results of the global analysis. Finger millet is one of the top candidate crops in calcium breeding, and the ability to use biofortification to breed staple food crops with greater calcium levels has shown encouraging results. This requires multiple nutrition enhancements with high-yielding backgrounds in future. With continued ICAR-IIMR research coordination and commitment, biofortified finger millet cultivars have the potential to significantly contribute to nutritional security and improve the quality of life for millet depended populations in India and globally.

References

Antony U, Chandra TS (1998) Antinutrient reduction and enhancement in protein, starch, and mineral availability in fermented flour of finger millet (Eleusine coracana). J Agric Food Chem 46(7):2578–2582

Arya C, Bisht A (2022) Small millets: path to food and nutrition security. In: Small millet grains: the superfoods in human diet. Springer Nature Singapore, pp 161–190

Ashik S, Morya S (2022) A review on finger millet properties, processing, health benefits, and applications. The Pharma Innovation Journal, 11(7), 3388-3395.

Babu BK, Agrawal PK, Pandey D, Kumar A (2014) Comparative genomics and association mapping approaches for opaque2 modifier genes in finger millet accessions using genic, genomic and candidate gene-based simple sequence repeat markers. Mol Breed 34:1261–1279

Backiyalakshmi C, Babu C, Deshpande S, Govindaraj M, Gupta R, Sudhagar R et al (2024) Characterization of finger millet global germplasm diversity panel for grain nutrients content for utilization in biofortification breeding. Crop Sci 64(5):2569–2588

Bailey-Serres J, Parker JE, Ainsworth EA, Oldroyd GE, Schroeder JI (2019) Genetic strategies for improving crop yields. Nature 575(7781):109–118

Beenish I, Omar Bashir I, Tashooq Ah Bhat I, Bazila Naseer I, Tahiya Qadri I, Nusrat Jan I et al (2018) Hidden hunger and its prevention by food processing: a review. Int J Unani Integr Med 2:1–10

Bhat BV, Hariprasanna K, Sooganna, Ratnavathi CV (2023) Global and Indian scenario of millets. Indian Farming 73:16–18

Bouis H (2018) Reducing mineral and vitamin deficiencies through biofortification: progress under HarvestPlus. World Rev Nutr Diet 118:112–122. https://doi.org/10.1159/0004843427
Bouis H, Saltzman A, Birol E (2019) Improving nutrition through biofortification. In: Shenggen F, Sivan Y, Rajul P-L (eds) Agriculture for improved nutrition: seizing the momentum. International Food Policy Research Institute, pp 47–57. https://doi.org/10.1079/9781786399311.0047
Ceasar SA, Babu BK (2023) Genetic and genomic resources for harnessing the health-related genes in finger millet. In: Compendium of crop genome designing for nutraceuticals. Springer Nature Singapore, pp 267–282
Ceasar SA, Ignacimuthu S (2011) Agrobacterium-mediated transformation of finger millet (Eleusine coracana (L.) Gaertn.) using shoot apex explants. Plant Cell Rep 30:1759–1770
Chandra AK, Pandey D, Tiwari A, Gururani K, Agarwal A, Dhasmana A, Kumar A (2021) Metal based nanoparticles trigger the differential expression of key regulatory genes which regulate iron and zinc homeostasis mechanism in finger millet. J Cereal Sci 100:103235
Chaudhary R (2024) Enhancement of iron and zinc bioavailability in finger millet (Eleusine coracana) using bacterial endophytes
Chaudhary R, Chaturvedi S, Sharma R, Tiwari S (2020) Global scenario of vitamin deficiency and human health. In: Advances in agri-food biotechnology. Springer Singapore, pp 199–220. https://doi.org/10.1007/978-981-15-2874-3_9
Chaudhary R, Kumar V, Gupta S, Naik B, Prasad R, Mishra S et al (2023) Finger millet (Eleusine coracana) plant–endophyte dynamics: plant growth, nutrient uptake, and zinc biofortification. Microorganisms 11(4):973
Chaudhary R, Sharma C, Kumar V, Rajput V, Naik B, Prasad R et al (2025) Millet biofortification for enhanced iron content: roadmap for combating hidden hunger. J Agric Food Res:101654
Collard BC, Mackill DJ (2008) Marker-assisted selection: an approach for precision plant breeding in the twenty-first century. Philos Trans R Soc B Biol Sci 363:557–572. https://doi.org/10.1098/rstb.2007.2170
Connorton JM, Jones ER, Rodríguez-Ramiro I, Fairweather-Tait S, Uauy C, Balk J (2017) Wheat vacuolar iron transporter TaVIT2 transports Fe and Mn and is effective for biofortification. Plant Physiol 174:2434–2444
Devi PB, Vijayabharathi R, Sathyabama S, Malleshi NG, Priyadarisini VB (2014) Health benefits of finger millet (Eleusine coracana L.) polyphenols and dietary fiber: a review. J Food Sci Technol 51:1021–1040
Díaz-Gómez J, Twyman RM, Zhu C, Farré G, Serrano JCE, Portero-Otín M, Muñoz P, Sandmann G, Capell T, Christou P (2017) Biofortification of crops with nutrients: factors affecting utilization and storage. Curr Opin Biotechnol 44:115–123
Dida MM, Srinivasachary, Ramakrishnan S, Bennetzen JL, Gale MD, Devos KM (2007) The genetic map of finger millet, Eleusine coracana. Theor Appl Genet 114:321–332
Dwivedi SL, Garcia-Oliveira AL, Govindaraj M, Ortiz R (2023) Biofortification to avoid malnutrition in humans in a changing climate: enhancing micronutrient bioavailability in seed, tuber, and storage roots. Front Plant Sci 14:1119148
FAO (2018) The state of food security and nutrition in the world: building climate resilience for food security and nutrition, Rome
Fred OH, Sheunda P, Kibuka J, Kumar A, Rathore A, Manyasa E, Ajaku D (2021) Characterization of finger millet germplasm for mineral contents: prospects for breeding. J Cereals Oilseeds 12(1):33–44
Ganapathy KN, Patro TSSK, Palanna KB, Das IK, Elangovan M, Prashant B et al (2021a) Development of improved mutants for grain yield and related traits in finger millet (Eleusine coracana L Gaertn.) through gamma mutagenesis. Int J Plant Soil Sci 33(18):225–235
Ganapathy KN, Hariprasanna K, Tonapi V (2021b) Breeding for enhanced productivity in millets. In: Millets and pseudo cereals. https://doi.org/10.1016/B978-0-12-820089-6.00005-7
Ganesh BT, Dilipbhai KA, Ragi S, Katral A (2024) Breeding finger millet (Eleusine coracana L. Gaertn) for improvement of quality characters and yield. In: Genetic improvement of small millets. Springer Nature Singapore, pp 213–224

Gebreyohannes A, Shimelis H, Mashilo J, Odeny DA, Tadesse T, Ojiewo CO (2024) Finger millet (Eleusine coracana) improvement: challenges and prospects—a review. Plant Breed 143(3):350–374

Gimode D, Odeny DA, de Villiers EP, Wanyonyi S, Dida MM, Mneney EE et al (2016) Identification of SNP and SSR markers in finger millet using next generation sequencing technologies. PLoS One 11(7):e0159437

Gödecke T, Stein AJ, Qaim M (2018) The global burden of chronic and hidden hunger: trends and determinants. Glob Food Sec 17:21–29. https://doi.org/10.1016/j.gfs.2018.03.004

Govindaraj M, Rai KN, Cherian B, Pfeiffer WH, Kanatti A, Shivade H (2019) Breeding biofortified pearl millet varieties and hybrids to enhance millet markets for human nutrition. Agriculture 9(5):106. https://doi.org/10.3390/agriculture9050106

Gregorio GB, Senadhira D, Htut H, Graham RD (2000) Breeding for trace mineral density in rice. Food Nutr Bull 21:382–386

Gupta SM, Arora S, Mirza N, Pande A, Lata C, Puranik S et al (2017) Finger millet: a "certain" crop for an "uncertain" future and a solution to food insecurity and hidden hunger under stressful environments. Front Plant Sci 8:643

Hariprasanna K (2023) Biofortified millets to alleviate micronutrient malnourishment. Indian Farming 73(1):84–87

Hema R, Vemanna RS, Sreeramulu S, Reddy CP, Senthil-Kumar M, Udayakumar M (2014) Stable expression of mtlD gene imparts multiple stress tolerance in finger millet. PLoS One 9(6):e99110

Hittalmani S, Mahesh HB, Shirke MD, Biradar H, Uday G, Aruna YR et al (2017) Genome and transcriptome sequence of finger millet (Eleusine coracana (L.) Gaertn.) provides insights into drought tolerance and nutraceutical properties. BMC Genomics 18:1–16

Ignacimuthu S, Ceasar SA (2012) Development of transgenic finger millet (Eleusine coracana (L.) Gaertn.) resistant to leaf blast disease. J Biosci 37:135–147

Ingle K, Thakur N, Moharil MP, Suprasanna P, Awio B, Narkhede G et al (2023) Current status and future prospects of molecular Marker Assisted Selection (MAS) in millets. In: Nutriomics of millet crops

Iyengar KG, Dorasami LS, Iyengar RS (1945) Ragi (Eleusine coracana Gaertn.). Mysore Agr J 24:33–49

Joshi P, Gupta SK, Ojulong H, Sharma R, Vetriventhan M, Kudapa H et al (2023) Finger millet improvement in post-genomic era: hundred years of breeding and moving forward. In: Smart plant breeding for field crops in post-genomics era. Springer Nature Singapore, Singapore, pp 221–253

Kiran K (2020) Advanced approaches for biofortification. In: advances in agri-food biotechnology. Springer, Singapore, pp 29–55

Kumar A, Metwal M, Kaur S, Gupta AK, Puranik S, Singh S et al (2016) Nutraceutical value of finger millet [Eleusine coracana (L.) Gaertn.], and their improvement using omics approaches. Front Plant Sci 7:934

Kumar S, Palve A, Joshi C, Srivastava RK (2019) Crop biofortification for iron (Fe), zinc (Zn), and vitamin A with transgenic approaches. Heliyon 5:e01914

Laborde D, Parent M, Smaller C (2020) Ending hunger, increasing incomes, and protecting the climate: what would it cost donors?

Lenaerts B, Collard BC, Demont M (2019) Improving global food security through accelerated plant breeding. Plant Sci 287:110207

Lowe NM (2021) The global challenge of hidden hunger: perspectives from the field. Proc Nutr Soc 80:283–289. https://doi.org/10.1017/S0029665121000902

Mandial A, Negi SC, Kaur P, Kaur N (2023) Performance of different rice-based cropping systems in the wet temperate zone of Himachal Pradesh. Environ Conserv J 24(4):248–253

Mane RS, Prasad BD, Sahni S, Quaiyum Z, Sharma VK (2024) Biotechnological studies towards improvement of finger millet using multi-omics approaches. Funct Integr Genomics 24(5):148

Mrudula G, Rani PS, Sreekanth B, Madhuri KV, Luther MM (2022) Influence of various methods of zinc fertilization on growth and yield of finger millet (Eleusine coracana) varieties. Agric Sci Dig - Res J 42(4):438–443

National Research Council (1996) Lost crops of Africa: Volume I: Grains, vol 1. National Academies Press

Olson R, Gavin-Smith B, Ferraboschi C, Kraemer K (2021) Food fortification: the advantages, disadvantages, and lessons from sight and life programs. Nutrients 13:1118. https://doi.org/10.3390/nu13041118

Ozkan H, Brandolini A, Torun A, Altintas S, Eker S, Kilian B, Braun HJ, Salamini F, Cakmak I (2007) Natural variation and identification of microelements content in seeds of einkorn wheat (Triticum monococcum). In: Buck HT, Nisi JE, Salomón N (eds) Wheat production in stressed environments. Springer, Dordrecht, pp 455–462

Paltridge NG, Palmer LJ, Milham PJ, Guild GE, Stangoulis JC (2012) Energy-dispersive X-ray fluorescence analysis of zinc and iron concentration in rice and pearl millet grain. Plant Soil 361:251–260

Pujar M, Govindaraj M, Gangaprasad S, Kanatti A, Shivade H (2020) Genetic variation and diversity for grain iron, zinc, protein and agronomic traits in advanced breeding lines of pearl millet [Pennisetum glaucum (L.) R. Br.] for biofortification breeding. Genet Resour Crop Evol 67:2009–2022

Puranik S, Kam J, Sahu PP, Yadav R, Srivastava RK, Ojulong H, Yadav R (2017) Harnessing finger millet to combat calcium deficiency in humans: challenges and prospects. Front Plant Sci 8:1311

Puranik S, Sahu PP, Beynon S, Srivastava RK, Sehgal D, Ojulong H, Yadav R (2020) Genome-wide association mapping and comparative genomics identifies genomic regions governing grain nutritional traits in finger millet (Eleusine coracana L. Gaertn). Plants, People, Planet 2(6):649–662

Rai KN, Govindaraj M, Rao AS (2012) Genetic enhancement of grain iron and zinc content in pearl millet. Qual Assur Saf Crops Foods 4:119–125

Ramashia SE, Gwata ET, Meddows-Taylor S, Anyasi TA, Jideani AIO (2018) Some physical and functional properties of finger millet (Eleusine coracana) obtained in sub-Saharan Africa. Food Res Int 104:110–118

Reddy VD, Rao KV, Reddy TP, Kishor PBK (2008) Finger millet. In: Kole C, Hall TC (eds) Compendium of transgenic crop plants: transgenic cereals and forage grasses. Blackwell, London, pp 191–198

Saint Ville A, Po JYT, Sen A, Bui A, Melgar-Quiñonez H (2019) Food security and the food insecurity experience scale (FIES): ensuring progress by 2030. Food Secur 11:483–491. https://doi.org/10.1007/s12571-019-00936-9

Saleh AS, Zhang Q, Chen J, Shen Q (2013) Millet grains: nutritional quality, processing, and potential health benefits. Compr Rev Food Sci Food Saf 12(3):281–295

Sharma S, Garg AP (2023) Effect of germination on the physicochemical and anti-nutritional properties of finger millet (Eleusine coracana), pearl millet (Pennisetum glaucum), and sorghum (Sorghum bicolor). Pharma Innov J 12:4763–4772

Sharma D, Jamra G, Singh UM, Sood S, Kumar A (2017) Calcium biofortification: three pronged molecular approaches for dissecting complex trait of calcium nutrition in finger millet (Eleusine coracana) for devising strategies of enrichment of food crops. Front Plant Sci 7:2028

Shobana S, Krishnaswamy K, Sudha V, Malleshi NG, Anjana RM, Palaniappan L, Mohan V (2013) Finger millet (Ragi, Eleusine coracana L.): a review of its nutritional properties, processing, and plausible health benefits. Adv Food Nutr Res 69:1–39

Shrestha J (2024) Crop biofortification of finger millet through agronomic approaches to enhance bioactive compounds for human health and nutrition. Agron J Nepal 8:237–249

Shrestha J, Subedi S, Shrestha S, Rai S, Poudel A, Adhikari S et al (2024) Crop biofortification of finger millet through agronomic approaches to enhance bioactive compounds for human health and nutrition. Agron J Nepal 8:237–249

Singh P, Srivastava S (2006) Nutrient composition of some new varieties of finger millet (Eleusine coracana). J Commun Mobiliz Sustain Dev 1(1and2):115–120

Singh UM, Metwal M, Singh M, Taj G, Kumar A (2015) Identification and characterization of calcium transporter gene family in finger millet in relation to grain calcium content. Gene 566(1):37–46

Sood S, Kumar A, Babu BK, Gaur VS, Pandey D, Kant L, Pattnayak A (2016) Gene discovery and advances in finger millet [Eleusine coracana (L.) Gaertn.] genomics—an important nutri-cereal of future. Front Plant Sci 7:1634

Taylor JR, Emmambux MN (2008) Gluten-free foods and beverages from millets. In: Gluten-free cereal products and beverages. Academic Press, pp 119–125

Udeh HO, Duodu KG, Jideani AIO (2017) Finger millet bioactive compounds, bioaccessibility, and potential health effects—a review. Czech J Food Sci 35(1):7

Velu G, Rai KN, Muralidharan V, Longvah T, Crossa J (2011) Gene effects and heterosis for grain iron and zinc density in pearl millet (Pennisetum glaucum (L.) R. Br). Euphytica 180:251–259

Vetriventhan M, Azevedo VC, Upadhyaya HD, Nirmalakumari A, Kane-Potaka J, Anitha S et al (2020) Genetic and genomic resources, and breeding for accelerating improvement of small millets: current status and future interventions. Nucleus 63:217–239

Vinoth A, Ravindhran R (2017) Biofortification in millets: a sustainable approach for nutritional security. Front Plant Sci 8:29

Wakeel A, Labuschagne MT (2021) Editorial: crop biofortification for food security in developing countries. Front Sustain Food Syst 5:756296. https://doi.org/10.3389/fsufs.2021.756296

Wakeel A, Farooq M, Bashir K, Ozturk L (2018) Micronutrient malnutrition and biofortification: recent advances and future perspectives. In: Hossain MA, Kamiya T, Burritt DJ, Tran LSP, Fujiwara T (eds) Plant micronutrient use efficiency: molecular and genomic perspectives in crop plants. Elsevier, pp 225–243. https://doi.org/10.1016/B978-0-12-812104-7.00017-4

Welch RM, Graham RD (2004) Breeding for micronutrients in staple food crops from a human nutrition perspective. J Exp Bot 55:353–364

Wijesinha-Bettoni R, Burlingame B (2013) Milk and dairy product composition. In: Milk and Dairy Products in Human Nutrition, eds. E. Muehlhoff, A. A. Bennett, and D. McMahon, 41–102. Rome: Food and Agriculture Organization of the United Nations

Zhu C, Naqvi S, Gomez-Galera S, Pelacho AM, Capell T, Christou P (2007) Transgenic strategies for the nutritional enhancement of plants. Trends Plant Sci 12(12):548–555. https://doi.org/10.1016/j.tplants.2007.09.007

Chapter 9
Marketing Biofortified Pearl Millet: Scaling Strategies from Seed to Market

Binu Cherian

Abstract Pearl millet, a climate-resilient crop, is an integral part of the food system for 2 billion people who are living in semi-arid and dryland regions of Asia and Africa. This chapter explores leveraging the value of biofortified pearl millet as a multifunctional crop for food, feed, and fodder. It highlights the strategies to create demand across diverse markets and to position it as a cost-effective solution to malnutrition. Malnutrition remains one of the largest public health challenges in the twenty-first century, with more than 2 billion people suffering from the insidious effects in South Asia and Africa. The chapter looks at the various options available to promote biofortified seeds and further value chain development at various market segments. Due to the rising incomes in urban areas, expanding urban markets, and demand for health and nutritious food options. It examines the marketing strategies and tactics required to position biofortified pearl millet in different market segments, from commercial to institutions' food programs, as cost-effective and effective solutions. It also examines the opportunities to leverage emerging market food trends and marketing tools to drive demand for biofortified crops. The chapter positions biofortified pearl millet as a nutritious food and its role in sustainable agriculture, public health, and climate resilience.

Keywords Biofortification · Seed production · Scaling strategy · Types of market · Biofortified pearl millet · Climate-resilient crops · Iron deficiency · Nutrition-sensitive agriculture · Seed systems

---

B. Cherian (✉)
HarvestPlus Program, International Centre for Tropical Agriculture (CIAT), Patancheru, Telangana, India
e-mail: binu-cherian@outlook.com

M. Govindaraj (ed.), *Breeding Biofortified Millets for Nutritional Security*, Sustainability Sciences in Asia and Africa,
https://doi.org/10.1007/978-981-95-9827-4_9

9.1 Introduction: The Millet Renaissance in the Twenty-First Century

Pearl millet (*Pennisetum glaucum*), a hardy cereal crop, is undergoing a much-needed renaissance, which is traditionally used for food, feed, and fodder and plays a key role in the food systems and climate resilience in the arid and semi-arid dry land areas of South Asia and Sub-Saharan Africa (FAO 2021). About 46% of the world's land areas are drying land inhabiting 2 billion people who depend on agriculture, livestock, and natural resources for their primary livelihood. Two billion people inhabit the dryland areas, which contribute about 46% of the world's land areas, and they mostly depend on agriculture, livestock, and natural resources for their primary livelihood (Global Environment Facility [GEF] 2006; Millennium Ecosystem Assessment 2005). Pearl millet is an integral part of their food system in these regions, serving their nutrition and energy demands. Pearl millet is a C4 plant with better climate resilience to face drought, heatwaves, and degradation of soils in agricultural ecosystems (ICRISAT 2022).

Pearl millet has demonstrated high potential for biofortification, which offers a food-based strategy to address iron deficiency among vulnerable populations. Micronutrient malnutrition, which is also known as hidden hunger, affects more than 2 billion people worldwide. Iron deficiency alone is responsible for nearly half of all anemia cases globally, affecting women of reproductive age and young children (World Health Organisation n.d.; WHO 2023; UNICEF–WHO 2023). Nutrition studies conducted with biofortified pearl millet have demonstrated measurable improvements in child and maternal health outcomes (Finkelstein et al. 2015; Kodkany et al. 2013). HarvestPlus, a program of the CGIAR, led an alliance consisting of ICRISAT, NARES, and seed companies to develop biofortified varieties of pearl millet. First pearl millet variety, Dhanashakti, has up to 70 mg/kg iron and 40 mg/kg zinc—substantially more than conventional varieties. Subsequently, 17 biofortified pearl millet cultivars have been developed by NARES in India, and one biofortified pearl millet variety has been developed and released in Africa.

This chapter explores the strategic approaches for leveraging the potential of biofortified pearl millet for long-term market sustainability by considering strengthening of seed systems, favorable public policies, commercial and institutional markets, nutrition evidence, and strategic marketing to create demand at scale across the value chain. The declaration of the International Year of Millets in 2023, by the United Nations, has further boosted the awareness of millets globally. These promotions have shifted the perception of millets from an "orphan crop" to "nutri-cereals." Biofortified pearl millet offers a sustainable and scalable solution to address the nutrition requirements in the semi-arid and drylands in Asia and Africa

9.2 Biofortified Pearl Millet

Biofortified pearl millet is specially bred for higher levels of micronutrients, such as iron, as compared to conventional varieties. The plant breeder, while developing biofortified varieties, takes into consideration the farmers' preferred traits, tolerance to heat, drought, resistance to disease, and agronomical superiority. Nutrition studies have established a measurable impact on health and well-being when biofortified varieties are consumed daily. Biofortification as an approach is considered cost-effective, sustainable, and rural in approach. Since it is an agricultural-based approach, it starts from the rural area.

A significant quantity of staple crops, such as pearl millet, is consumed by producers on the farm. The surplus production of biofortified pearl millet after meeting farm consumption requirements finds its way to markets. The initial cost of biofortification is front-loaded to meet the cost of breeding and initial seed production. The further cost declines once these varieties are integrated into the seed production system. As compared to fortification, where the annual recurring investment is required, biofortification needs a one-time initial investment to develop varieties. Once these varieties are scaled, the cost of further adoption tapers off. Therefore, biofortification as an approach is considered cost-effective, sustainable, and rural in approach.

For example: Breeding standard set by HarvestPlus for biofortified pearl millet

- Iron target: 77 mg/kg or more of iron (compared to ~30–40 mg/kg in conventional varieties). Delivering about 25–50% of the Estimated Average Requirement (EAR) for young children and women of reproductive age.
- Zinc target: Zinc levels are targeted around 32 mg/kg or more.

9.3 Markets' Potential and the Opportunities for Biofortified Pearl Millet

Due to the unique ability to address the challenges in nutrition, climate resilience, and rural livelihoods, the millets are gaining prominence. Biofortified pearl millet can be a strategic fit for traditional and modern food systems. Africa and Asia together have over 95% of the global pearl millet area of cultivation. Africa contributes approximately 14–16 million hectares (60–64%), whereas Asia contributes around 7–9 million hectares (34–38%) of the total millet area. The leading pearl millet-producing countries are India, Nigeria, Niger, Mali, and Sudan (Satyavathi et al. 2021; FAO 2023; Directorate of Millets Development 2019; NRC 1996; Smart Food Initiative 2022).

In India, there is a growing demand for pearl millet from the urban consumer. Market research and behavioral studies have identified increased willingness to pay among urban and peri-urban consumers, especially women, for nutrient-rich millets

(Banerji et al. 2016). In India, pearl millet is increasingly perceived as a food solution for addressing lifestyle diseases. Pearl millet is being increasingly used as a nutritious ingredient for various consumer-based product categories such as energy bars, breakfast cereals, multigrain flour mixes, and beverages marketed as gluten-free, low-GI, and rich in iron and fiber. The current Indian millet market is US dollars 2.5 billion, with projected annual growth above 20 per cent in ready-to-eat and health-food segments (FICCI 2022). In West Africa, millet value chains are valued more than US dollars 1 billion annually, with high potential for growth in the food processing sector (AGRA 2020).

Urban demand for millet-based products is increasing in countries such as Nigeria, Kenya, Ghana, and Uganda, where millet-based flours, porridges, snacks, and beverages are entering premium and mid-tier urban retail markets. This trend is driven by rising incomes and changing dietary preferences, especially among youth, rising health awareness, the convenience factor, and the expansion of packaged food industries. The rapid growth of millet-based snacks that are gluten-free, healthy, and high in fiber in urban supermarkets shows a strong consumer response to nutrient-dense, traditional grain innovations in Africa's emerging health-conscious food segment (Muskan 2025) (Table 9.1).

The value chain presents diverse market opportunities for biofortified pearl millet from seed to the end consumer. The seed companies have to develop biofortified pearl millet varieties that are rich in iron and undertake seed production. While farmers grow and consume them, they benefit from the additional nutrition and crop yield. There is a willingness to pay premium prices by health-conscious urban consumers, which can be leveraged to position the iron-rich pearl millets product in the markets. The government institution's food programs can be a potential channel to incorporate nutrition millets into public nutrition programs, which can create additional demand. The export markets, food processing companies, and feed producers can create commercial market opportunities across the entire value chain.

Table 9.1 Potential target segments across the value chain for biofortified pearl millet

Market segment	Characteristics
Seed companies	A new product line that offers better productivity and enhanced nutrition, for added value and increased profit potential.
Farmers	Cultivate and consume biofortified cultivars, benefit from improved productivity and income, along with health improvement.
Consumers	Rising income, changing dietary preferences, seeking better and nutritious alternatives to rice/wheat, and a willingness to pay a premium.
Government	Incorporating biofortified varieties in the public food programs for children, women, and vulnerable groups.
Exporters	Ethenic/traditional food demand- overseas markets.
Processors	Millet-based snacks, beverages, and gluten-free products.
Feed industry	Increasing use of millets for animal feed.

9.4 Policy Integration and Global Micronutrient Standards

Polices can be an important lever to scale up biofortified pearl millet that aligns agricultural research with public health nutrition. The Indian Council of Agricultural Research (ICAR), through its All India Coordinated Research Project on Pearl Millet (AICRP-PM), has developed minimum iron and zinc standards as benchmarks to be integrated into its national variety release process. As per this policy, all new pearl millet cultivars must meet at least the following: iron—42 mg/kg and above; zinc—32 mg/kg and above. HarvestPlus, in collaboration with the British Standards Institute (BSI), published Publicly Available Standard, PAS 234:2021, "Iron Enriched Bean and Pearl Millet Grain – Specification." (Fig. 9.1)—https://www.bsigroup.com/siteassets/pdf/en/insights-and-media/insights/brochures/pas_234.pdf. This standard sets global minimum iron thresholds for classifying iron-enriched pearl millet ready for food systems and trade (Table 9.2).

- Class I criteria (≥ 77 mg/kg) are considered optimal, as they deliver approximately >80 % of the daily iron needs for young children and non-pregnant women when eaten regularly as part of a typical diet.
- The global baseline iron level for conventional pearl millet is set at 47 mg/kg, indicating that Class I varieties must contain at least 30 mg/kg more iron than standard grain.
- PAS 234 also outlines essential sampling protocols, analytical testing methods (e.g., ICP–OES, XRF), and labeling requirements to support transparent procurement and trade of nutritionally enhanced grains.

These micronutrient standards help to inform national food safety and labeling policies and support the inclusion of biofortified crops in social safety nets, school meals, and, in the future, public food procurement guidelines (IFPRI 2021). Crop micronutrient standards for micronutrients for the development of new cultivars of pearl millet will attract investments in research and development from the private and public sectors in mainstreaming biofortified pearl millet. The Publicly Available Standard (PAS) plays a key role in benefiting the value chain actors for maintaining the quality control and quality assurance for the product across the value chains and maintaining the traceability. It will require coordination among key stakeholders to make a compelling case for the implementation of these standards.

9.5 Marketing Strategies for Biofortified Pearl Millet Market Development

Biofortification in pearl millet represents a cutting-edge innovation that is evidence-based for addressing hidden hunger and better rural livelihood and prosperity to the communities in the semi-arid and drylands of Asia and Africa. However, its transformative potential remains largely untapped due to limited market development

Fig. 9.1 Publicly Available Standard (PAS 234: 2021 Cereals and cereal products)

and weak demand-side integration. To scale and create sustainable demand for biofortified pearl millet, differentiated marketing strategies can be created for three critical potential market segments:

Table 9.2 PAS defines the classification thresholds[a]

Class	Iron content (mg/kg)	Nutrition impact
I	≥ 77	Highest impact; recommended as a whole grain
II	69 –< 77	Moderate impact
III	62 –< 69	Lower but still enriched compared to baseline

[a]Measured in mg/kg iron in grain, using AOAC 999.10 analysis

9.5.1 *Urban Consumer Markets*

In the survey conducted between 2023–2024, Urban households in India accounted for nearly 40% of total household spending in food expenditure (MoSPI). Due to the increase in urbanization, there is a shift in consumer behavior, making urban markets key drivers of growth in health-focused lifestyle products (Reuters 2024). The urban consumer market that caters to health-conscious individuals, professionals, and fitness enthusiasts is increasingly looking for nutritious, gluten-free, and low-glycemic alternatives to rice and wheat. In India, non-communicable lifestyle diseases such as diabetes, hypertension, and obesity are rising at an alarming rate, accounting for over 67% of all deaths annually (Indian Council of Medical Research [ICMR] 2023).

The health and wellness food market of India, mostly driven by rising consumer interest in functional and fortified foods, has been valued at approximately USD 35.9 million in 2024, with a projected CAGR of 18.5% through 2032 (Data Bridget Market Report 2024). Capturing this market segment will require strategic branding, consumer awareness, attractive packaging, digital marketing, and product innovation (e.g., ready-to-eat snacks, breakfast cereals, millet-based noodles) to incorporate biofortified pearl millet. Influencer campaigns, the right messaging, and training the retail outlets can help in educating consumers and building trust.

9.5.2 *Bottom-of-the-Pyramid (BoP) Markets*

Serving bottom-of-the-pyramid (BoP) markets comes with great potential and unique challenges. These markets represent the poorest socio-economic group, which requires strategic products that are cost-effective and with marketing strategies that resonate well with their needs, behavior, and aspirations. This segment includes rural women, smallholder farmers, tribal populations, urban poor, and adolescent girls and children in resource-poor settings. These groups suffer the highest burden of micronutrient malnutrition, particularly iron-deficiency anemia. Awareness campaigns targeting women on the benefits for them and their family, using local languages, street plays, and recipe demonstrations, are required to overcome myths, taste barriers, and initial adoption hesitancy.

9.5.3 Institutional Markets

In India, there is a large opportunity to integrate pearl millet with the institutional food programs. Pearl millet still has not been mainstreamed in public food programs in all the states in India, unlike rice and wheat. Due to the growing awareness about the benefits of nutrient and climate resilience of millets, some states such as Odisha and Karnataka have taken progressive steps to integrate millet grains into the Public Food Distribution Scheme (PDS). Including biofortified millet can significantly strengthen the nutrition component of the existing food program that can benefit the most vulnerable population and improve the public health outcomes. The large public food program includes the subsidized Public Distribution Scheme (PDS), mid-day meals, Integrated Child Development Services (ICDS), and Poshan Abhiyaan (Nutrition Mission). These programs serve children, pregnant and lactating mothers, and marginalized communities—making them ideal entry points for scaling nutrition impact.

Biofortified pearl millet can be integrated into school meals, with recipes such as traditional foods, ready to eat porridge mixes, or snack packs. Success depends on centralized procurement mechanisms, policy alignment with nutrition goals (e.g., Anaemia Mukt Bharat), pilot success stories, and partnerships with ministries, UN agencies (e.g., WFP, UNICEF), and state nutrition missions. Uptake from these programs will ensure sustained demand and nutrition outcomes. Implementing iron-rich pearl millet into the various national food programs will require a strong policy framework, collaboration with the Food Corporation of India (FCI), and other civil supply ministries. A well-structured procurement plan needs to be established to maintain the robust traceability and quality control of the iron-rich pearl millet raw materials. For some of the programs that provide hot cooked meals, such as Integrated Child Development Scheme (ICDS) and Mid-Day Meal, appropriate recipes need to be developed that are appealing and accepted locally are essential for nutrition impact on all the beneficiaries. Each of these segments will require unique tactics for positioning, messaging, and the final delivery to unlock both nutritional impact and commercial feasibility (Tables 9.3 and 9.4).

The BoP market strategy for biofortified pearl millet targets rural and peri-urban consumers, smallholder farmers, tribal communities, children, and mothers with the value proposition: "Stronger children, healthier mothers." Messaging highlights its role in fighting anemia, supporting child growth, and thriving in drought-prone areas—framed as "Shakti ka dana" (grain of strength).

Products offered may include whole grain, flour, porridge mixes, and fortified school meals, distributed through channels dedicated to rural retail outlets, SHG networks, and village retail centers. Grassroots tools such as community theatre, radio jingles, peer educators, and mother group recipe demos can be leveraged.

Table 9.3 Consumer market strategy for biofortified pearl millet

Component	Strategy
Target audience	Urban youth, working professionals, health-conscious individuals/families, fitness, and senior citizen communities.
Value proposition	Attractive value proposition that resonates well with the consumer groups, such as iron-rich super grain biofortified millets that enhance immunity, strength, and mental performance.
Messaging	Functional food, climate-smart crop, ancestral grain with modern relevance, energy and fitness, women's and children's health, clean label, and organic.
Product offering	Breakfast cereals, protein bars, instant mixes (idli/dosa), pearl millet-based pasta/noodles, roasted snacks, flour blends.
Channels	Modern trade/supermarkets, specialized gourmet and health stores, e-commerce platforms (Amazon, BigBasket, Jumia), nutrition blogs, fitness influencers.
Promotions	Social media storytelling, QR-code nutrition labels, celebrity endorsements, recipe contests, and sustainability certifications, traceability assurance.
Partnerships	FMCG brands, health-tech platforms, nutritionists, chefs, and fitness centers.
Case example	Urban India: millet food brands in India—Slurrp Farm, Manna, ITC Foods, and TruMillets promote millet blends for children's nutrition, health, and well-being, and diabetic diets

Table 9.4 Bottom-of-the pyramid (BoP) market strategy for biofortified pearl millet

Component	Strategy
Target audience	Rural consumers, smallholder farmers, tribal communities, children and adolescents, pregnant/lactating mothers
Value proposition	"Stronger children, healthier mothers"—Biofortified pearl millet as a nutrition shield against anemia and malnutrition, strength, and mental performance
Messaging	"Shakti ka Dana" (grain of strength), better school performance, healthier babies, resilient crop for drought-prone farms
Product offering	Whole grain, basic flour (bajra atta), porridge mixes, school meals, mid-day meals, healthy snacks
Channels--marketing	Anganwadi workers, SHG retail points, rural fairs, rural and peri-urban retailers
Promotions	Community nutrition campaigns, drama/street theatre, posters, radio jingles, peer educators, recipe demos in mother groups
Partnerships	State nutrition missions, NGOs, FPOs, women's SHGs, Agri Krishi Vigyan Kendras (KVKs), local ASHAs/health workers, NRLMs

9.5.4 Institutional Markets

Institutional market strategy targets biofortification as a tool for combating anemia and malnutrition (Table 9.5). A targeted strategy that focuses on key government stakeholders such as the Ministries of Women and Child Development, Education, Health, State Nutrition Missions, and public nutrition programs like Mid-Day Meals and ICDS. The core value proposition positions pearl millet as a cost-effective,

Table 9.5 Institutional market strategy for biofortified pearl millet

Component	Strategy
Target audience	Government ministries (WCD, Education, Health), State Nutrition Missions, Mid-Day Meal Authorities, ICDS functionaries, Public Health Institutions, NGOs
Value proposition	"Cost-effective, iron- and zinc-rich staple for fighting anaemia and malnutrition at scale"—aligns with SDGs, Anaemia Mukt Bharat, and Poshan Abhiyaan
Messaging	Nutrition equity, child cognitive development, maternal health, climate resilience, Make-in-India nutrition innovation, millet revival under IYM 2023 legacy
Product	Raw grains/flour for school kitchens; ready-to-cook premixes (idli, upma); fortified snacks; blends with rice/wheat; porridge mixes for ICDS
Procurement channels-raw materials (Grain)	Integration via centralized procurement (e.g., AP Food Corporation, Odisha Millet Mission); tie-ups with FPOs, SHGs, WCD nutrition kitchens; Tech platforms for supply chain traceability, Private foundations—Akshaya Patra Foundation
Promotions	Evidence briefs, pilot success stories, endorsement by nutrition councils, and inclusion in school nutrition curriculum
Partnerships	State departments (Agriculture/Women and Child Development/Health), (World Food Program/UNICEF), JEEViKA, Odisha Millet Mission, local FPOs, and SHGs
Case example	Odisha Millet Mission, including millets in the Karnataka State Public Distribution Scheme (PDS)

iron- and zinc-rich staple aligned with national priorities, including *Anaemia Mukt Bharat*, *Poshan Abhiyaan*, and the *Sustainable Development Goals (SDGs)*.

Procurement strategies need to be well structured and coordinated between FCI and the state Civil Supply department, partnerships with FPOs, SHGs, and tech-enabled traceability tools. Promotional efforts include evidence briefs, pilot case studies, and endorsements by nutrition councils, along with curriculum integration in school health programs. The initiative can be further strengthened through multi-sectoral partnerships with government departments, development organizations, and NGOs such as Akshaya Patra Foundation. Successful models like the Odisha Millet Mission and Karnataka's inclusion of millets in the PDS illustrate scalable pathways for integration (Table 9.6).

Biofortified pearl millet serves different roles across various markets, such as consumer, BoP, and institutional, each of which is driven by unique drivers that will require targeted strategies. The consumer markets are driven by demand from health and lifestyle. This market will require consistent branding and messaging, with products that resonate well with the customers' preferences, and premium retail outlets for sales. The BoP market requires a combination of affordable, need-based product innovation and low-cost products with strategic marketing to build trust and loyalty. There is a large market opportunity at the same time, and various challenges to meet the needs of the segment. The institutional markets are driven by different dynamics like government policies, public nutrition, and health mandates, for the

Table 9.6 Key differences and synergies across market segments

Factors	Consumer market	BoP market	Institutional market
Drivers	Wellness, fitness, sustainable food choices	Combating malnutrition, maternal-child health	Public health objectives, nutrition targets, policy alignment
Consumer behavior	Influenced by branding, packaging, and convenience	Influenced by peer networks, trust, and government provisioning	Driven by evidence, mandates, and program guidelines
Product formats	Branded, processed, value-added	Whole or semi-processed, basic staples	Bulk grain/flour, ready mixes for kitchens, centrally procured
Value chain partners	Retailers, FMCGs, food innovators	Government agencies, SHGs, NGOs, rural cooperatives	Ministries, ICDS/MDM schemes, food corporations, WFP, NGOs
Levers for success	Storytelling, packaging, traceability, and sustainability	Behavior change, inclusion in public schemes, and affordability	Policy advocacy, pilot success, procurement integration
Impact focus	Demand-side growth, price premium, lifestyle food	Supply side scale-up, anemia reduction, food access	System-level nutrition outcomes, sustained institutional uptake

respective government and flagship programs. Together, these segments offer complementary pathways to scale, combining demand-side growth, nutrition access, and systemic impact (Table 9.7).

High uptake of biofortified pearl millet is constrained by low consumer awareness, poor flour shelf life, weak market infrastructure for segregation, lack of consistent branding, and strong nutrition messaging. Investment for R&D is required for improving the shelf life of pearl millet flour and product innovation for urban and specialty overseas markets. However, major opportunities to scale biofortified pear millet exist through growing consumer demand for functional and gluten-free foods, development of supportive policies to integrate into India's public food program, community-led FPOs, and SHGs (Self Help Groups). There is strong support for biofortification from Indian Government such as the Indian Council of Agricultural Research (ICAR). The Honorable Prime Minister of India endorsed biofortified crops in 2018. Yet, threats remain from competing with subsidized staples like rice and wheat, weak procurement systems for pearl millet, fragmentation of policies, and lack of convergence of various interlinked ministries such as agriculture, health, and nutrition. Public confusion with GMOs will require strong advocacy efforts,

Table 9.7 SWOT analysis of the biofortified pearl Millet market

Strengths	**Weaknesses**
Rich in iron, zinc, and dietary fiber, non-gluten supports addressing anemia and micronutrient gaps	Low awareness among consumers about biofortification and health benefits
Drought-tolerant, climate-resilient—ideal for semi-arid regions of Asia and Africa	Poor shelf life of pearl Millet flour without specialized packaging
Culturally familiar—part of traditional diets (e.g., *bajra roti), easy to integrate into the consumer market*	Weak infrastructure for aggregation, processing, and market access
Multipurpose usage—food + fodder + feed—supports both human and livestock nutrition	Lack of strong branding, nutritional messaging, and quality control standards in the market
Multiple high-yielding biofortified crop products are available for cultivation in various agro-ecological zones	Limited R&D and promotion for Millet-based products in urban and export markets
Opportunities	**Threats**
Growing demand for iron-rich, gluten--free, functional foods in domestic and export markets	Competition from subsidized staples (e.g., rice and wheat) in India
Indian Government support for Millets--Millet Mission programs in states, IYM2023	Absence of a Minimum Support Price (MSP), guaranteed procurement channels for farmers/no premium price
Community/Women-led FPOs/SHGs for processing and microenterprise development	Policy neglect, lack of convergence across various ministries—health, ag, and nutrition ministries
Availability of food-processing infrastructure, scope for value addition. Product category—Snacks, porridge mixes, instant foods	Misperceptions regarding biofortification (e.g., confusion with GMOs)
QR code labeling and digital marketplaces, nutrition messaging for creating demand	Iron and zinc are invisible traits—They require robust traceability protocols and strong segregation infrastructure to maintain identity preservation across the value chain

9.6 Unique Selling Proposition (USP) of Biofortified Pearl Millet

Biofortified pearl millet is more than just a staple crop—it is a strategic solution at the intersection of nutrition, climate resilience, and rural livelihoods. Its unique agronomic and nutritional profile positions it as a "smart grain" for the twenty-first century, particularly in regions facing the triple burden of malnutrition, poverty, and climate stress. The following table highlights the core Unique Selling Propositions (USPs)—of biofortified pearl millet across five key dimensions: health, environment, market adaptability, inclusivity, and global relevance (Table 9.8).

The unique selling propositions (USPs) of biofortified pearl millet offer a powerful foundation for differentiated marketing strategies across diverse segments. Its

Table 9.8 Unique selling propositions (USPs) of biofortified pearl millet

USP category	Key features	Impact highlights
Nutritional superiority	Rich in iron, zinc, fiber, and antioxidants, biofortified varieties improve hemoglobin and cognitive health; suitable for diabetes and obesity management.	Up to 80% daily iron needs met; significant reduction in anemia (Birol et al. 2019; Finkelstein et al. 2015).
Climate resilience	C4 plant with high water-use efficiency; thrives in drought-prone, low-fertility soils; short maturity period (65–85 days); supports CSA practices.	Resilient under extreme climates; ideal for dryland farming in Asia and Africa (ICRISAT 2022; FAO 2021).
Versatility in food applications	Adaptable to traditional foods (roti, porridges) and modern formats (energy bars, cereals, gluten-free mixes); suitable across demographics.	Enhances nutrition accessibility and consumer acceptance through familiar and innovative products
Support for smallholders and women's empowerment	Supports livelihoods of smallholders, especially women; promotes SHG- and FPO-led value chains in India, Nigeria, and Kenya.	Empowers women economically; strengthens local economies through value addition (IFAD and ICRISAT 2022).
Alignment with global goals and ethical consumption	Contributes to SDGs: Zero hunger, good health, gender equality, climate action; appeals to ESG investors and ethical brands.	Supports public health, climate adaptation, and inclusive development goals via sustainable food systems.

nutritional superiority can be positioned to attract health-conscious urban consumers and public health programs. The climate resilience narrative resonates strongly with policymakers, ESG investors, and sustainability-driven brands. Product versatility in innovation and applications enables bridging between traditional diets and modern convenience. Meanwhile, its role in empowering smallholders and women appeals to inclusive business models and social enterprises. Alignment with Sustainable Development Goals (SDGs) provides legitimacy for public procurement from Governments and donor-backed programs. Together, these USPs can help in shaping tailored messaging, create category leadership, and unlock multi-channel growth—from consumer retail to institutional demand and export opportunities. LINK Excel.SheetBinaryMacroEnabled.12 C:\\Users\\Binu\\Downloads\\USP_of_Biofortified_Pearl_Millet.csv USP_of_Biofortified_Pearl_Mille!R1C1:R6C3 \a \f 4 \h * MERGEFORMAT

9.7 Branding, Messaging, and Media Mix

A well-crafted branding and communication strategy is central to scaling the adoption of millets. Effective messaging not only informs but also shifts consumer perceptions—positioning millets as aspirational, nutritious, and climate-smart staples. Branding strategies should emphasize health and nutrition by highlighting the biofortified nature of millet varieties with labels such as "Iron-Rich Pearl Millet." At the same time, celebrating cultural roots by connecting millets to indigenous food traditions and regional pride can build emotional resonance (FAO 2021). Millets can also be reimagined for modern audiences as superfoods through terms like "smart carbs," "gluten-free power grain," or "Clean label," which particularly appeal to urban millennials (FICCI 2022). Sustainability storytelling, highlighting the role of millets in climate resilience and eco-friendly agriculture, adds credibility among environmentally conscious consumers and institutions (ICRISAT 2022).

Messaging should follow key principles: clarity through jargon-free, relatable language; credibility via endorsements from nutritionists, chefs, and farmers; consistency across all touchpoints from packaging to digital platforms; and customization tailored to different audiences—urban, rural, institutional buyers, or youth. A diverse media mix is crucial. Digital channels like Instagram, YouTube, WhatsApp, and Facebook are ideal for influencer campaigns, nutrition storytelling, and recipe sharing. In rural and semi-urban areas, traditional media such as radio jingles, TV features, and print remain powerful. On-ground activations like millet food festivals, cooking demos, and school nutrition drives create immersive experiences. Additionally, retail and public-facing spaces—such as point-of-sale (POS) displays, Kirana (neighborhood grocery shop), branding—public transports, and branded APMC stalls—help reinforce visibility and drive trust at the last mile.

9.8 Seed Systems and Input Marketing for Biofortified Pearl Millet

9.8.1 Seed System Models and Community Engagement in India

There is a robust seed infrastructure for production and distribution, with strong investment and active participation from the public and private sectors. Over 70% of the pearl millet area in India is covered by hybrid varieties, with the private sector contributing significantly to seed supply (Mula et al. 2014). The private sector companies have vertically integrated infrastructure from Research and Development (R&D) to distribution and sales. Apart from the private sector, the public sector research program on pearl millet is led by the Indian Council of Agriculture Research (ICAR), Indian Institute of Millets Research (IIMR), which coordinates national-level research programs like the All India Coordinated Research Project (AICRP) on Pearl Millet, the International Centre for Research for the Semi-Tropics (ICRISAT), and State Agriculture Universities (SAU). ICRISAT has played an important role in the pearl millet research and development program in India, which has helped India to double the productivity in the last 20 years.

The private sector can also commercialize their research product as truthfully labeled (TFL) seeds without active participation in the national testing trials (NTTs).

This also helps to reduce the time to market. The pearl millet hybrid seeds are marketed in 1.5 kg packs, whereas the open-pollinated varieties (OPVs) are marketed in 1.5 kg and 3 kg packs. Private sector seed companies distribute their pearl millet hybrids through well-established distribution and retail outlet channels to cater to markets. The sales and product managers engage the farmers with training and capacity-building programs to improve the best practices on the farm. The private sector incentivizes the retailers and distributors, as they play an important role in making purchase decisions of products and knowledge dissemination.

The public sector seed companies, such as National Seed Corporation (NSC), State Seed Corporations, State Agriculture Universities SAUs), can play an important role in the production and distribution of public sector biofortified pearl millet hybrids. Krishi Vigyan Kendras (KVKs) conduct product demonstrations and training, frontline demonstrations (FLDs) and capacity building of farmers to create awareness. Participatory Varietal Selection (PVS) has proven effective in identifying locally adapted, nutritionally beneficial varieties while ensuring accessibility for smallholders (Bioversity International 2018). Biofortified pearl millet can leverage the infrastructure for private and public sectors to scale seed production and distribution to provide access to farmers.

9.8.2 Effective Seed Marketing for Biofortified Pearl Millet: A Strategic Approach

Mainstreaming biofortified pearl millet hinges on an effective seed marketing strategy that showcases not only high-yielding, climate-resilient agronomic traits but also the nutritional advantages of iron- and zinc-rich varieties. Unlike traditional seed sales, this requires engaging stakeholders across the entire seed value chain—farmers, agri-retailers, distributors, SHGs, and public institutions—through a mix of pre-sowing and post-marketing strategies.

Pre-sowing marketing strategies focus on awareness, positioning, and first-use incentives. Engaging farmers to communicate the key value proposition, such as agronomic features—“high yielding,” “early maturing,” and “drought-tolerant.” Developing nutrition messaging, branding, and packaging that labels a compelling message about the product—iron-rich pearl millet, health benefits to their family. Mini-kit distribution for farmer trials, exclusive retailer promotions, farmer engagement programs demonstrations, and positioning within government subsidy schemes.

Post-sowing marketing strategies are built on the experiences of the grower to influence and deepen market penetration. These include field demonstrations, farmer feedback from progressive farmers, and showcasing performance data on yield. Consumers can be engaged through recipe demonstrations, food festivals, school meals, and community nutrition days. Repeat purchase incentives, loyalty schemes for retailers, and storytelling via local champion farmers and farm advocates can help build a sustainable market. A critical pillar to consider during marketing planning is training and capacity building. This includes equipping agri-retailers, last-mile distributors, SHGs, and FPOs with the knowledge and tools to promote biofortified seeds. Training needs to cover seed handling, nutritional messaging, demo plot management, pricing strategies, and digital sales tools. Such capacity-building ensures consistency, credibility, and quality assurance throughout the supply chain. Combining good quality seeds, marketing strategies, and systematic stakeholder engagement can help to mainstream biofortified pearl millet. Biofortification as an approach can be a market-driven solution that delivers on both nutrition and livelihood outcomes.

9.8.3 Strategic Priorities for Scalable Seed Systems for Biofortified Pearl Millet

To ensure the widespread adoption and long-term impact of biofortified pearl millet, seed systems must be designed to be scalable, farmer-centric, and nutrition-sensitive. A foundational priority is investing in research and development (R&D) to breed farmer-preferred varieties that combine high iron and zinc content with essential agronomic traits such as early maturity, high grain yield, drought tolerance, disease resistance, and consumer-desired cooking quality. These multi-trait cultivars will enhance marketability and field performance, supporting both nutrition and livelihood goals.

Public–private partnerships (PPPs) must be established to scale seed production, efficient last-mile delivery, and support co-branding and marketing campaigns. Such partnerships ensure that seed innovations move beyond research plots into farmers' fields through trusted, efficient supply channels. In parallel, community seed banks managed by self-help groups (SHGs) and farmer-producer organizations (FPOs) should be strengthened. These local institutions can be trained in seed saving, bulking, and quality assurance—building community ownership and enhancing resilience in remote areas that are not served by seed companies.

Maintenance of genetic purity and varietal integrity across the seed value chain is critical for success. This demands robust quality control (QC) and quality assurance (QA) systems—including field inspections, germination testing, moisture control, and labeling protocols, micronutrient testing, as per national seed certification standards. Good quality seed is the foundation of farmer trust and credibility. Institutionalizing biofortified trait standards in seed release policies is essential. Nutritional criteria should be embedded into varietal release protocols, public procurement standards (e.g., PDS, ICDS, mid-day meals). This policy alignment will anchor biofortified seeds and help in the mainstream, enabling sustainability.

Fig. 9.2 Dhanashakti—World First Iron Rich Pearl Millet Variety

9.9 Case Study: Scaling Biofortified Pearl Millet in India—A Seed-to-Market Strategy by HarvestPlus

HarvestPlus collaborated with Nirmal Seed in 2013 to commercialize Dhanashakti (Fig. 9.2) with the first biofortified pearl millet variety. Nirmal Seeds is a vertically integrated seed company with strong R&D, large-scale seed production capacity, and a national distribution network. Known for its diverse product portfolio across cereals, field crops, and vegetables, Nirmal marketed *Dhanashakti* primarily in 1 kg and 3 kg packs, suitable for both smallholders and larger farms. HarvestPlus market research revealed a strong positive response from end users. Over 73 per cent of farmers who planted *Dhanashakti* were willing to replant it, citing deeper grain colour, nutty taste, and family health benefits—particularly for children. Sensory studies confirmed high acceptability of products like rotis, porridges, and snacks in both rural and peri-urban households (Birol et al. 2019). To reduce the initial business risk for Nirmal Seed, the program introduced de-risking support mechanisms such as early-stage seed production, co-branded awareness campaigns, joint seed forecasting, joint training, and capacity building for farmers' engagement to create awareness. HarvestPlus field teams provided on-ground support to build confidence among stakeholders unfamiliar with biofortification as a trait.

In India, private seed companies are allowed to market their own research varieties as "truthfully labeled seeds" (TLS)—a regulatory provision under the Seed Act. This allows companies like Nirmal to introduce new varieties directly into the market without undergoing the lengthy national varietal release process. By leveraging TLS, the best commercial geographies were targeted for the test performance, gather rapid farmer feedback, and reduce the time-to-market—a crucial factor in scaling innovations like *Dhanashakti*. Recognizing the trust farmers place in agri-retailers and input dealers, HarvestPlus prioritized point-of-sale engagement, farmers's engagement and mobile van campaigns in remote areas, to create awareness and educate about the benefits of biofortified variety. Retailers and distributors were

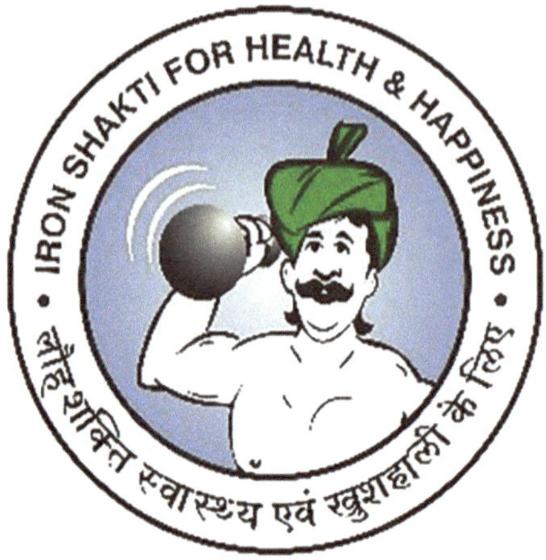

Fig. 9.3 Unique logo that endorses iron-rich pearl millet

1:2

Fig. 9.4 Design of the marketing collateral—Hoarding

trained to explain the seed's nutritional value and benefits, field performance, supported by marketing materials such as a unique logo to endorse iron-rich pearl millet (Fig. 9.3), Hoarding (Fig. 9.4), Dealer gate (Fig. 9.5), mobile marketing campaign on a van to reach farmers in remote areas (Fig. 9.6), seed package design with logos (Fig. 9.7), shop displays, and demonstration videos. To drive early adoption, the program provided incentives to retailers and promotional kits in high-potential districts.

In India, private seed companies are allowed to market their research varieties as "truthfully labeled seeds" (TLS)—a regulatory provision under the Seed Act. This allows companies like Nirmal to introduce new varieties directly into the market without undergoing the lengthy national varietal release process. By leveraging

Fig. 9.5 Design of the marketing collateral—Dealer Gate

TLS, companies can test performance in their best commercial geographies, gather rapid farmer feedback, and accelerate time-to-market—a crucial factor in scaling innovations like *Dhanashakti*.

Retailers reported high levels of repeat purchase, driven by both farmer satisfaction and positive consumer word-of-mouth. The campaign successfully positioned *Dhanashakti* as a "grain of strength," proving that biofortified crops can thrive commercially when supported by trusted seed systems, targeted branding, and local market champions. After the official release of Dhanashakti variety in 2014, to scale

Fig. 9.6 Design of the marketing campaigns on the van to reach farmers in remote areas

Fig. 9.7 Design of the seed pack design with logo endorsing iron–rich pearl millet

its adoption in Maharashtra and Karnataka, HarvestPlus partnered with Mahabeej (Maharashtra State Seed Corporation) and the Karnataka State Seed Corporation. Under the broader Alliance established by HarvestPlus, similar models were implemented to scale biofortified hybrids. HarvestPlus continues to play a catalytic role in mainstreaming biofortified pearl millet through strategic partnerships. As a result of this integrated strategy, under the partnership between 2014 and 2024, over 4.1 million kilograms of biofortified pearl millet seed were produced, enabling cultivation across 1.9 million acres and benefitting 1.8 million farming households.

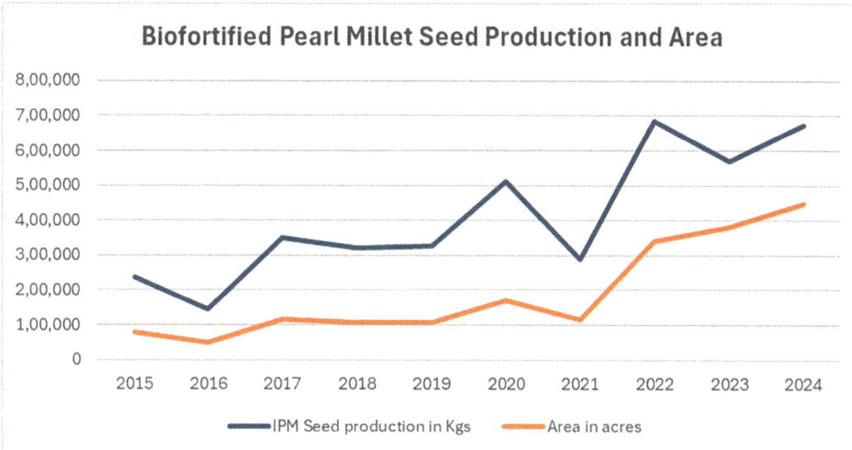

Fig. 9.8 Biofortified pearl millet–Seed production (in kg) and area (in acres) from 2015 to 2024

Fig. 9.8 shows the annual seed production (in kg) and area (in acres) of biofortified pearl millet cultivars from 2015 to 2024. This case provides valuable insights for scaling other biofortified crops. It highlights the importance of building variety, identity, and visibility at the seed level, empowering retailers as nutrition messengers, and de-risking early market entry through shared incentives and training. Ultimately, *Dhanashakti* demonstrates how nutrition and agribusiness goals can align, especially when innovation is embedded in multi-stakeholder partnerships spanning seed development to last-mile delivery.

9.10 Conclusion: Toward a Millet-Forward Future

We are confronted by the reality of malnutrition, climate volatility, and rural livelihood distress, and biofortified pearl millets offer solutions. Pearl millets are undergoing a period of renaissance from an "orphan crop" to "nutri-cereals" recognized for their climate resilience, rich nutritional profile, to strengthen the food systems in the dryland and semi-arid regions of South Asia and Africa. Biofortified pearl millet as an innovation stands out to address the iron deficiencies. Biofortification is recognized for its scientific credibility and is valued for its cost-effectiveness, sustainability, and benefit to rural communities directly. Promoting biofortified pearl millet is strengthening the food systems for the future. Strategic marketing, robust seed systems, and establishing inclusive value chains, biofortified pearl millet can enable transformational benefits such as improved nutrition outcomes and diversified diets for farmers. Consumer awareness campaigns and behavioral nudges are needed to reposition millets as aspirational, healthy, and climate-smart staples. At the same time, governments, donors, and philanthropic partners must help build capacity at the last mile. Leveraging Government Institutional food programs offers a powerful lever to create demand and scale of biofortified pearl millet and to target. This will require deliberate and coordinated action. Supportive public policies—such as micronutrient standards, targeted procurement, and crop incentives—must align with private sector investments in seed multiplication, value-added product development, and distribution networks. Biofortified millets offer a powerful intersection of nutrition-sensitive agriculture and climate-smart development. They can serve as cornerstone crops for achieving multiple Sustainable Development Goals—especially Zero Hunger (SDG 2), Good Health and Wellbeing (SDG 3), Climate Action (SDG 13), and Reduced Inequality (SDG 10). It will require collaboration across a wide spectrum of actors—governments, researchers, agri-businesses, farmers, development agencies, and empowered consumers. Only through such partnerships can we scale biofortified pearl millet from pilot projects to national programs, from niche markets to mainstream consumption. In doing so, we will build food systems that are resilient, equitable, and future-ready.

References

AGRA (2020) Africa agriculture status report: feeding Africa's cities: opportunities, challenges, and policies for linking African farmers with growing urban food markets. Alliance for a Green Revolution in Africa

Banerji A, Birol E, Karandikar B, Roy D (2016) Information, branding, certification, and consumer willingness to pay for high-iron pearl millet: evidence from experimental auctions in Maharashtra, India. Food Policy 62:133–141

Bioversity International (2018) Community seed banks: origins, evolution and prospects. Bioversity International

Birol E, Roy D, Torheim LE, Fatima F (2019) Developing country consumers' demand for nutritious foods: are biofortified staple crops a viable option? Glob Food Sec 23:100350

Data Bridget Market Report (2024) India health and wellness food market—industry trends, forecast, and competitive analysis. Data Bridget Market Research

Directorate of Millets Development (2019) Status paper on millets. Ministry of Agriculture & Farmers Welfare, Government of India

FAO (2021) The state of food and agriculture: making Agrifood systems more resilient to shocks and stresses. Food and Agriculture Organization of the United Nations

FAO (2023) FAOSTAT: crops and livestock products. Food and Agriculture Organization of the United Nations. https://www.fao.org/faostat/en/

FICCI (2022) Millets market in India: trends and opportunities. Federation of Indian Chambers of Commerce & Industry

Finkelstein JL, Mehta S, Udipi SA, Ghugre PS, Luna SV, Wenger MJ et al (2015) A randomized trial of iron-biofortified pearl millet in school children in India. Am J Clin Nutr 101(4):991–996

GEF (2006) Land degradation as a global environmental issue: a synthesis of three studies commissioned by the Global Environment Facility to strengthen the knowledge base to support the land degradation focal area. Global Environment Facility

ICMR (2023) India: health of the nation's states—the India state-level disease burden initiative. Indian Council of Medical Research

ICRISAT (2022) Annual Report 2022: science and innovation for a food secure future. International Crops Research Institute for the Semi-Arid Tropics

IFAD & ICRISAT (2022) Enhancing the livelihoods of smallholder farmers through climate-resilient agriculture. International Fund for Agricultural Development & ICRISAT

IFPRI (2021) Standards and certification for biofortified crops: a global scan. International Food Policy Research Institute

Kodkany BS, Bellad RM, Mahantshetti NS, Westcott JE, Krebs NF, Kemp JF et al (2013) Biofortification of pearl millet with iron and zinc in India: a randomized controlled trial. Nutr Rev 71(1):110–116

Millennium Ecosystem Assessment (2005) Ecosystems and human well-being: desertification synthesis. World Resources Institute

Mula MG, Gupta SK, Rai KN (2014) Adoption and impact of improved cultivars of pearl millet in India. International Crops Research Institute for the Semi-Arid Tropics (ICRISAT). https://www.icrisat.org/adoption-impact-pearl-millet-india.pdf

Muskan (2025) Urban retail trends and millet-based product innovation in West Africa

NRC (1996) Lost crops of Africa: volume I—grains. National Research Council. National Academies Press

Reuters (2024) India consumer spending trends rise with urbanization. Reuters News Service

Satyavathi CT, Khandelwal V, Bhattacharjee R (2021) Pearl millet: genetic improvement for productivity and nutritional quality. In: Gupta MK (ed) Millets and pseudo cereals. Springer, pp 103–123

Smart Food Initiative (2022) Smart food for smart farming: market insights report. ICRISAT and the Smart Food Initiative. 5. Top of Form

UNICEF – State of the World's Children (2023) https://www.unicef.org/reports/state-of-worlds-children-2023

WHO—Micronutrient Deficiencies (n.d.) https://www.who.int/health-topics/micronutrients

WHO—Anaemia Fact Sheet (2023) https://www.who.int/news-room/fact-sheets/detail/anaemia

Chapter 10
Biofortification of Sorghum Quality Traits Through Agronomic Manipulation

M. Joseph, M. Hemalatha, N. Vinitha, G. Guru, and Mahalingam Govindaraj

Abstract Sorghum (*Sorghum bicolor* L.) is a nutri- cereal crop, widely cultivated for its resilience to harsh environmental conditions and its role in ensuring global nutritional security. However, improving its productivity and quality parameters such as grain yield, protein content, digestibility, and micronutrient composition is essential to maximize its nutritional benefits. This chapter explores agronomic strategies to enhance sorghum quality through optimized crop management practices, soil fertility enhancement, precision agriculture, and stress mitigation techniques. Key interventions include balanced fertilization, biofortification, water-efficient irrigation methods, and the adoption of improved cultivars with superior nutritional traits. Additionally, integrated pest and disease management, along with sustainable farming practices such as intercropping and conservation tillage, contribute to better grain quality and higher productivity. Leveraging modern balanced agronomic advancements can significantly enhance sorghum grain nutrition in addressing malnutrition and food security challenges, particularly in arid and semi-arid regions. Future research should focus on refining these agronomic strategies to achieve both high yields and improved nutritional profiles in sorghum production, particularly in drylands.

Keywords Agronomic manipulation · Biofortification · Food security · Quality traits · Sorghum

---

M. Joseph · G. Guru
Agricultural Research Station (AICRPDA Main Centre), Tamil Nadu Agricultural University, Kovilpatti, TN, India

M. Hemalatha (✉) · N. Vinitha
VOC Agricultural College and Research Institute, Tamil Nadu Agricultural University, Killikulam, Vallanadu, Tuticorin, TN, USA
e-mail: hemalatha.m@tnau.ac.in

M. Govindaraj
Alliance of Bioversity International and the International Centre for Tropical Agriculture (CIAT), Cali, Colombia

M. Govindaraj (ed.), *Breeding Biofortified Millets for Nutritional Security*, Sustainability Sciences in Asia and Africa,
https://doi.org/10.1007/978-981-95-9827-4_10

10.1 Introduction

Malnutrition results from an unbalanced diet, while undernourishment refers to inadequate nutrient intake, leading to insufficient energy and impaired body tissue repair. In India, malnutrition and undernourishment are major concerns, contributing to nearly 45% of deaths among children under the age of 5. The issue is especially severe in Southeast Asia, where 30.1% of children under five experience stunting and 14.5% suffer from wasting, significantly higher than the global averages of 22% and 6.7%, respectively (WHO 2022). Globally, over 2 billion people, approximately one in three, are affected by micronutrient deficiencies, often referred to as "hidden hunger" (Prom-U-Thai et al. 2020). These deficiencies are widespread, even in highly developed countries, and are particularly common among children, pregnant and lactating women, athletes, and manual laborers. Zinc (Zn), selenium (Se), and iron (Fe) are the most associated micronutrients with this form of malnutrition worldwide.

With a rapidly growing population and shifting dietary patterns, food demand is rising significantly, while agricultural land remains limited. This has led to the intensive use of natural resources and often lower-quality produce. Implementing agricultural strategies to enhance the nutritional value of edible and forage crops for livestock could play a crucial role in improving human nutrition, particularly in developing countries. Biofortification is an effective strategy to reduce micronutrient deficiencies by enhancing the nutritional value of edible crops through selective breeding, mineral fertilization, or advanced transgenic techniques. Alongside improving nutrient content, biofortification can also enhance crop quality, reduce post-harvest losses, improve taste, shorten cooking time, increase yield, and boost pest resistance (Fig. 10.1). This approach is part of OneHealth that primarily benefits rural populations who rely heavily on staple crops for their nutrition. Since rural communities have limited access to processed fortified foods or supplements, biofortification helps address nutrient deficiencies in their daily diets.

10.2 Area, Production, and Productivity of Sorghum

The United States is the world's largest sorghum producer, followed by India, Ethiopia, Nigeria, and Australia. Global sorghum production reaches approximately 63 million metric tons annually, covering around 44 million hectares, with an average yield of 4.3 tons per hectare. Africa leads in sorghum production, followed by the Americas. About 47.5% of global output is used for food, while 34.7% is allocated for animal feed. Developing countries account for approximately 90% of the global sorghum cultivation area and 70% of total production, with Africa and Asia each contributing around 25–30% of global output. India ranks fourth in global sorghum production, producing 4.4 million metric tons over 3.97 million hectares in 2023–24, with the majority cultivated during the rabi season. Kharif sorghum

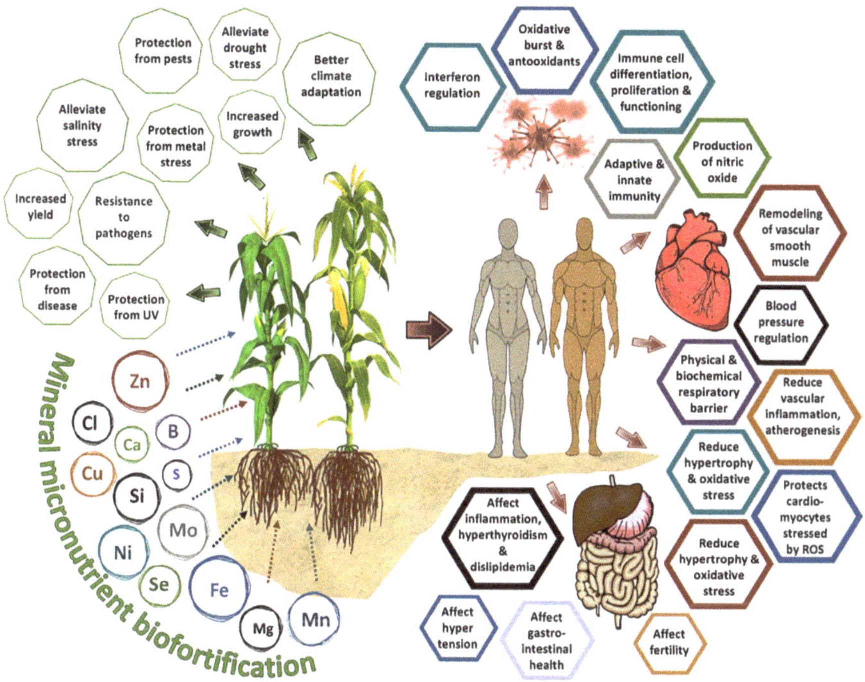

Fig. 10.1 Impact of mineral micronutrient biofortification on plant-physiological processes and its connection to human health and immunity. (Courtesy: Bhardwaj et al. (2022))

(36.6%) is primarily grown in Rajasthan (43.3%), Uttar Pradesh (15.6%), Haryana (10.1%), and Madhya Pradesh (9.8%), whereas rabi sorghum (63.4%) is concentrated in Maharashtra (63.5%), Karnataka (22%), Tamil Nadu (7.9%), and Andhra Pradesh (3.2%). Maharashtra and Karnataka are the leading sorghum-producing states in India. India's sorghum yield, at 840 kg per hectare, is lower than the global average of 1435 kg per hectare. According to the Government of India's third advance estimates for 2023–24, sorghum production was projected at 47.42 lakh tonnes, an increase from 38.14 lakh tonnes in 2022–23.

10.3 Sorghum: A Key Crop for Nutritional Security

Sorghum (*Sorghum bicolor* L. Moench) is known as "Jowar" in North India and "Cholam" in other regions. It is a carbohydrate-rich staple crop ranked as the world's fifth most important cereal after wheat, rice, maize, and barley (Karthikeyan et al. 2017). Its high-yielding potential and adaptability to diverse climates, along with strong heat resistance, have earned it the title "Camel crop." Predominantly grown in tropical and subtropical regions, particularly in Africa and Asia. Before the Green Revolution, sorghum was India's second-most cultivated grain, though it has since dropped to third place in area and production (Jain and Patel 2012). However, its importance is resurging due to its coarse texture, distinct grain flavor, and dual use for grain and fodder (Rao et al. 2015). Like rice, sorghum is consumed in various forms, including cooked food, malt, and flour for dosai, chapati, and roti. It can also be popped, while its fodder remains succulent, palatable, and free from toxins (Ayub et al. 2010). Sorghum is increasingly valued for its nutritional benefits, containing dietary fiber (7.6–9.2%), starch (72.6%), protein (10–12%), minerals (1.6%), and fat (1.9%). It is also rich in essential amino acids like lysine, thiamine, and folic acid. Approximately 19.5% of the global population and 27% of India's population rely on sorghum as a staple (Shinde et al. 2018). Known as the "King of Millets," it plays a crucial role in food security, particularly in semi-arid regions, due to its resilience to low inputs and minimal rainfall (Hossain et al. 2016). Packed with fiber, antioxidants, and essential nutrients, sorghum offers numerous health benefits, making it a key component in combating malnutrition and hunger.

10.4 Types of Sorghum

Sorghum exists in several varieties, each serving distinct purposes. Grain sorghum, a type of grass, is used as livestock feed and processed into flour for human consumption. It comes in shades of white, tan, orange, red, bronze, and black. The red, orange, and bronze varieties are highly versatile, used for applications ranging from animal feed to biofuel. Tan, cream, and white sorghum are primarily milled into flour for the food industry. Meanwhile, burgundy and black sorghum are notable for their high levels of antioxidants.

10.5 Nutritional Profile of Sorghum

Whole sorghum has a nutritional profile and chemical composition comparable to rice, wheat and maize. Its energy content ranges from 296.1 to 356.0 kcal per 100 grams, with its primary components being polysaccharides (both starch and non-starch), proteins, and lipids. According to Mkandawire et al. (2013), sorghum has the lowest starch digestibility among carbohydrate-rich crops due to the strong interactions between starch granules, proteins and tannins (Table 10.1).

10.6 Sorghum Health Benefits

Sorghum is packed with essential nutrients, including B vitamins, which support metabolism, nerve cell growth, and the health of hair and skin. It is also an excellent source of magnesium, a vital mineral that aids bone formation, heart function, and

Table 10.1 Nutritional status of sorghum

Components	Range	Components	Range
Protein (%)	4.40–21.10	Tannin (%)	0.1–7.22
Water-soluble protein (%)	0.30–0.90	Phytic acid (mg/100 g)	875.00–2211.90
Lysine (%)	1.06–3.64	Energy value (kcal/100 g)	296.1–356.0
Starch (%)	55.60–75.20	Non-starch polysaccharides (g/100 g)	6.0–15.0
Amylose (%)	21.20–30.20	Insoluble fibers (%)	75.0–90.0
Soluble sugar (%)	0.70–4.20	Soluble fibers (%)	10.0–25.0
Reducing sugar (%)	0.05–0.53	α-kafirins (%)	66–84
Crude fiber (%)	1.00–3.40	β-kafirins (%)	8–13
Fat (%)	2.10–7.60	γ-kafirins (%)	9–21
Ash (%)	1.30–3.30	Reduced lipid (g/100 g)	1.24–3.07
Calcium (mg/100 g)	11.00–586.00	Unsaturated fatty acids (%)	83–88
Phosphorus (mg/100 g)	167.00–751.00	Linoleic acid (%)	45.6–51.1
Iron (mg/100 g)	0.90–20.00	Oleic acid (%)	32.2–42.0
Thiamine (mg/100 g)	0.24–0.54	Palmitic acid (%)	12.4–16.0
Niacin (mg/100 g)	2.90–6.40	Palmitic acid (%)	12.4–16.0
Riboflavin (mg/100 g)	0.10–0.20	Linolenic acid (%)	1.4–2.8

Source: Hossain et al. (2022)

more than 600 biochemical processes, such as energy production and protein metabolism. Additionally, sorghum is rich in antioxidants like flavonoids, phenolic acids, and tannins, which help reduce oxidative stress and inflammation. Half a cup of sorghum contains over 7 grams of fiber, around 25% of the recommended daily intake. A high-fiber diet supports weight management, lowers cholesterol, regulates blood sugar, and prevents constipation. With the increasing demand for gluten-free products, sorghum serves as a nutritious substitute for wheat and other gluten-containing grains. It can be used in baking, as a flour replacement in bread and desserts, or as a hearty side dish.

10.7 Production Technologies for Sorghum

Several methods are available to assess yield potential, though each has its limitations. The most reliable approach is to analyze long-term yield data specific to the producer, as it accounts for the natural productivity of the environment, the impact of agronomic practices like soil preparation, plant spacing, fertilization, weed and pest control, and the producer's management skills.

10.7.1 Soil Requirements

Sorghum is primarily cultivated in low-quality, shallow soils with high clay content, which are generally unsuitable for maize farming. It tends to struggle in sandy soil unless there is a heavy-textured subsoil. Compared to other grain crops, sorghum is more tolerant of alkaline salts and can thrive in soil with a pH ranging from 5.5 to 8.5. It also withstands short periods of waterlogging better than maize. The ideal soil for sorghum cultivation contains between 10% and 30% clay.

10.7.2 Climatic Requirements

The climatic conditions necessary for sorghum production are categorized into temperature and rainfall requirements.

10.7.2.1 Temperature

Sorghum thrives in warm climates, requiring high temperatures for germination and growth. Germination begins at 7–10 °C, with optimal rates at 15 °C, while the ideal soil temperature for planting is at least 15 °C at a 10 cm depth. Post-germination, growth is best between 27 and 30 °C, though it remains unaffected down to 21 °C, whereas extreme heat can reduce yield. Rising temperatures delay flowering and primordia development. Plants with 4–6 leaves exposed to temperatures below 18 °C may develop lateral shoots, while those with 8+ leaves maintain apical dominance. Freezing temperatures can be fatal, with young plants surviving brief exposure to −5 °C but dying at −7 °C, while older plants are vulnerable at 0 °C.

10.7.2.2 Rainfall Requirements

Sorghum typically needs 400 to 600 mm (16 to 24 inches) of rainfall during its growing cycle. However, it can still grow in drier regions if irrigation is available. Rainfall should ideally be evenly distributed throughout the growing season. Even with low rainfall, sorghum can still produce a harvest, though yields may be lower. Moderate rainfall or irrigation is necessary for seedling establishment and early growth, whereas moisture is crucial during flowering and grain filling. A dry spell during these stages can severely affect yield. Excessive rain or waterlogging can promote diseases such as root rot and downy mildew, making proper drainage essential in high-rainfall areas.

10.7.3 Field Preparation

The field should be prepared well in advance, making use of early rainfall. Farmyard manure (FYM) at a rate of 12.5 t/ha should be applied during the final ploughing.

1. The soil should be chiseled using a hand pan or shallow plough once every 3 years.
2. Land-shaping techniques should be chosen based on rainfall and soil type to conserve moisture.

In black soil regions with high rainfall, broad bed and furrow methods should be used, whereas compartmental bunding can be implemented under areas with low rainfall, or seeds can be sown on a flat bed, followed by furrow formation between crops during inter-cultivation.

10.7.4 Seed Rate

- Sorghum can be cultivated under both rainfed and irrigated conditions, either by direct sowing or by transplanting. The recommended seed rate for irrigated sorghum is 7.5 kg ha^{-1} for transplanted crops and 10 kg ha^{-1} for direct sowing, whereas under rainfed conditions, a seed rate of 15 kg ha^{-1} is recommended for direct sowing (Fig. 10.2). Transplanting offers several benefits:

  (a) It shortens the crop's duration in the main field by 10 days.
  (b) Shoot fly can be effectively and economically managed in the nursery.
  (c) A healthy and optimal plant population is ensured, as only strong seedlings are transplanted.
  (d) The seed rate requirement is reduced by 2.5 kg/ha.

10.7.5 Nursery Practices for Transplanted Sorghum

Nursery preparation Choose a 300 m^2 (7.5 cents) area near a water source that has good drainage to prevent water stagnation.

Application of FYM to the Nursery Apply 750 kg of FYM or compost initially, followed by an additional 500 kg of compost or FYM to cover the seeds after sowing.

Laying the Nursery Provide three distinct, 2 m × 1.5 m irrigation units with a 30-cm gap between each plot and surrounding the unit. Dig out soil from these gaps and the surrounding area to a depth of 15 cm to create channels, then distribute the excavated soil evenly over the beds and level them. Water the beds immediately after sowing, followed by a second irrigation on the third day. Seedlings should be 17–18 days old at the time of planting.

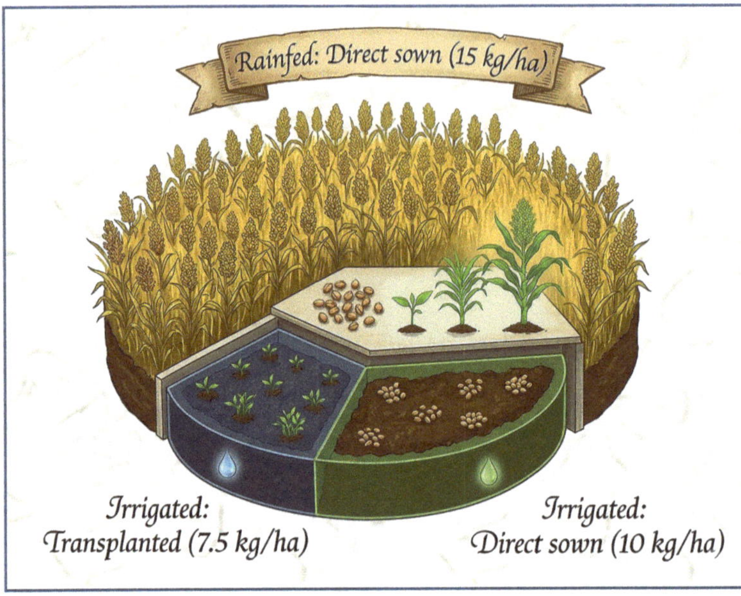

Fig. 10.2 Seed rate at various sowing methods

Seed Treatment Soak the seeds 24 hours before planting with Carbendazim, Captan, or Thiram at a rate of 2 g per kg of seed. Additionally, coat the seeds with three packets (600 g/ha) of Azospirillum and three packets (600 g) of phosphobacteria, or alternatively, six packets (1200 g) of Azophos, using rice gruel as a binding agent.

Sowing and Covering the Seeds Create shallow grooves no deeper than 1 cm on the bed by running your fingers vertically over it. Evenly distribute 7.5 kg of treated seeds across the beds. Gently cover the seeds by smoothing out the grooves with your hand.

Water Management Adjust the irrigation frequency based on soil type: for red soils, water every 4–5 days, while for black soils, maintain a 5–6-day interval.

10.7.6 In a Rainfed Situation

10.7.6.1 Seed Hardening

Seed hardening enhances germination by pre-soaking seeds in a 2% potassium dihydrogen phosphate solution or 500 ppm CCC (Cycocel) (1 ml per liter of water) for 6 hours, followed by shade drying for 5 hours and restoring the original moisture content before sowing.

Alternatively: (i) Seeds were soaked for 16 hours in a 1% (w/v) aqueous fresh leaf extract prepared from Prosopis juliflora and Pongamia pinnata in a 1:1 ratio, using a seed-to-solution ratio of 1:0.6 (w/v). Subsequently, the seeds were dried and coated with Pongamia leaf powder at 300 g kg^{-1} using gruel as an adhesive. (ii) Halogenation involved treating the seeds with a mixture of CaOCl, $CaCO_3$, and Arappu leaf powder in a 5:4:1 ratio, or with an iodine-based formulation (2 mg iodine mixed with 3 g $CaCO_3$) applied at 3 g kg^{-1}. The treated seeds were then stored in a poly-lined cloth bag, which maintained seed viability for more than 10 months. (iii) Seeds are treated with three packets (600 g) each of Azospirillum and phosphobacteria, or with six packets (1200 g ha^{-1}) of Azophos. (iv) Seed pelleting involves coating the seeds with 15 g of chlorpyrifos using 150 ml of gum as an adhesive, followed by shade drying.

10.7.6.2 Nutripriming

Nutripriming, or seed priming, involves soaking seeds in a nutrient-rich solution before planting (Farooq et al. 2019; Raj and Raj 2019). This technique is primarily used to improve germination, root development, seedling establishment, and yield, with some studies also reporting enhanced grain nutrient content. Micronutrient seed priming is both environmentally friendly and effective in boosting crop yield and nutrient levels. Nano-priming, a more advanced method, offers greater efficiency by enhancing surface response and electron exchange in plant cells. It also promotes hydroxyl radical generation, facilitating starch breakdown, antioxidant activity, and nanopore formation in shoots for better water absorption (Nile et al. 2022). However, nutripriming techniques remain relatively unfamiliar to farmers due to their technical complexity.

10.7.7 Main Field Preparation for Both Direct and Transplanted Crops

Plough the field once or twice with an iron plough; a finely tilled seedbed is unnecessary and may hinder germination and yield in direct-sown sorghum. In Alfisols (deep red soils) with a subsoil hard pan, chiseling at 0.5 m intervals to a depth of 40 cm in both directions improves yield. Incorporate 12.5 t/ha of farmyard manure (FYM) or compost into the unploughed field, along with 2 kg each of Azospirillum, Phosphobacteria, or Azophos. Applying 5 t/ha of well-decomposed poultry manure further boosts grain yield and soil health. For rainfed sorghum, prepare the field early to utilize initial rainfall. In hardpan soils, chisel ploughing at 0.5 m intervals every 3 years improves productivity. Combining chiseling with 12.5 t/ha of FYM or composted coir pith can enhance yields by 30%. A field experiment on the sorghum variety K 12, conducted under the soil moisture conservation method of ridges and furrows, recorded the highest average yield of 1393 kg/ha, a benefit–cost ratio of 1.18, and a rainwater use efficiency (RWUE) of 4.9 kg/ha-mm. In comparison, compartmental bunding resulted in a lower yield of 1116 kg/ha, a benefit–cost ratio of 1.08, and a slightly higher RWUE of 4.95 kg/ha-mm (Annual Report of ARS, Kovilpatti 2023).

10.7.8 Soil-Less Cultivation

Soil-less cultivation is a modern crop-growing technique that relies on inert organic, inorganic, or liquid media combined with precisely formulated nutrients. This method is becoming increasingly significant in human nutrition, especially for microgreens, by enhancing them with micronutrient-rich growing media. By carefully regulating environmental factors such as temperature, light, and nutrient levels, plant growth is maximized, ensuring optimal nutrient absorption through continuous root contact. The constant exposure to fertilizer solutions improves nutrient uptake, ultimately enhancing the crops' nutritional value. Furthermore, soilless farming allows for year-round cultivation. Despite its efficiency and environmental advantages, high infrastructure costs may restrict its widespread adoption in certain areas.

10.7.9 Sowing

Sow the seeds at a depth of 5 cm just before the onset of the monsoon, using a seed drill, seed-cum-fertilizer drill, or traditional plough. Prior to sowing, coat the seeds with a mixture of 15 g chlorpyrifos and 150 ml gum, and allow them to shade dry. The major cultivated varieties and hybrids in India are listed in Table 10.2.

Pre-monsoon Sowing/Dry Seeding Sowing should be done one to two weeks before the onset of the monsoon. For instance, the optimal sowing period for PMS in different districts of Tamil Nadu is as follows:

Coimbatore: 37th–38th week (second to third week of September)
Erode: 38th week (third week of September)
Sivaganga and Ramanathapuram: 40th week (first week of October)
Thoothukudi and Tirunelveli: 39th–40th week (last week of Sept to the first week of Oct)

10.7.10 Thinning

This is a crucial practice in sorghum cultivation. Thinning should be carried out 10 to 15 days after emergence, ensuring that only one plant remains per hill.

Table 10.2 Important varieties and hybrids in India

Sl. No	State name	Season	Popular hybrids/varieties
1	Haryana	Kharif	CSH-16
2	Rajasthan	Kharif	CSH-13, CSH-16, CSH-17, CSH-18, MLSH-296, ICI-501, SPH-837, CSH-23
3	Uttar Pradesh	Kharif	CSH-13, CSH-16, CSH-18, CSH-23
4	Bihar	Kharif	CSH-16MP Chari, Pusa Chari-1, CSH-1, CSH-2, CSH-3, CSH-5, Swarna hybrid-, BR-1, BR-2
5	Madhya Pradesh	Kharif	CSH-13, CSH-16, ICI-501, CSH-17, CSH-18, CSH-21, CSH-22 SS, CSH-23JK-22, JK-741, JK-938, JK-1041, SPV-15, CSV-15, SPV-1022, CSH-5, CSH-16, CSH-9, CSH-14
6	Gujarat	Rabi	CSH-13 R, CSH-19 R GJ-37, GJ-39, GJ-40, GJ-41, JF-4, CSH-5, CSH-6
7	Gujarat	Sweet sorghum	GJ-37, GJ-39, GJ-40, GJ-41, JF-4, CSH-5, CSH-6
8	Maharashtra	Rabi	CSH-15 R, CSH-19 R, Parbhani Shakti (ICSR 14001)
9	Andhra Pradesh	Kharif	PSV-1, CSV-14 R, PJ-890, SJ-092, 122, 2169, Palem-2, N-13, N-14, APS-1, ICMV-221, SPV-462, CSH-5, CSH-6, CSH-9, CSH-16; ASH-1, CSH-13, MLSH-296, CSH-16, JKSH-22, PSH-1, CSH-18, ASH-1, CSH-21, CSH-23
10	Andhra Pradesh	Rabi	CSH-12 R, CSH-13 R, CSH-15 R, CSH-19 RPSV-1, CSV-14 R, PJ-890, SJ-092, 122, 2169, Palem-2, N-13, N-14, APS-1, ICMV-221, SPV-462, CSH-5, CSH-6, CSH-9, CSH-16; ASH-1, ICMH-451, ICTP-8203
11	Karnataka	Kharif	CSH-13, MLSH-296, CSH-216, CSH-21, CSH-23
12	Karnataka	Rabi	CSH-15 R, CSH-19-R, DSH-4 R
13	Tamil Nadu	Kharif	CSH-13, MLSH-296, CSH-16, ICI-501, CSH-17CO(S) -28, CO(S)-26, APK-1, Payur-1, Payur-2, CO-10, CO-18, CO-19, CO-21, CO-25, CO-26, K-4, K-5, K-6, K-7, K-8, K-9, K-10, K-11, K Tall, BSR-1, IS 3541, CSH-1, CSH-5, CSH-6, COH-3, COH-4
14	Tamil Nadu	Rabi	CSH-19 R CO(S)-28, CO(S)-26, APK-1, Payur-1, Payur-2, CO-10, CO-18, CO-19, CO-21, CO-25, CO-26, K-4, K-5, K-6, K-7, K-8, K-9, K-10, K-11, K Tall, BSR-1, IS 3541, CSH-1, CSH-5, CSH-6, COH-3, COH-4
15	Tamil Nadu	Summer	CO(S)-28, CO(S)-26, APK-1, Payur-1, Payur-2, CO-10, CO-18, CO-19, CO-21, CO-25, CO-26, K-4, K-5, K-6, K-7, K-8, K-9, K-10, K-11, K Tall, BSR-1, IS 3541, CSH-1, CSH-5, CSH-6, COH-3, COH-4

Source: https://nutricereals.dac.gov.in/sorghum.aspx

10.7.11 Maintenance of Plant Geometry for Sorghum

Plant density is a crucial factor in crop production, directly impacting grain yield and key agronomic traits. Maintaining optimal density is challenging, as low densities promote weed infestation, while excessive density causes lodging, reduced light penetration, and lower yields (Lemerle et al. 2004, 2006).

Studies show that higher plant densities often reduce individual plant productivity but can enhance dry matter production in crops like sorghum due to improved light interception. A 2023 experiment assessing row spacings of 25, 45, and 76 cm with seeding rates of 210,000 to 300,000 plants per hectare found significant variations in moisture content, protein levels, plant height, and yield. The highest protein content (12.61%) and yield (10.59 t/ha) were observed at 210,000 plants per hectare with the narrowest 25 cm row spacing (Szemerits et al. 2024).

10.7.12 Nutrient Management for Sorghum

Effective nutrient management in sorghum is essential for maximizing yield, preserving soil health, and promoting sustainable farming practices. As a drought-resistant and adaptable cereal crop, sorghum thrives in diverse environments, including areas with scarce water resources. Despite its resilience, maintaining an optimal nutrient balance is vital for ensuring high productivity and superior crop quality. Below are some key reasons highlighting the importance of nutrient management in sorghum cultivation:

Maximizing Yield Potential Sorghum thrives on a well-balanced supply of essential nutrients, particularly nitrogen (N), phosphorus (P), potassium (K), and various micronutrients. To support optimal growth, farmers should apply fertilizers in the correct amounts and at the appropriate times. This ensures that sorghum receives sufficient nutrients during key growth phases such as flowering and grain filling, when nutrient requirements are at their peak.

Improving Crop Quality Nutrient management impacts both the yield and its quality. For instance, nitrogen plays a crucial role in protein synthesis within grains, directly affecting their nutritional value. Phosphorus supports root growth and energy transfer, enhancing overall plant health and quality.

Preventing Soil Depletion Over time, poor nutrient management can exhaust soil resources, negatively impacting future crop production. Like other crops, sorghum depletes soil nutrients as it grows, making proper fertilization essential for maintaining long-term soil fertility. Using a mix of organic materials, such as compost or manure, along with synthetic fertilizers, helps sustain or even enhance soil health. This balanced approach improves soil structure, boosts water retention, and promotes microbial activity, all of which support healthy sorghum growth.

Enhancing Drought Tolerance Sorghum is recognized for its resilience to drought, but well-nourished plants are more capable of withstanding environmental stress. Adequate nutrition helps plants sustain essential physiological functions under challenging conditions, such as optimizing stomatal regulation and improving water-use efficiency. Specific nutrients, like potassium, play a crucial role in managing water stress by maintaining osmotic balance and preserving cell turgor.

Minimizing Environmental Impact Effective nutrient management reduces nutrient runoff and leaching into water sources, preventing environmental contamination. Overuse of synthetic fertilizers can disrupt nutrient balance and contribute to pollution. Implementing precision agriculture enhances nutrient application efficiency while minimizing waste. This strategy integrates organic, biological, and mineral fertilizers to improve nutrient utilization and mitigate environmental impact.

Economic Benefits for Farmers Effective nutrient management enables farmers to prevent excessive fertilization, which can be expensive. By precisely evaluating soil nutrient levels and crop requirements, they can minimize input costs while maximizing yields. Additionally, in some cases, it provides more fodder for livestock, enhancing overall farm productivity.

Improving Soil Microbial Activity Effective nutrient management improves soil microbial diversity and activity. A thriving microbial community plays a crucial role in nutrient cycling, enhancing nutrient availability for sorghum and future crops. Incorporating organic fertilizers and crop residues into the soil supports beneficial microbial processes such as nitrogen fixation and organic matter breakdown.

10.7.12.1 Agronomic Biofortification

Agronomic biofortification is a simple and effective method to enhance crop nutrition by applying nutrients through soil, foliar sprays, or seed priming. Its primary goal is to address dietary deficiencies by increasing the micronutrient content of food crops, particularly staple grains, legumes, and tubers, which are essential in global diets. Fertilizer application at different growth stages significantly boosts mineral content in grains and fruits, making biofortification a vital strategy for populations with limited access to diverse diets, supplements, or fortified foods. Recent advancements in specialty fertilizers such as nano-fertilizers, chelated fertilizers, biofertilizers, and water-soluble fertilizers have improved nutrient absorption and efficiency. However, the high cost of mineral fertilizers raises the price of biofortified crops, making them less accessible to low-income communities. Farmers also play a crucial role, as they may choose not to apply fertilizers if it is not profitable. Additionally, frequent fertilizer use can lead to toxic accumulation, and increased demand for certain minerals such as selenium may have environmental consequences.

Application of Fertilizers

Transplanted Crop

If soil test recommendations are unavailable, follow a general guideline of applying 90 kg N, 45 kg P_2O_5, and 45 kg K_2O per hectare. Distribute nitrogen application in a 50:25:25 ratio at 0, 15, and 30 days after sowing (DAS), while the full dose of phosphorus (P_2O_5) and potassium (K_2O) should be incorporated into the soil before planting. For sulphur-deficient soils, apply 30 kg of sulphur at the basal stage. Additionally, the Soil Test Crop Response-based Integrated Plant Nutrition System (STCR–IPNS) can be used to determine fertilizer application rates for achieving specific yield targets.

Direct Sown Crop

Apply NPK fertilizers based on soil test recommendations whenever possible. If soil test data are unavailable, follow the standard recommendation (similarly to a transplanted crop). If basal application is not feasible, these nutrients should be top-dressed within 24 hours. For bed-planted crops, create furrows 5 cm deep and 45 cm apart, placing the fertilizer mixture at a depth of 5 cm along the lines. Cover the lines with soil, leaving 2 cm at the top before sowing. When growing sorghum as an intercrop with pulses, open furrows 30 cm apart to a depth of 5 cm. Apply fertilizer in two rows designated for sorghum, covering it with 2 cm of soil. Leave the third row for the pulse crop without fertilization, then apply fertilizer in the next two rows and cover with soil up to 2 cm. For biofertilizer application, if Azospirillum is used, apply only 75% of the recommended nitrogen dosage for irrigated sorghum.

Rainfed Crop

Apply 12.5 tons per hectare of composted coir pith along with NPK at a ratio of 40:20:0. Additionally, incorporate enriched farmyard manure (FYM) at a rate of 750 kg per hectare. For rainfed sorghum, the standard recommendation of 40 kg nitrogen and 20 kg P_2O_5 per hectare can be reduced by half if FYM is applied at 5 tons per hectare.

Deficiency Symptoms of Sorghum

- *Zinc Deficiency:* Symptoms initially emerge in young leaves around 20 to 30 days after growth. Older leaves develop yellow streaks or chlorotic striping between the veins.
- *Iron Deficiency:* Interveinal chlorosis is the first noticeable sign. If the deficiency persists, the entire leaf, including the veins, may turn chlorotic. Newly developed leaves also display chlorotic symptoms. In severe cases, the entire crop may appear bleached, dry out, and eventually perish.

Application of Micronutrients

Transplanted Crop

Mix 12.5 kg/ha of the micronutrient mixture developed by the Department of Agriculture, Tamil Nadu, with sufficient sand to reach a total weight of 50 kg, then apply it over the furrows and the upper one-third of the ridges. Alternatively, for irrigated crops, use 12.5 kg/ha of the TNAU MN mixture, while for rainfed crops, apply 7.5 kg/ha as enriched FYM. If the micronutrient mixture is unavailable, mix 25 kg of zinc sulphate with sand to reach a total of 50 kg and apply it similarly over the furrows and the top one-third of the ridges.

Direct Sown Crop

Like a transplanted crop, however, for soils deficient in zinc, apply 25 kg of $ZnSO_4$ per hectare as a basal dose, or alternatively, use 12.5 kg of $ZnSO_4$ along with 12.5 tons of farmyard manure per hectare. In the case of iron deficiency, apply 50 kg of $FeSO_4$ per hectare along with 12.5 tons of FYM. If nutrient deficiencies appear in plants, conduct foliar spraying with 0.5% $ZnSO_4$ and 1.0% $FeSO_4$ mixed with 0.1% citric acid, applying the solution three times at 30, 40, and 50 days after sowing (DAS).

10.7.12.2 Importance of Organic Manures

Organic manures enhance soil fertility by ensuring the availability of essential macro and micronutrients, addressing deficiencies, and maintaining the carbon-to-nitrogen (C:N) ratio. They improve soil structure, texture, and water-holding capacity while acting as mulch to reduce evaporation losses. Despite their benefits, organic manures face challenges like bulkiness, slow nutrient release, and limited supply (Yadav et al. 2019). Among them, poultry manure is the most effective, providing higher nutrient content and rapid release, improving yield and soil fertility. Likewise, farmyard manure gradually enhances soil properties, reduces bulk density, increases porosity, and stimulates microbial activity. Applying FYM at 10 t ha^{-1} significantly improves plant growth and sorghum yield. Vermicompost further aids plant development by supplying growth hormones, vitamins, enzymes, and beneficial microbes. Rich in enzymes like protease and amylase, it accelerates organic matter decomposition, ensuring continuous nutrient availability. Research suggests organic nutrient sources enhance crop growth by supplying essential nutrients throughout the growth period, stimulating amino acid synthesis, and promoting overall plant development.

Organic Manures on Quality Parameters of Sorghum

The application of poultry manure at a rate of 3.0 t ha^{-1} resulted in increased crude protein and crude fiber content in sorghum, likely due to the ample supply of essential nutrients in poultry manure. Similarly, Agbede et al. (2008) reported improved protein and fiber levels following poultry manure application to soil. Nandhagopal et al. (2003) and Karwasra and Anil Kumar (2006) observed a rise in crude protein content in dual-purpose sorghum due to vermicompost application. Likewise, Kumari and Ushakumari (2002) recorded enhanced protein and nutrient levels in cowpea when vermicompost was applied at 3.0 t ha^{-1}. The application of pressmud also improved nutrient uptake and grain and fodder quality, as noted by Bhosale et al. (2012). A steady increase in N, P, K, and Ca content in sorghum plant tissues was observed with pressmud application at 12.5 t ha^{-1}. Kalibhavi et al. (2001) also reported higher N, P, K, and Zn levels in sorghum when pressmud was used. According to Kumar et al. (2017a), incorporating pressmud into the soil enhances nutrient availability and plant uptake by fostering a favorable soil environment. Vermicompost at 5 t ha^{-1} also performed comparably to poultry manure. However, FYM at 12.5 t ha^{-1} limited nitrogen uptake, subsequently affecting organic nitrogen synthesis and leading to lower quality parameters.

10.7.12.3 Inorganic Fertilizers on Sorghum

Nitrogen, phosphorus, and potassium are crucial macronutrients required for plant growth and development. A deficiency in these elements can lead to significant yield losses. Inorganic fertilizers are essential for crop production as they supply readily available nutrients for rapid plant uptake. Increasing nitrogen levels from 0 to 150 kg ha^{-1} has been associated with enhanced plant characteristics such as height, stem thickness, and leaf count. In sorghum, applying 60 kg N ha^{-1} led to an increase in the leaf area index. A deficiency in nitrogen results in reduced dry matter accumulation, lower quality parameters, and decreased yield. Mishra et al. (2019) confirmed that increasing fertilizer application levels led to improved plant growth attributes such as plant height, dry matter production (DMP), leaf area index (LAI), SPAD values and crop growth rate (CGR), and also promoted cell division and development, accelerating protoplasm activity, which ultimately enhanced leaf count, stem diameter, and other growth characteristics. Valadabadi and Farahani (2010) found that increasing fertilizer application from 100% to 125% RDF led to a 10% rise in dry weight and a 12.3% increase in LAI, RGR, and CGR. Nitrogen concentration, ranging from 0 to 120 kg ha^{-1}, strongly influenced chlorophyll content, with maximum levels observed at 120 kg N ha^{-1}, emphasizing nitrogen's role in chlorophyll production.

Research by Roy et al. (2018) indicated that higher phosphorus levels led to increased sorghum plant height, while Shivran et al. (2013) reported that the tallest plants were recorded at 75 kg P ha^{-1} compared to lower levels. Similarly, Dambale et al. (2018) found that phosphorus application up to 50 kg ha^{-1} on vertisols improved plant height, LAI, and dry matter accumulation compared to 0 or 20 kg ha^{-1}. Higher potassium doses enhanced LAI, improved nutrient translocation from source to sink, and optimized nutrient absorption, leading to better yield parameters. The highest sorghum growth parameters were observed at 125% RDF across different growth stages, though results were comparable to 100% RDF.

Levels of Inorganic Fertilizers on Quality Parameters in Sorghum

The crude protein and crude fiber content in fodder sorghum were significantly influenced by nitrogen application, with the highest quality parameters observed at 100% RDF compared to lower levels. Increasing nitrogen levels resulted in maximum crude protein and crude fiber content at 125% RDN, outperforming 0, 50, 75, and 100% RDN. The lowest-quality parameters were recorded under absolute control conditions, likely due to insufficient nutrient availability to support crop growth (Damame et al. 2013). Kumar et al. (2017b) also reported that higher nitrogen levels enhanced succulence, palatability, crude protein, and crude fiber content in fodder sorghum. Similarly, Singh and Chauhan (2016) found that applying 125% RDF led to higher crude protein, crude fiber, and malic acid content in dry fodder compared to 50, 75, and 100% RDF.

10.7.12.4 Effect of Fortification of Quality Parameters of Sorghum

While soil application of fertilizers remains the most common method, it has limitations in terms of nutrient availability to plants. Since inorganic nutrients are highly soluble in soil, they can be lost due to leaching from rainfall or irrigation (Alshaal and El-Ramady 2017). Foliar application provides an effective alternative by addressing these challenges. It is the fastest way to correct nutrient deficiencies, enhance crop yield, and improve product quality while reducing soil fertilizer use, minimizing environmental pollution, and increasing nutrient efficiency. Research has shown that foliar application of micronutrients like zinc (Zn), iron (Fe), and manganese (Mn) significantly boosts crop yield, even when applied in small quantities. Furthermore, a study by Yvin and Dufils (2010) indicated that foliar fertilization directly enhances plant quality.

According to Malakouti (2008), using micronutrient-enriched NPK fertilizers improves the nutritional quality of harvested grains by increasing protein content and micronutrient concentrations. Intensive farming, the use of high-purity NPK fertilizers, and limited zinc application have contributed to widespread zinc deficiency in sorghum. Zinc is a crucial nutrient for plant growth and reproduction, acting as a structural component or regulatory cofactor for various enzymes in key biochemical pathways. To prevent Zn deficiency and support cereal grain biofortification, zinc fertilizers are widely used.

Foliar application of Fe and Zn has been found to increase the total carbohydrate, starch, indole acetic acid (IAA), and protein content in grains. Research conducted on Vertisol soil using the K12 sorghum variety demonstrated that applying a fertilizer regime of 50:25:25 kg NPK ha^{-1} along with foliar sprays of 0.5% $ZnSO_4$ and 0.2% $FeSO_4$ at 15, 30, and 45 DAS significantly improved crude protein and crude fiber content. Additionally, the inclusion of potash fertilizers in rainfed sorghum cultivation enhanced crude protein and crude fiber content compared to sorghum grown with only the recommended dose of fertilizer (RDF), where potash was not applied. Foliar spraying is also an effective strategy for mitigating midseason drought and improving productivity. Applying water-soluble fertilizer (19:19:19) at 0.5% concentration during dry spells resulted in the highest sorghum grain yield (1448 kg/ha), a benefit-cost (BC) ratio of 1.16, and a rainwater use efficiency (RWUE) of 3.85 kg/ha-mm. Foliar application of the micronutrients Mn and Zn can enhance grain Zn content (Aziz et al. 2019). Tamoor et al. 2021 found that all micronutrients (Zn, Fe and Se) improved the growth, quality and uptake of nutrients. However, Se produced the highest plant height, stem diameter, and 1000-grain weight, while Zn recorded the maximum protein, oil, and starch contents in the tested sorghum accessions.

The finding of Markole et al. 2020 revealed that variety CSV–31 with soil application of zinc sulphate @ 50 kg ha^{-1} + foliar spray of ZnSO4 @ 0.2% at knee high stage and at flowering (F5) registered significantly maximum zinc concentration (grain and stover) and uptake by grain and stover as compared to other treatments. Cipriano et al. (2022) assessed the selenium biofortification in sorghum through soil by various selenium doses and sources. The sources SeB (methyl hydroxy-selenide and SeC (hydroxy-selenide) at 240 μg dm^{-3}, along with SeD (sodium selenate) at 60 μg dm^{-3}, showed the most promise. SeD led to the highest root absorption, Se recovery, and translocation efficiency, resulting in the greatest Se accumulation in shoots and grains. The study found that applying 20 kg $ZnSO_4$ and 20 kg $FeSO_4$ per hectare as a basal dose, along with a 0.5% foliar spray of $ZnSO_4$ and $FeSO_4$ at 45 DAS, improved the green fodder yield and quality (crude protein yield, protein content, crude fiber) of maize and sorghum by enhancing crude protein content (Adesh et al. 2021).

Choudhary et al. (2017) found that applying micronutrients along with N, P, and K fertilizers boosted yield by 25.33% and enhanced grain nutritional quality. Combined soil and foliar application enhanced N, P, and K uptake by 14.12% and 11.75%, 13.98% and 10.86%, and 12.33% & 9.57% compared to soil and foliar application alone. The application of RDF with Fe, Zn, and B further increased N, P, and K uptake by 33.25%, 33.14%, and 18.51% over the control. Additionally, micronutrient uptake improved significantly, with Zn increasing by 47.90% & 18.32%, Fe by 30.20% & 16.11%, and B by 25.60% & 19.75% under combined soil + foliar application and (Fe + Zn + B) treatment compared to the control (Choudhary et al. 2017).

Zahida et al. (2024) found that applying 5 kg of zinc and 2 kg of boron per hectare, combined with 10 ppm triacontanol foliar spray 30 days after sowing, resulted in the highest yields of green fodder (493.2 q/ha), dry matter (125.6 q/ha), and crude protein (10.80 q/ha). The study revealed that combining soil and foliar applications of zinc (Zn) and iron (Fe) significantly enhanced their uptake and content in sorghum grain and straw. Treatment 0.5% $FeSO_4$ + 0.5% $ZnSO_4$ foliar spray at 30 and 45 DAS was the most effective, achieving the highest Zn uptake (grain: 99.90 g ha^{-1}, straw: 107.66 g ha^{-1}), Fe uptake (grain: 181.11 g ha^{-1}, straw: 568.23 g ha^{-1}), and maximum grain yield (3496 kg ha^{-1}). This method offers a practical strategy to boost sorghum production, improve crop nutrition, and mitigate micronutrient deficiencies (Krushiv et al. 2024).

10.7.12.5 A Concept of Integrated Nutrient Management

Enhancing grain and fodder quality requires organic amendments to boost mineral nutrient content, while integrated nutrient management (INM) lowers production costs, improves economic returns, and reduces environmental pollution. INM also enhances crude protein and fiber content in fodder, while inorganic fertilizers accelerate plant growth. Combining organic manures with inorganic fertilizers optimizes economic yield while preserving soil properties (Krishnaprabu 2014). INM improves water-holding capacity, enabling roots to access deeper moisture and sustain high relative water content under drought. Organic matter further enhances nutrient uptake, improving overall grain quality.

Organic Manures and Inorganic Fertilizers on Growth, Yield Attributes, Yield, and Quality Parameters of Sorghum

The application of poultry manure at 5 t ha^{-1} combined with 100% RDF on medium black soil resulted in the highest root mass density, root volume, and total dry matter accumulation in sorghum compared to using 100% RDF alone. Additionally, a 25% yield increase was recorded with vermicompost application compared to 100% RDF alone. The integration of organic manures with inorganic fertilizers also enhanced the quality parameters of sorghum, whereas the absolute control plot exhibited lower nutrient uptake. Bayu et al. (2006) reported that incorporating organic nutrients along with varied fertilizer levels resulted in higher grain protein yield. The addition of poultry manures along with 125% RDF improved quality parameters in maize, both in grain and fodder (Boateng et al. 2009). Specifically, the combination of poultry manure at 5 t ha^{-1} with 125% RDF resulted in higher crude protein, crude fiber, and crude fat content. This improvement in quality was attributed to increased nitrogen absorption from both organic and inorganic sources, as nitrogen is a key component of amino acids, ultimately enhancing fodder quality.

10.7.13 Weed Management in Sorghum

Weeds pose a major challenge to sorghum productivity, significantly reducing yields by competing with crops for water, nutrients, sunlight, and space. In many cases, their impact is severe, with sorghum grain yield losses ranging from 30% to 50% due to weed interference, higher than in most other grain crops. Several studies have highlighted the detrimental effects of prolonged weed competition, with unweeded sorghum fields experiencing yield reductions of 40% to 80% (Thakur et al. 2016). The most critical period for weed competition in sorghum occurs around 45 days after sowing (DAS). Managing weeds in sorghum cultivation is particularly difficult for farmers due to limited access to herbicides and labor, restrictions on rotational crops after using multiple herbicides approved for sorghum, and the increasing prevalence of herbicide-resistant weeds (Fromme et al. 2012). Common weed species that dominate in sorghum fields include *Cynodon dactylon*, *Dactyloctenium aegyptium*, *Euphorbia hirta*, *Phyllanthus niruri*, *Cyperus rotundus*, *Solanum nigrum*, *Celosia argentea*, and *Leucas aspera*. Traditionally, Indian farmers have relied on manual weeding, which has been effective against annual weeds but less so against perennial grasses due to their ability to regenerate. Mechanical weeding serves as a more efficient alternative to hand weeding, especially during peak growing periods, as it saves time and effort. Both manual and mechanical methods are most effective when applied before crop emergence. The development of herbicides has revolutionized weed management by offering a more efficient, cost-effective, and time-saving solution. Pre-emergence herbicides are particularly valuable as they control weeds from the early stages, while post-emergence herbicides help manage weeds that appear later in the crop cycle (Deshmukh and Usadadia 2017).

10.7.13.1 Herbicides Recommended for Sorghum

Atrazine (0.25 kg), metribuzin (1–2 kg), and terbutryn (1–2 kg) are common pre-emergence herbicides in sorghum, applied 3–5 days after sowing (DAS) or transplanting (DAT), followed by hand weeding or hoeing. Post-emergence herbicides like halosulfuron (30–40 g) and prosulfuron (15–30 g) are also effective. In pulse intercropping, atrazine is avoided, and pendimethalin (1.0 kg) is preferred. For sorghum + soybean, alachlor (2.0 kg) or pendimethalin (1.0 kg) is recommended. Weed-free conditions should be maintained for five weeks after germination. Research shows atrazine (3000 g a.i. ha^{-1}) provides 73.17% weed control, with metolachlor + atrazine (1.0 kg/ha) achieving superior results. Weed management improves sorghum quality, with the highest crude protein content (7.93%) observed in hand-weeded plots, followed by atrazine (0.375 kg a.i. ha^{-1}) and pendimethalin (0.75 kg a.i. ha^{-1}). Unweeded conditions resulted in the lowest protein content due to weed interference (Singh and Pandey 2019). The efficiency of weed control was observed to be highest at 73.17% with atrazine at 3000 g a.i. ha^{-1}, followed by 68.12% with atrazine at 2000 g a.i. ha^{-1}. Atrazine effectively controlled most broadleaf weeds and some grasses (Pimentel et al. 2019). Shakoor et al. (2000) reported that pre-emergence application of metolachlor + atrazine at 1.0 kg/ha resulted in superior weed control, with a lower weed density of 20 plants/m^2 and reduced weed dry matter of 533 kg/ha. Similarly, Paswan et al. (2017) found that 2,4-D and metsulfuron-methyl, either alone or in combination, outperformed isoproturon and carfentrazone-ethyl in weed suppression.

10.7.13.2 Sorghum: Parasitic Weeds

Parasitic weeds can be classified as either total or partial parasites. Parasitic weeds that infest plant roots are known as root parasites. In sorghum, *Striga lutea* acts as a partial root parasite.

Control Measures for Striga in Sorghum
- Apply post-emergence 2,4-D Na salt at 2.0 kg/ha between 25 and 30 DAS.
- Practice intercropping with redgram.
- Rotate crops with trap crops such as cotton, sunflower, groundnut, and cowpea, which stimulate weed seed germination but are not affected by parasitism.
- Apply high doses of nitrogen (N) and farmyard manure (FYM).
- Flood the field to suppress weed growth.
- Spray a 10% urea solution at 25–30 DAS.
- Utilize germination stimulants like Strigol and Ethylene gas.

10.7.14 Water Management in Sorghum

The water requirement for achieving maximum sorghum yield is not a fixed value, as it depends on factors such as temperature, relative humidity, wind speed, and soil moisture, all of which influence evaporation and plant transpiration. In favorable seasons, the crop may require as little as 400–450 mm of water, whereas in hot, dry conditions, this demand can rise to 700–850 mm. Under Tamil Nadu conditions, sorghum requires approximately 450–500 mm of water, with irrigation at 50% depletion of available soil moisture or maintaining an IW/CPE ratio of 0.6 being adequate as an irrigation interval based on a climatological approach. However, irrigation is required at the critical stage (blue line), particularly from stage three (35 DAS) to stage six (65 DAS), which coincides with head initiation to flowering, which has to be maintained for optimal moisture that leads to optimal productivity. (Fig. 10.3 and Table 10.3).

Critical stages: (1) Seedling, (2) vegetative, (3) flowering, (iv) dough stage.

The period one week before and after flowering is highly critical for moisture availability. In moisture-stressed conditions, five irrigations are adequate, whereas under normal conditions, eight irrigations should be provided. The recommended schedule includes irrigation on day 1, day 4, day 15, day 28, day 40, day 53, day 64, day 76, and day 88. Irrigation should be discontinued after 88 to 90 DAS.

10.7.15 Harvesting and Threshing

Most high-yielding sorghum varieties and hybrids reach maturity within 100 to 115 days. The ideal time for harvesting is when the grains become hard and contain less than 25% moisture. It is not necessary to wait for the stubble and leaves to dry,

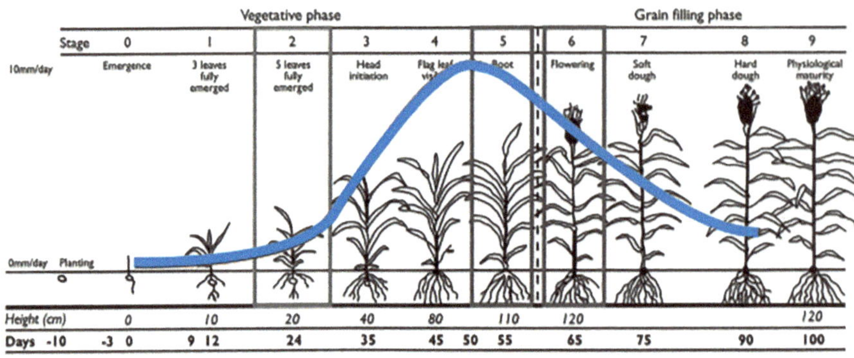

Fig. 10.3 Grain sorghum water use pattern and critical growth stages

Table 10.3 Potential reduction in yield from moisture stress

Growth stage	% yield reduction
Seedling emergence to boot	10–15
Boot to flowering	30–50
Soft dough to maturity	10–20

Source: Kelley and Faske (2024)

as hybrid sorghum often remains green even after reaching maturity. Harvesting can be carried out at physiological maturity by either cutting the entire plant or first removing the earheads and then cutting down the plants, allowing them to dry for 2 to 5 days. Threshing is performed using a thresher or by manually beating the ear heads. The threshed grains are then sun-dried for about a week to reduce the moisture content to 10–12% for safe storage.

10.7.16 Yield

Grain yield ranges from 2 to 3 tons per hectare under rainfed conditions, while dry stover production is approximately 8 to 10 tons per hectare.

10.8 Bad Qualities of Sorghum

- Sorghum contains a high amount of niacin, which interferes with the synthesis of tryptophan, a precursor to produce indole-3-acetic acid (IAA).
- "Pellagra" is a nutritional disorder caused by a high leucine-to-isoleucine ratio (3.4). Reducing this ratio can lower the yield. This condition is commonly observed in Africa.
- Sorghum contains a significant amount of oxalic acid, which hampers calcium absorption and metabolism.

- Due to the presence of high oxalic acid, phytin phosphorus remains unavailable, and iron uptake is also negatively affected.
- Low digestibility and poor palatability of sorghum are attributed to the presence of phenolic compounds, glucosides, tannins, and lignin.
- Sorghum contains a cyanogenic glucoside known as "Dhurin," which converts into hydrogen cyanide (HCN) in the stomachs of ruminants, leading to bloating and reduced oxygen transfer in the bloodstream, potentially causing death. This condition is known as "sorghum poisoning" or the "sorghum effect." HCN levels exceed 100 ppm in the early growth stage, while the critical level is 50 ppm, typically occurring between 60 and 65 days after sowing or at the heading stage. If harvested earlier, sorghum should be dried before being fed to cattle.
- "Sorghum injury" occurs due to the high carbon-to-nitrogen (C:N) ratio (50:1) in sorghum stubble and roots, leading to low nitrogen availability. Soil microbes utilize soil nitrogen for decomposition instead of drawing it from the sorghum residue, temporarily immobilizing nitrogen. As a result, the succeeding crop experiences early-stage nitrogen deficiency, necessitating higher nitrogen supplementation.

10.9 Crop Protection

10.9.1 Pest Management

- *To protect the nursery,* apply one of the following insecticides in 6 liters of water on the 7th and 14th day after sowing:
  - Methyl demeton 25EC at 12 ml.
  - Dimethoate 30EC at 12 ml.
- Plough the field immediately after harvest, remove and destroy the stubble.
- Treat seeds before sowing with either chlorpyrifos 20EC, phosalone 35EC (4 ml/kg), imidacloprid 48FS, imidacloprid 70WS, or thiamethoxam 30FS (10 g/kg).
- Avoid repeated application of the same insecticides to prevent pest resurgence.
- Complete sorghum sowing within a short period to minimize continuous flowering, which promotes grain midge and earhead bug multiplication.
- Use light traps until midnight to monitor, attract, and eliminate stem borer, grain midge, and earhead caterpillars (Table 10.4).

10.9.2 Disease Management

10.9.2.1 Nursery Practices

Seed treatment: Treat the seeds 24 hours before sowing with either carbendazim, captan, or thiram at a rate of 2 g/kg of seeds or use metalaxyl at 6 g/kg of seeds (Table 10.5).

Table 10.4 Control measures for important pests

Shoot fly *Atherigona soccata*	Sow sorghum early, immediately after receiving the Southwest or Northeast monsoon, to reduce shoot fly infestation. For direct seeding, use a higher seed rate of up to 12.5 kg/ha and remove shoot fly-damaged seedlings during thinning. In transplanted crops, ensure only healthy seedlings are used. Spray dimethoate 30EC (12 ml) for every 120 m^2 nursery area. Install fish meal traps at a rate of 12 per hectare until the crop reaches 30 days of age. Plough the field immediately after harvest, remove, and destroy stubble. Apply any one of the following per hectare: Carbofuran 3CG at 33.3 kg (at sowing) Dimethoate 30EC at 500 ml Neem seed kernel extract 5% Quinalphos 25EC at 1500 ml
Stem borer *Chilo partellus; Sesamia inferens*	Intercrop sorghum with lab or cowpea in a 4:1 ratio to reduce stem borer infestation. Apply carbofuran 3CG at 17 kg/ha, mixed with sand to reach a total of 50 kg/ha, and apply it in the leaf whorls.
Grain midge, *Contarinia sorgicola*	Apply any of the following per hectare on the 3rd and 18th day after panicle emergence: Dimethoate 30EC at 1650 ml Malathion 50EC at 1600 ml Malathion 5D at 25 kg Neem seed kernel extract 5% Phosalone 35EC at 1150 ml Phosalone 4D at 25 kg
Earhead bug, *Calocoris angustatus*	Apply any of the following per hectare on the 3rd and 18th day after panicle emergence: Malathion 50EC at 1000 ml Malathion 5D at 25 kg Neem seed kernel extract 5% Quinalphos 1.5DP at 25 kg

Table 10.5 Control measures for important diseases

Name of the disease	Recommendations
Rust: *Puccinia purpurea*	Apply mancozeb at a rate of 1 kg/ha and repeat the fungicide treatment after 10 days.
Ergot or sugary disease: *Sphacelia sorghi*	Modify the sowing schedule to avoid flowering during the rainy and winter seasons. Apply mancozeb at 1000 g/ha or propiconazole at 500 ml/ha during the 5–10% flowering stage and again at 50% flowering. If required, repeat the spray after one week.
Downy mildew: *Peronosclerospora sorghi*	Remove infected plants up to 45 days after sowing. Apply metalaxyl + mancozeb at 500 g/ha or mancozeb at 1000 g/ha after symptoms appear (as per CIB recommendation). Treat seeds with metalaxyl-M 31.8% ES at 2 ml/kg or use slurry seed treatment with metalaxyl 35% WS at 2 g/kg of seed.
Charcoal rot: *Macrophomina phaseolina*	Coat the seeds with Pseudomonas fluorescens at a rate of 10 g/kg or Trichoderma viride at 4 g/kg of seed.
Grain smut: *Sphacelotheca sorghi*	As per the CIB recommendation, treat the seeds with sulfur 80% WP at a rate of 3–4 g per kg of seed.

References

Adesh RP, Singh S, Pal RP, Verma D (2021) Effect of micronutrient application on growth, yield, and quality of fodder maize (Ziya maize L.) and fodder sorghum (Sorghum bicolor L.). Pharma Innov J 10(10):2356–2361

Agbede TM, Ojeniyi SO, Adeyemo AJ (2008) Effect of poultry manure on soil physical and chemical properties, growth and grain yield of sorghum in southwest, Nigeria. Am-Eurasian J Sustain Agric 2(1):72–77

Alshaal T, El-Ramady H (2017) Foliar application: from plant nutrition to biofortification. Environ Biodiver Soil Secur 1(2017):71–83

Annual Report of Agricultural Research Station, Kovilpatti, Tamil Nadu (2023)

Ayub M, Nadeem MA, Tahir M, Ghafoor A, Ahmed Z, Naeem M (2010) Comparative studies on the growth, forage yield and quality of sorghum (Sorghum bicolor L.) varieties under irrigated conditions of Faisalabad. Pak J Life Soc Sci 8(2):94–97

Aziz MZ, Yaseen M, Abbas T, Naveed M, Mustafa A, Hamid Y et al (2019) Foliar application of micronutrients enhances crop stand, yield and the biofortification essential for human health of different wheat cultivars. J Integr Agric 18(6):1369–1378

Bayu W, Rethman NFG, Hammes PS, Alemu G (2006) Effects of farmyard manure and inorganic fertilizers on sorghum growth, yield, and nitrogen use in a semi-arid area of Ethiopia. J Plant Nutr 29(2):391–407

Bhardwaj AK, Chejara S, Malik K, Kumar R, Kumar A, Yadav RK (2022) Agronomic biofortification of food crops: an emerging opportunity for global food and nutritional security. Front Plant Sci 13:1055278

Bhosale PR, Chonde SG, Nakade DB, Raut PD (2012) Studies on physico-chemical characteristics of waxed and dewaxed pressmud and its effect on water holding capacity of soil. ISCA J Biol Sci 1(1):35–41
Boateng S, Zickermann J, Kornahrens M (2009) Poultry manure effect on growth and yield of maize. West Afr J Appl Ecol 9(1). https://doi.org/10.4314/wajae.v9i1.45682
Choudhary SK, Jat MK, Mathur AK (2017) Effect of micronutrient on yield and nutrient uptake in sorghum. J Pharmacogn Phytochem 6(2):105–108
Cipriano PE, da Silva RF, de Lima FRD, de Oliveira C, de Lima AB, Celante G et al (2022) Selenium biofortification via soil and its effect on plant metabolism and mineral content of sorghum plants. J Food Compos Anal 109:104505
Damame SV, Bhingarde RN, Pathan SH (2013) Effect of different nitrogen levels on nutritional quality and nitrate nitrogen accumulation in forage pearl millet genotypes grown under rainfed conditions. Forage Res 39(2):93–95
Dambale AS, Ghotmukale AK, Khandekar SD, Suryawanshi SB, Suryararshi VP, Shinde RS (2018) Influence of integrated nutrient management on growth, yield, quality and economics of sunflower (Helianthus annuus L.). Int J Curr Microbiol App Sci 6:1226–1233
Deshmukh SP, Usadadia VP (2017) Weed management influence on crop-weed competition in sorghum under South Gujarat conditions. Indian J Weed Sci 49(3):272–275
Farooq M, Usman M, Nadeem F, ur Rehman H, Wahid A, Basra SM, Siddique KH (2019) Seed priming in field crops: potential benefits, adoption and challenges. Crop Past Sci 70(9):731–771
Fromme DD, Dotray PA, Grichar WJ, Fernandez CJ (2012) Weed control and grain sorghum (Sorghum bicolor) tolerance to pyrasulfotole plus bromoxynil. Int J Agron 2012(1):951454
Hossain AKMZ, Mahmud MS, Islam MS, Tajkia JE, Sagar A, Khaton MA (2016) Effect of moisture stress on morphological and yield attributes of four sorghum varieties. Progress Agric 27(3):265–271
Hossain MS, Islam MN, Rahman MM, Mostofa MG, Khan MAR (2022) Sorghum: a prospective crop for climatic vulnerability, food and nutritional security. J Agric Food Res 8:100300. https://nutricereals.dac.gov.in/sorghum.aspx
Jain SK, Patel PR (2012) Genetic variability in land races of forage sorghum {Sorghum bicolor (l) moench} collected from different geographical origin of India. Int J Agric Sci 4(2):182
Kalibhavi CM, Kachapur MD, Patil RH (2001) Performance of rabi sorghum under integrated nutrient management system. Indian J Dryland Agric Res Develop 16(1):45–50
Karthikeyan BJ, Babu C, Amalraj JJ (2017) Nutritive value and fodder potential of different sorghum (Sorghum bicolor L. Moench) cultivars. Int J Curr Microbiol Appl Sci 6(8):898–911
Karwasra RS, Anil Kumar AK (2006) Effect of nitrogen and phosphorus on fodder sorghum production. cabidigitallibrary.org
Kelley J, Faske T (2024) Arkansas corn and grain sorghum research studies 2023. Arkansas Agricultural Experiment Station Research Series. Retrieved from https://scholarworks.uark.edu/aaesser/ 225
Krishnaprabu S (2014) Evaluation of vermicompost and biofertilizers on soil properties and yield of sweet sorghum. Int J Curr Res Life Sci 3(3):191–192
Krushiv P, Leva RL, Patel HK, Genan H, Modh MB, Bhavsar A, Patel SP (2024) Enhancing sorghum nutrition through agronomic fortification with zinc and iron. Int J Res Agron SP-7(11):479–484
Kumar S, Meena RS, Jinger D, Jatav HS, Banjara T (2017a) Use of pressmud compost for improving crop productivity and soil health. Int J Chem Stud 5(2):384–389
Kumar R, Kumawat N, Kumar S, Singh AK, Bohra JS (2017b) Effect of NPKS and Zn fertilization on, growth, yield and quality of baby corn-a review. Int J Curr Microbiol App Sci 6(3):1392–1428
Kumari MS, Ushakumari K (2002) Effect of vermicompost enriched with rock phosphate on the yield and uptake of nutrients in cowpea (Vigna unguiculata [L.] Walp). J Trop Agric 40:27–30
Lemerle D, Cousens RD, Gill GS, Peltzer SJ, Moerkerk M, Murphy CE et al (2004) Reliability of higher seeding rates of wheat for increased competitiveness with weeds in low rainfall environments. J Agric Sci 142(4):395–409

Lemerle D, Verbeek B, Diffey S (2006) Influences of field pea (Pisum sativum) density on grain yield and competitiveness with annual ryegrass (Lolium rigidum) in south-eastern Australia. Aust J Exp Agric 46(11):1465–1472

Malakouti MJ (2008) The effect of micronutrients in ensuring efficient use of macronutrients. Turk J Agric For 32(3):215–220

Markole SK, Suresh K, Madhavi A, Shanti M (2020) Agronomic biofortification and nutrient uptake of sorghum (Sorghum biochor L.) as influenced by fertilization strategies. Int J Chem Stud 8:2107–2110

Mishra JS, Rao SS, Das IK (2019) Effect of tillage and nutrient management on sorghum (Sorghum bicolor) productivity in Alfisols of semi-arid tropical India. Indian J Agric Sci 89(7):1133–1139

Mkandawire NL, Kaufman RC, Bean SR, Weller CL, Jackson DS, Rose DJ (2013) Effects of sorghum (Sorghum bicolor (L.) Moench) tannins on α-amylase activity and in vitro digestibility of starch in raw and processed flours. J Agric Food Chem 61(18):4448–4454

Nandhagopal A, Subramanian KS, Jayakumar R, Balasubramanian A (2003) Integrated nutrient management for hybrid sunflower (Helianthus annus L.). Madras Agric J 90:1

Nile SH, Thiruvengadam M, Wang Y, Samynathan R, Shariati MA, Rebezov M et al (2022) Nanopriming as emerging seed priming technology for sustainable agriculture—recent developments and future perspectives. J Nanobiotechnol 20(1):254

Paswan AK, Mandal D, Kumawat N, Kumar J, Kumar R, Singh AK, Kumar A (2017) Efficacy of separate and premix formulation of metsulfuron-methyl and carfentrazone-ethyl on weeds in wheat: a review. Int J Curr Microbiol App Sci 6(7):2439–2453

Pimentel GV, Guimarães DF, Moreira SG, Ávila MOT, Martins IA, Bruzi AT (2019) Selectivity and effectiveness of herbicides in the grain sorghum crop. Planta Daninha 37:e019187771

Prom-U-Thai C, Rashid A, Ram H, Zou C, Guilherme LRG, Corguinha APB et al (2020) Simultaneous biofortification of rice with zinc, iodine, iron and selenium through foliar treatment of a micronutrient cocktail in five countries. Front Plant Sci 11:589835

Raj AB, Raj SK (2019) Seed priming: an approach towards agricultural sustainability. J Appl Nat Sci 11(1):227

Rao PS, Prakasham RS, Rao PP, Chopra S, Jose S (2015) Sorghum as a sustainable feedstock for biofuels. In: Biomass and biofuels: advanced biorefineries for sustainable production and distribution. CRC Press, pp 27–47

Roy P, Basu S, Chatterjee J, Goswami A, Barik S (2018) Response of various yield of sweet sorghum crop (Sorghum bicolour L. Moench) to different levels of fertilizer in lateritic soil of Birbhum, West Bengal. Int J Agric Environ Biotechnol 11(1):25–32

Shakoor A, Islam S, Naeem M (2000) Efficacy of different herbicides for control of weeds in sorghum (Sorghum bicolor L.) under rainfed conditions. Pak J Biol Sci 3:463–465

Shinde NK, Nemade SM, Ghorade RB, Mohod NB (2018) Effect of integrated nutrient management on yield and quality of parching sorghum varieties. Int J Curr Microbiol App Sci 6(12):4268–4272

Shivran RK, Kumar R, Kumari A (2013) Influence of sulphur, phosphorus and farm yard manure on yield attributes and productivity of maize (Zea mays L.) in humid south eastern plains of Rajasthan. Agric Sci Dig 33(1):9–14

Singh SB, Chauhan SK (2016) Effect of integrated nutrient management on pearl millet crop grown in semi-arid climate. J Multidisc Adv Res 5(2):54–57

Singh A, Pandey IB (2019) Effect of crop establishment methods, nutrient levels and weed management on yield of hybrid rice. J Pharmacogn Phytochem 8(4):91–95

Szemerits B, Kukorelli G, Molnár Z (2024) Influence of seed rate and row spacing across on two different maturity groups of sorghum grain yield and quality characteristics. In: BIO Web of Conferences, vol 125. EDP Sciences, p 01009

Tamoor QM, Faizan AM, Nasir I (2021) Agronomic bio-fortification of iron, zinc and selenium enhance growth, quality and uptake of different sorghum accessions, pp 549–557

Thakur NS, Kushwaha BB, Girothia OP, Sinha NK, Mishra JS (2016) Effect of integrated weed management on growth and yields of rainy-season sorghum (Sorghum bicolor). Indian J Agron 61(2):217–222

Valadabadi SA, Farahani HA (2010) Effects of planting density and pattern on physiological growth indices in maize (Zea mays L.) under nitrogenous fertilizer application. J Agric Extens Rural Develop 2(3):40–47
World Health Organization (WHO). World health statistics (2022) Monitoring health for SDGs, sustainable development goals. World Health Organization, Geneva
Yadav S, Lal M, Naresh RK, Yadav RB, Yadav AK, Yadav KG, Rajput P (2019) Effect of organic and inorganic nutrient sources on productivity, grain quality of rice and soil health in north-west IGP: a review. Int J Curr Microbiol App Sci 8(12):2488–2514
Yvin JC, Dufils A (2010) Incidences of Fertileader Elite® foliar spray applications on the improvement of Fruits quality and their conservation. Trials realized on Pink Lady® Cripps Pink Cov. J Horticult For Biotechnol 14(3):1–4
Zahida R, Agarwal RK, Naveen K, Shekara BG, Prasad S, Sabina N et al (2024) Optimizing sorghum yield and quality: A study on the role of plant growth regulators and micronutrient management strategies. J Sci Res Reports 30(11):208–222

Chapter 11
Organic Approach for Sorghum Cultivation to Achieve Better Nutrition

M. Hemalatha, M. Joseph, M. Aswin Ajay Raj, N. Vinitha, and Mahalingam Govindaraj

Abstract The growing demand for sustainable agricultural practices has highlighted the importance of organic farming in enhancing soil health, increasing biodiversity, and improving crop quality. Sorghum, a drought-resistant cereal crop, offers significant potential for organic cultivation, particularly in regions facing environmental stressors. This study explores an organic approach to sorghum cultivation aimed at achieving better nutritional outcomes. By utilizing natural fertilizers, crop rotation, and integrated pest management, the research assesses the impact of organic farming techniques on sorghum yield, nutritional content, and environmental sustainability. The findings suggest that organic cultivation not only enhances the nutrient profile of sorghum, increasing levels of essential vitamins, minerals, and antioxidants, but also supports soil fertility and biodiversity. This approach presents a viable solution for promoting sustainable agriculture while addressing the nutritional needs of growing populations, making organic sorghum an integral component of a healthy, sustainable food system. The advent of improved organic farming techniques, policy support, and enhanced market infrastructure can unlock the full potential of organic sorghum cultivation and its native grain's nutritional quality in India and Africa, where it's feed several millions.

Keywords Sorghum · Organic cultivation · biofortification · quality traits · food security

M. Hemalatha · M. Aswin Ajay Raj · N. Vinitha
VOC Agricultural College and Research Institute, Tamil Nadu Agricultural University, Thoothukudi, Tamil Nadu, India

M. Joseph (✉)
Agricultural Research Station (AICRPDA Main Centre), Tamil Nadu Agricultural University, Kovilpatti, Tamil Nadu, India
e-mail: joseph.m@tnau.ac.in

M. Govindaraj
Alliance of Bioversity International and the International Centre for Tropical Agriculture (CIAT), Cali, Colombia

M. Govindaraj (ed.), *Breeding Biofortified Millets for Nutritional Security*, Sustainability Sciences in Asia and Africa,
https://doi.org/10.1007/978-981-95-9827-4_11

11.1 Introduction

Sorghum (*Sorghum bicolor*) is known for its ability to withstand drought and its important contribution to world food security. Sorghum grown organically promotes biodiversity and soil health while also supporting sustainable farming methods. Organic sorghum growing reduces environmental effects, enhances farmer livelihoods, and promotes safer food production by eschewing synthetic fertilizers and pesticides (Mwamahonje et al. 2024). This diverse method of growing sorghum emphasizes how crucial it is to solve the problems of food security and climate change (Tsehaye et al. 2024). Sorghum, a highly resilient and nutritious cereal, has gained increasing attention in organic farming due to its sustainability, climate adaptability, and health benefits. As global consumers shift toward organic and healthy diets, the demand for organically grown sorghum is expected to rise significantly. This report explores the projected demand for organic sorghum in India and worldwide, driven by health-conscious consumers, sustainable agriculture initiatives, and government support for organic farming. The Indian organic food market, valued at around USD 1 billion in 2023, is projected to grow at a CAGR of 20–25% by 2030, with organic sorghum playing a crucial role in this expansion. Organic sorghum is gaining popularity in gluten-free and diabetic-friendly diets, increasing its market scope for the health and wellness industry. The global organic food market was valued at USD 180 billion in 2023 and is expected to surpass USD 400 billion by 2032, driven by consumer demand for chemical-free, sustainable food. India, a leading producer of organic grains, has significant export opportunities for organic sorghum, especially in Europe, North America, and the Middle East, where demand for organic and gluten-free grains is rising.

11.2 Importance of Organic Cultivation of Sorghum

11.2.1 Environmental Benefits of Organic Sorghum Cultivation

Organic sorghum farming promotes sustainability by using natural fertilizers and insect control methods, as well as promoting soil health and biodiversity. By avoiding dangerous pesticides and fertilizers, it ensures safer consumption while minimizing the environmental impact by lowering the use of synthetic chemicals that damage water and soil (Takola et al. 2025; Abobatta and Fouad 2024). The deep roots of sorghum improve soil stability and structure, which are essential for reducing erosion during periods of intense precipitation. Sorghum's root structure helps to increase aggregate stability, which further reduces soil erodibility (Ola et al. 2015). By improving soil organic carbon storage, organic sorghum farming aids in carbon sequestration. Research shows that using organic fertilizers like manure instead of synthetic ones can lower greenhouse gas emissions and enhance agricultural systems' carbon budget (Yang et al. 2025; Tirkey 2024). Organic farming practices like mulching and composting enhance soil structure and water retention, increasing the drought resistance of crops like sorghum. Maintaining production in climatically changing environments requires this (Tirkey 2024; Varma et al. 2024). Because organic farming does not use synthetic fertilizers and pesticides, contamination of the land and groundwater is avoided. By encouraging beneficial microbes, this technique not only safeguards human health but also improves soil vitality (Varma et al. 2024).

11.2.2 Nutritional and Health Benefits

Organic sorghum contains high levels of phenolic compounds, which correlate strongly with antioxidant capacity (D'Almeida et al. 2025) These compounds, including flavonoids and tannins, are linked to reduced oxidative stress and potential neuroprotective effects (Rezaee et al. 2024) Sorghum is rich in essential minerals such as iron, zinc, and magnesium, which are crucial for overall health Its nutrient density supports dietary needs, particularly in food-insecure regions Sorghum serves as an excellent gluten-free grain, making it suitable for individuals with celiac disease or gluten intolerance (Fig. 11.1). Organic sorghum often exhibits

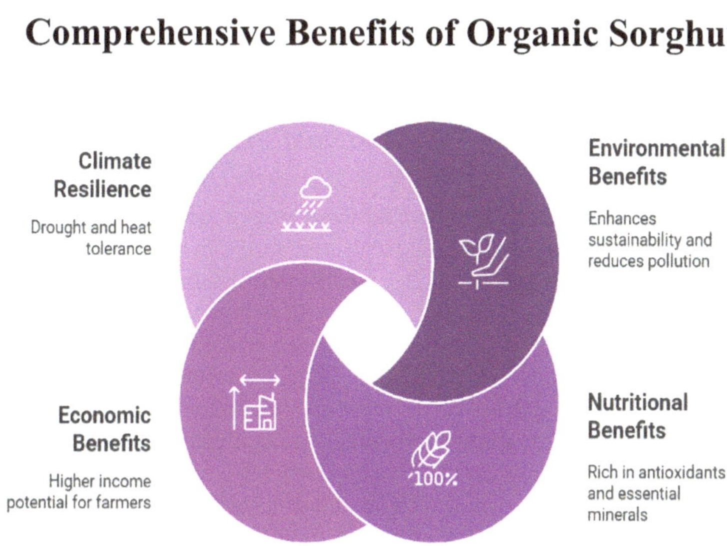

Fig. 11.1 Intersection of benefits in organic sorghum
This figure presents a Venn diagram highlighting four key advantages of organic sorghum *Climate Resilience*: Sorghum exhibits drought and heat tolerance, making it suitable for cultivation in arid and semi-arid regions *Environmental Benefits*: Its sustainable cultivation reduces pollution and enhances ecological balance *Economic Benefits*: Sorghum offers a higher income potential for farmers due to its adaptability and market demand *Nutritional Benefits*: The crop is rich in antioxidants and essential minerals, making it a valuable addition to a nutritious diet

a richer taste and aroma due to its natural growth processes, which can enhance culinary applications (Rajeshkumar et al. 2024) The low glycemic index of organic sorghum makes it beneficial for managing diabetes and obesity, as it helps regulate blood sugar levels (El Mahdi et al. 2024) The absence of synthetic chemicals in organic sorghum enhances gut health and reduces allergic reactions, promoting better digestibility (Wiryawan et al. 2024)

11.2.3 Economic and Social Benefits

Farmers may now command higher rates for organic sorghum due to the increased demand for organic products, increasing their potential revenue. Smallholder farmers are empowered by organic farming methods since they improve market access and lower the financial risks connected to chemical farming (Varma et al. 2024). This price increase is a result of growing customer awareness of the advantages for the environment and health (Kumar et al. 2024a). Because organic farming eliminates the need for pricey synthetic pesticides and fertilizers, production costs are reduced overall. This diversification fosters ecological equilibrium and lowers the hazards associated with monoculture farming. Majhi (2024). Crop rotation and the use of natural fertilizers are two methods that increase soil health while lowering costs. Making the switch to organic practices can improve resilience and long-term economic viability (Rani and Johar 2024).

11.2.4 Climate Resilience and Adaptability

Sorghum exhibits remarkable drought tolerance through mechanisms such as early flowering and osmotic adjustment, allowing it to thrive in extreme conditions Its cultivation can conserve water resources, which is critical in the face of increasing drought conditions globally (Gidi 2023). Sorghum requires significantly less water compared to traditional crops like rice and wheat, making it a more sustainable option in water-scarce regions (Kirton 2024). Genetic improvements and biotechnological advancements are enhancing sorghum's resilience to heat and drought stress, making it a model crop for studying abiotic stress responses (Mwamahonje et al. 2024; Gidi 2023). Rotating sorghum with legumes and nitrogen-fixing crops improves soil fertility and reduces reliance on chemical fertilizers, promoting sustainable agricultural practices. This practice not only enhances soil health but also contributes to increased crop yields and resilience against climate variability. Organic farming practices associated with sorghum cultivation enhance soil moisture retention, mitigating the impacts of unpredictable weather patterns (Dar et al. 2024). This adaptability allows it to serve as a reliable food source in regions where other crops may fail due to poor soil quality (Getachew et al. 2023) (Fig. 11.2).

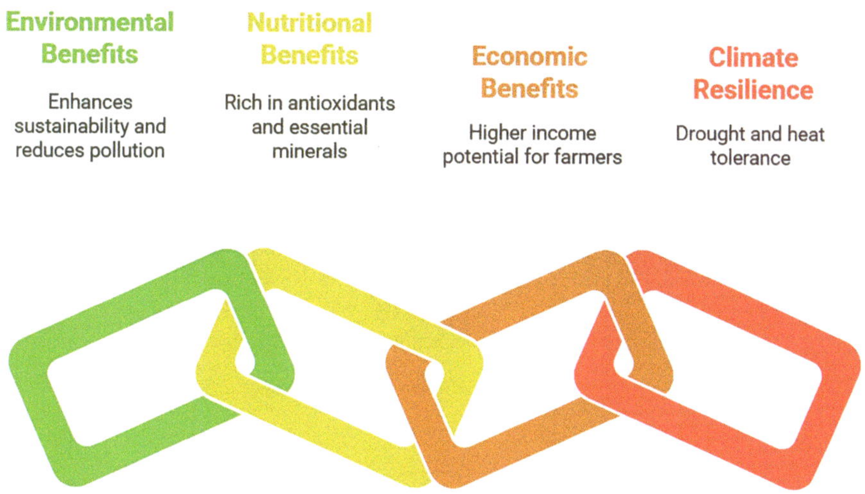

Fig. 11.2 Linked benefits of organic sorghum
This figure uses a chain-link design to illustrate the interconnectedness of the four benefits of organic sorghum: Green link, environmental benefits, yellow link, nutritional benefits, orange link, economic benefits, red link, climate resilience

11.3 Advantages of Organic Sorghum

11.3.1 Nutritional and Health Benefits of Organic Sorghum Grain

Sorghum is a good source of essential minerals such as iron, magnesium, phosphorus, and zinc, which are vital for various metabolic functions and reducing malnutrition risks (Tanwar et al. 2023; Dhar et al. 2024). It has a lot of dietary fiber, which helps with digestion and may help with weight control. In total, 10–12% of sorghum's protein helps maintain and develop muscle (Tanwar et al. 2023; Abbas et al. 2024). Flavonoids and phenolic substances found in the grain have antioxidant qualities that lower inflammation and oxidative stress (Pal and Gandhi 2024; Dhar et al. 2024). Sorghum is appropriate for diabetic diets because of its low-glycemic index, which helps control blood sugar levels. Sorghum is a healthy substitute for wheat-based products and is a gluten-free grain that can be used by those with celiac disease and gluten intolerance (McGinnis and Webster 2024).

11.3.2 Benefits of Organic Sorghum as Fodder

The use of organic sorghum aligns with sustainable agricultural practices, enhancing soil health and carbon sequestration (Sathiyabama et al. 2024). Sorghum's adaptability to harsh climates ensures consistent fodder availability, crucial for

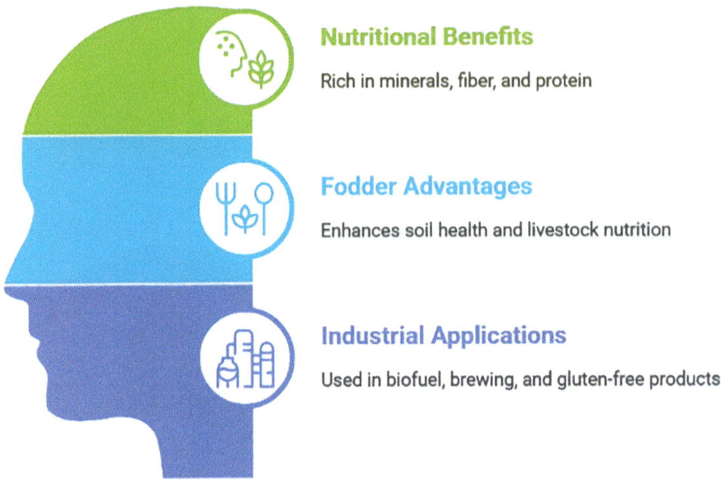

Fig. 11.3 Exploring organic sorghum benefits

livestock in arid regions (Mohamed et al. 2022). Organic sorghum eliminates harmful chemical residues, ensuring safer consumption of animal products. Sorghum contains 9–13% protein, essential for growth and metabolic functions (Tanwar et al. 2023). Enhanced protein levels contribute to improved milk production in dairy cattle. Studies show that sorghum's digestibility rates for dry matter and organic matter range from 76.29% to 82.16% (Liman et al. 2023) (Fig. 11.3).

11.3.3 Industrial and Value-Added Product Applications

Organic sorghum is a key feedstock for bioethanol, with studies indicating its effective conversion into fuel through bioconversion methods (Hu et al. 2022; Nasidi et al. 2016). Whole crop residues, including stalks and leaves, enhance ethanol yield, promoting sustainable biofuel production without competing with food supplies (Nasidi et al. 2016; Dahlberg 2019). The grain is increasingly used in craft brewing, particularly for gluten-free beers, capitalizing on its nutritional profile (Alavi et al. 2018). Organic sorghum flour is gaining traction in health-conscious markets, utilized in baking, pasta, and snacks, appealing to consumers seeking gluten-free options (Alavi et al. 2018). Sorghum serves as a sustainable alternative to maize and wheat in animal feed, particularly in poultry and aquaculture, providing essential protein sources (Alavi et al. 2018; Dahlberg 2019).

11.4 Qualitative Parameters of Organic Cultivated Sorghum: Crude Protein, Fiber, and Grain Quality

11.4.1 Crude Protein Content in Organic Sorghum

Organic sorghum generally exhibits a crude protein content ranging from 9% to 12% (Tanwar et al. 2023). Variability in protein levels can be attributed to genetic diversity and environmental factors, with some tropical accessions showing values as high as 15% (Cuevas et al. 2023). Organically grown sorghum has higher levels of essential amino acids, particularly lysine and methionine, which are crucial for muscle growth and metabolic functions (Tanwar et al. 2023). Organic sorghum typically contains 6–12% dietary fiber, varying by variety (Tanwar et al. 2023). The lower tannin content in organic sorghum enhances protein digestibility, making it easier for the body to absorb nutrients (Gallo et al. 2024).

11.4.2 Fiber in Organic Sorghum

Organic sorghum offers high fiber and resistant starch, promoting digestive health and reducing chronic disease risks. It supports gut microbiota and may lower colon cancer risk. However, limited global availability hinders widespread adoption (Stefoska-Needham 2024).

11.4.3 Grain Quality in Organic Sorghum

- *Lower Mycotoxin Contamination:* The absence of synthetic fertilizers in organic farming reduces the risk of fungal infections, leading to lower mycotoxin contamination. This results in safer grain quality, making organic sorghum a healthier choice for consumers (Jaćimović et al. 2023).
- *Enhanced Polyphenol Content:* Organic sorghum contains higher levels of polyphenols, which are known for their antioxidant properties. These compounds help mitigate oxidative stress and may lower the risk of chronic diseases (Dhar et al. 2024; Desta et al. 2023).
- *Better Grain Hardness and Milling Quality*: The density and uniformity of organic sorghum grains enhance their milling quality, resulting in better flour yield and texture. This is particularly beneficial for food processing applications (Pontieri et al. 2022). Conversely, while organic sorghum offers numerous

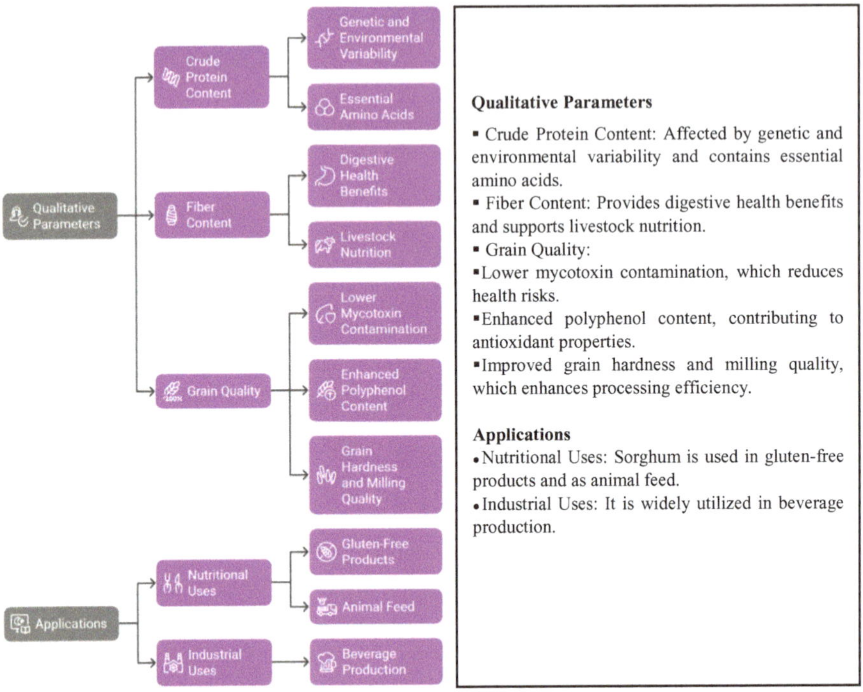

Fig. 11.4 Breakdown of qualitative parameters and applications of organic sorghum

advantages, conventional sorghum may still provide benefits in terms of yield and cost-effectiveness, which are critical for large-scale food production.

This hierarchical diagram (Fig. 11.4) categorizes the qualitative parameters and applications of organic sorghum into different aspects.

11.4.4 Industrial and Nutritional Applications of High-Quality Organic Sorghum

Organic sorghum flour is increasingly utilized in gluten-free baking, porridge, and whole-grain cereals. Its high antioxidant content and favorable glycemic index make it a healthy alternative to wheat, catering to the growing demand for gluten-free products (Slanev 2024; McGinnis and Webster 2024). Sorghum's rich fiber and protein content make it an ideal feed for dairy, poultry, and swine. Studies indicate that sorghum-based diets can match or exceed the performance of traditional feeds like maize, while also potentially reducing methane emissions in ruminants (Rugwete et al. 2024). In the beverage sector, organic sorghum is used in craft brewing and non-alcoholic malt drinks. Its unique flavor profile and lower glycemic

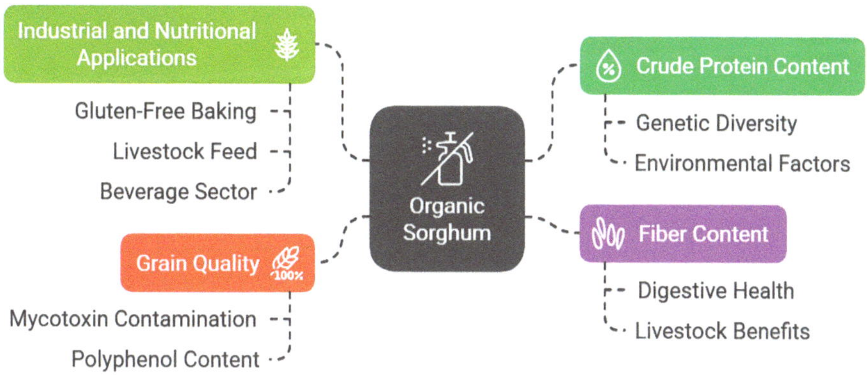

Fig. 11.5 Key attributes and applications of organic sorghum (Courtesy: McGinnis and Webster 2024)
This infographic presents the primary characteristics and applications of organic sorghum through a central concept map Crude Protein Content: Influenced by genetic diversity and environmental factors Fiber Content: Supports digestive health and livestock benefits Grain Quality: Affects mycotoxin contamination and polyphenol contentIndustrial and Nutritional Applications: Used in gluten-free baking Utilized as livestock feed Serves as an ingredient in the beverage sector

index make it a suitable ingredient for innovative beverages, including probiotic drinks (McGinnis and Webster 2024) (Fig. 11.5).

11.5 Management Practices for Organic Cultivation

11.5.1 Weed Management

- *Manual Weeding & Inter-Cultivation:* Techniques like hoeing and using tools such as tine weeders and rotary hoes effectively remove weeds during early growth stages (Aparna et al. 2024; Shrestha et al. 2024).
- *Crop Rotation:* Incorporating legumes like cowpea or pigeon pea before sorghum enhances soil fertility and suppresses weed growth (Patel et al. 2024).
- *Cover Cropping:* Fast-growing cover crops such as buckwheat and hairy vetch can outcompete weeds by shading them and inhibiting germination (Sharawat 2025).
- *High-Density Planting:* Utilizing narrow row spacing and increasing plant populations limits the space available for weeds to thrive (Kumar et al. 2024b).
- *Allelopathic Effects of Sorghum:* Sorghum residues effectively reduce weed density and enhance crop yields. Research indicates that sorghum surface mulching can achieve weed control efficiencies comparable to hand weeding (Patra et al. 2025; Tonni et al. 2024).

11.5.2 Nutrient Management

Compost and Farmyard Manure (FYM): These amendments enhance soil organic matter, microbial diversity, and nutrient availability, leading to improved soil structure and fertility (Singh et al. 2024; Khadim et al. 2024).

Biofertilizers: Inoculants such as Azospirillum and Mycorrhiza promote nitrogen fixation and phosphorus solubilization, significantly boosting nutrient uptake and soil health (Sathiyabama et al. 2024; Singh et al. 2024).

Green Manuring: Incorporating legumes like Crotalaria improves nitrogen content and soil structure, fostering a healthier growing environment for sorghum (Patel et al. 2024).

Foliar Sprays: Applications of seaweed extracts and compost tea enhance nutrient absorption and drought resilience, contributing to overall plant health (Sathiyabama et al. 2024).

Biochar: This amendment retains moisture and enhances microbial activity, increasing soil cation exchange capacity (CEC) and nutrient retention (Singh et al. 2024).

11.5.3 Pest Management

Parasitic Wasps: Species such as *Trichogramma chilonis* and *Telenomus remus* effectively target sorghum stem borers and fall armyworms, contributing to pest suppression (Khalid et al. 2024). *Fungal Bio-Pesticides:* Beauveria bassiana and Metarhizium anisopliae are utilized to control aphids, shoot flies, and cutworms, showcasing the efficacy of microbial agents in pest management (Tiwari 2024). *Neem Oil:* Extracted from *Azadirachta indica*, neem oil is effective against aphids, shoot flies, and sorghum midge, providing a natural alternative to synthetic pesticides (Bhall et al. 2024). *Diversity and Monitoring:* Maintaining insect diversity, including natural enemies, is crucial for effective pest management. Studies indicate that increased natural enemy populations can lead to reduced pest abundance (Bradicich et al. 2024). Intercropping sorghum with legumes like pigeon pea and cowpea attracts predatory insects, enhancing natural pest control and fostering a diverse ecosystem (Uyi and Toews 2024). While effective, these organic methods need further research to optimize their integration into pest management frameworks amid evolving challenges. Companion planting with marigold (*Tagetes* spp.) and basil (*Ocimum* spp.) also helps repel sorghum pests by releasing deterrent compounds, offering additional crop protection (Okosun et al. 2021).

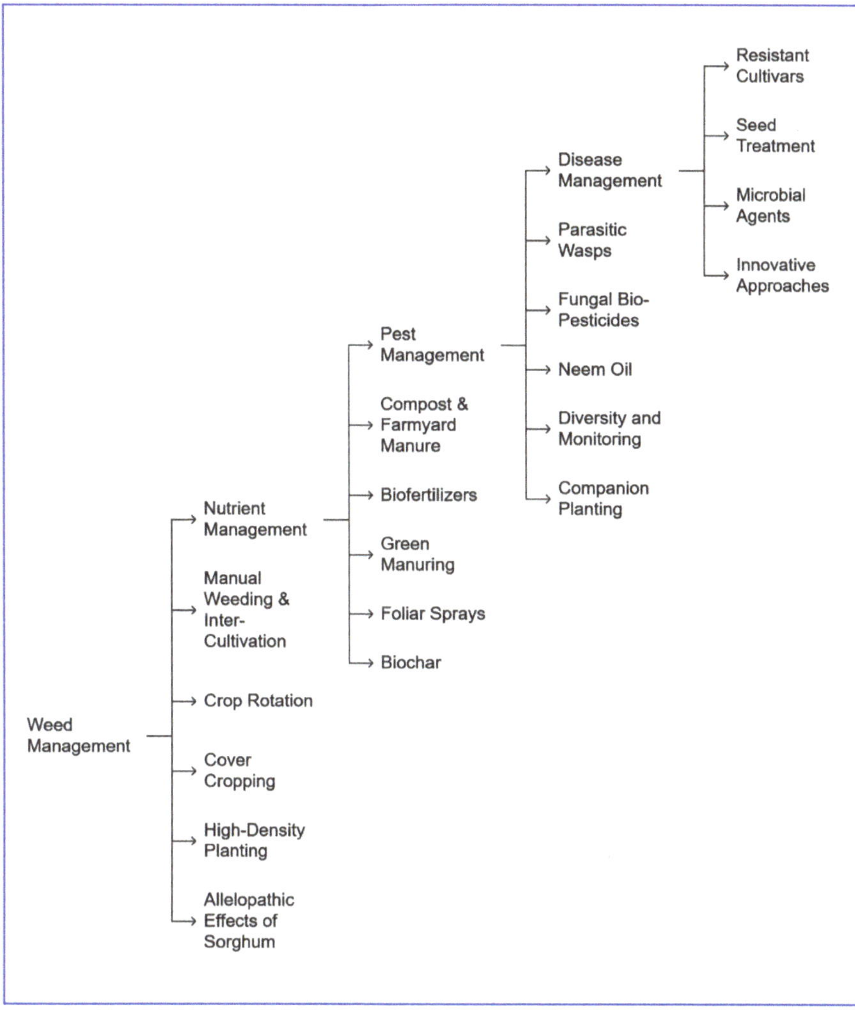

Fig. 11.6 Integrated organic sorghum management

11.5.4 Disease Management

Resistant cultivars like CSV 15, Phule Vasudha, and M 35–1 offer protection against major fungal diseases, ensuring yield stability (Aragaw and Terefe 2024; Anjum et al. 2024). Seed treatment with *Trichoderma viride* safeguards seedlings from blights and soil-borne fungi, promoting healthier plant establishment (Aragaw and Terefe 2024). Microbial agents such as *Bacillus subtilis* and *Pseudomonas fluorescens* effectively combat sorghum rust and anthracnose, providing a reliable biological control option (Anjum et al. 2024; Prom et al. 2024). Spraying cow urine extract at a 5% concentration helps reduce grain mold incidence, offering an innovative organic approach (Anjum et al. 2024) (Fig. 11.6).

11.6 Consequences and Limitations in Organic Cultivated Sorghum

Positive Consequences of Organic Sorghum Cultivation: By increasing organic matter and microbial activity, organic techniques like composting and cover crops enhance soil fertility and structure (Bhaumik et al. 2024). By avoiding artificial fertilizers and pesticides, ecosystem resilience is increased, and soil and water contamination is decreased (Yang et al. 2025). According to Takola et al. (2025), these systems help sustain soil bacteria and beneficial insects, resulting in a balanced agroecosystem that is necessary for sustainable farming. The lower yields of organic sorghum compared to conventional ways may deter farmers from implementing these approaches (Sharma and Bipana 2024). Negative consequences of organic sorghum cultivation are lower yields. Organic sorghum yields are typically 10–30% lower than conventional farming during the initial years. This is primarily due to the time required for soil nutrient balancing and the absence of synthetic fertilizers (Jiménez et al. 2025; Patel et al. 2024) (Figs. 11.7 and 11.8).

11.6.1 Pest and Weed Management Challenges

Weed Management Challenges: Cultural and Mechanical Approaches: Techniques such as surface mulching with sorghum residues have shown promise in reducing weed density and improving crop yield, achieving up to 80.9% weed control

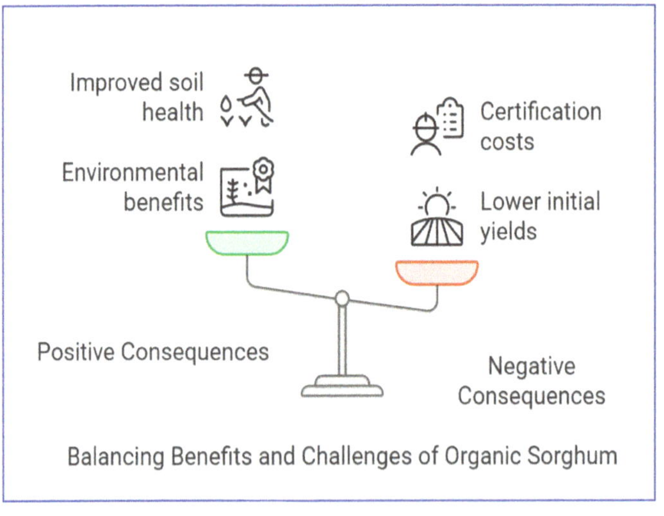

Fig. 11.7 Consequences and limitations in organic cultivated sorghum

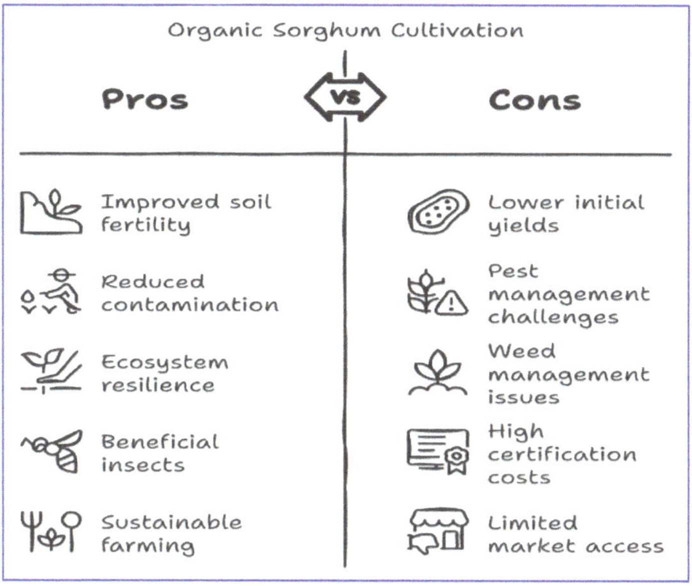

Fig. 11.8 Advantages and challenges of organic sorghum cultivation (Courtesy: Sharma and Bipana 2024)
This figure presents a comparative analysis of the pros and cons of organic sorghum cultivation, highlighting its benefits and limitations.

efficiency (Patra et al. 2025). Allelopathy: Sorghum's allelopathic properties can suppress weed growth, enhancing soil quality and reducing herbicide reliance (Patra et al. 2025). Utilizing variable seeding rates can optimize weed suppression and improve net returns, demonstrating a reduction in weed biomass by an average of 10 kg/ha (Loewen and Maxwell 2024).

Pest Management Strategies: Crop Rotation and Beneficial Insects: Organic farming relies on these methods to manage pests, although they can be unpredictable and require careful monitoring (Mansoor et al. 2024). Integrated Pest Management (IPM): Combining non-chemical methods with limited chemical use can enhance pest control effectiveness while minimizing environmental impact (Matzrafi et al. 2024).

11.6.2 Soil Microbial Activity

The nutrient cycling in organic systems relies heavily on microbial decomposition, which is influenced by environmental factors such as climate and soil moisture (Malhi and Brandt 2014). Adverse weather conditions can disrupt microbial processes, leading to reduced nutrient availability. For example, drought-stressed organic soils exhibited a 25% decrease in microbial activity, further limiting nutrient access for sorghum (Isha et al. 2024).

11.6.3 Slow Nutrient Release from Organic Fertilizers

Organic fertilizers, such as compost and manure, release nutrients gradually, which can delay crop uptake compared to synthetic fertilizers. A study indicated that the application of farmyard manure (FYM) alone resulted in lower soil nutrient levels compared to combined applications with inorganic fertilizers (Kisinyo et al. 2019).

11.6.4 Soil Acidity and Nutrient Deficiencies

Acidic soils, common in regions like Kenya, often exhibit low levels of nitrogen (N) and phosphorus (P), which are critical for sorghum growth. The application of organic amendments such as farmyard manure can improve soil pH but may not sufficiently address nutrient deficiencies compared to inorganic fertilizers (Kisinyo et al. 2019). Co-application of organic materials like charcoal and sago bark ash with inorganic fertilizers has been shown to enhance nitrogen availability and uptake, yet it can lead to reduced soil nitrogen levels over time (Hamidi et al. 2023). Long-term studies indicate that while organic amendments like manure and compost can improve soil health and increase sorghum yields, they may not fully replace the need for inorganic fertilizers to achieve optimal productivity (Soma et al. 2018).

11.6.5 Market Access and Certification Costs

The high costs associated with obtaining organic certification can deter farmers from entering the market, limiting their access to consumers willing to pay premium prices for organic products (Saini and Meena 2024). Limited market demand and inadequate marketing facilities further exacerbate the challenges faced by organic farmers, as highlighted in various studies (Chavan and Koshti 2023; Babajani et al. 2023). Government support programs, while beneficial, often fall short in addressing the comprehensive needs of organic farmers, particularly in terms of subsidies and market promotion (Kumar et al. 2024a, b).

11.6.6 Market Demand and Infrastructure

The organic sorghum market is niche, limiting sales opportunities for farmers (Kydyrbaeva et al. 2024). Poorly developed supply chains and storage facilities lead to post-harvest losses (Panday et al. 2024). A study indicated that organic grain farmers face a 20% market fluctuation risk due to inconsistent demand (Sandhiya and Balakrishnan 2024).

- *Land Expansion Requirements:* Organic farming typically requires 50% more land to match conventional yields (Sandhiya and Balakrishnan 2024). Extensive land use can threaten biodiversity if not managed properly.
- *Climatic and Environmental Limitations:* Organic sorghum is highly dependent on natural rainfall, making it vulnerable to climate change (Kumar and Kumari 2024). Research shows organic yields fluctuate 25% more than conventional yields under extreme weather conditions (Zhao et al. 2024).

Pros (Advantages)

- Improved Soil Fertility: Organic farming enhances soil structure and nutrient content.
- Reduced Contamination: Avoids synthetic pesticides and fertilizers, leading to cleaner produce and environment.
- Ecosystem Resilience: Supports biodiversity and soil health.
- Beneficial Insects: Encourages pollinators and natural pest predators.
- Sustainable Farming: Reduces dependency on chemical inputs and promotes long-term agricultural sustainability.

Cons (Challenges)

- Lower Initial Yields: Organic sorghum may produce lower yields initially compared to conventional farming.
- Pest Management Challenges: Without chemical pesticides, pest control relies on natural methods, which can be less effective.
- Weed Management Issues: Weeds compete with sorghum for nutrients, requiring labor-intensive or alternative management practices.
- High Certification Costs: Organic certification processes can be costly and time-consuming.
- Limited Market Access: Organic farmers may face challenges in reaching profitable markets due to niche demand.

Figure 11.9 explains that the organic approaches would enhance the soil fertility with fewer external inputs with gain of 1.5% soil organic matter retention. Productivity per unit area of sorghum can be increased by 2% or more through inorganic inputs; however, the long-term soil fertility would deteriorate, thus future yield plates are expected. Nutrition content and quality of organic sorghum for household consumption is recommended since the protein, fibres, and minerals are higher by 2% in research trials, which can be more achieved through a biofortification approach that suite for organic cultivations and targets for 50% higher nutrition than regular grains. By doing so, the cost of organic sorghum fetches higher price in markets (by 200USD/ha). At the end, environmental impact such as carbon footprint (400 kg CO_2 lesser) and water use efficiency (save 2000 liters of water) are highly supportive by promoting organic way of cultivations.

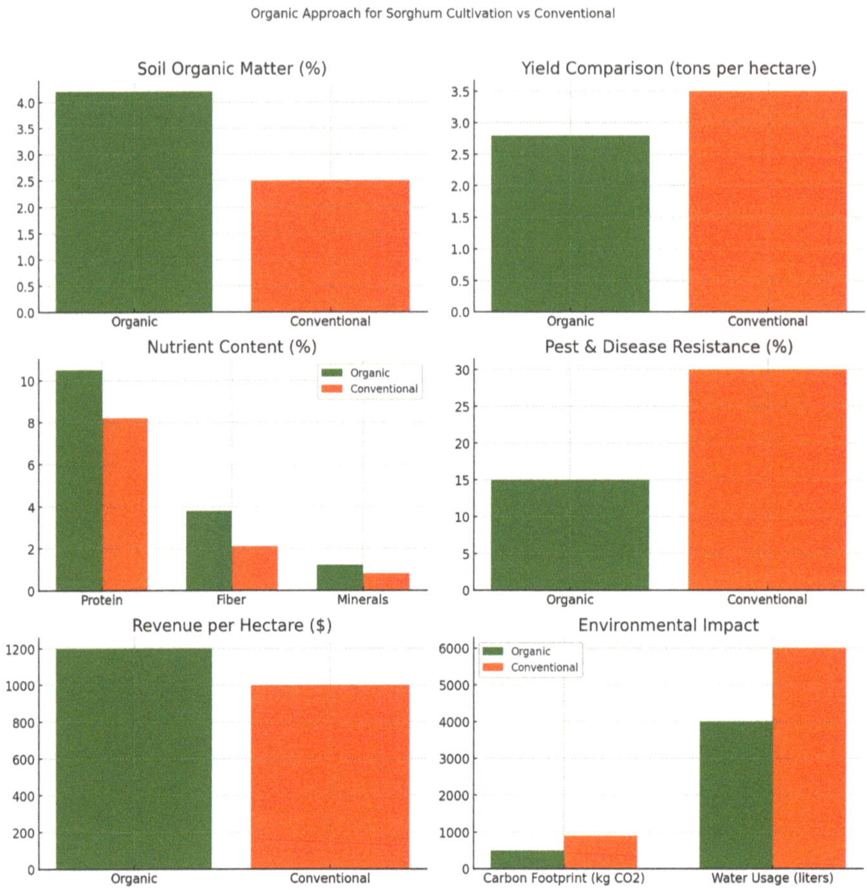

Fig. 11.9 Comparison of different aspects of an organic approach influences better sorghum cultivation and its impact on quality grains for nutritious food

11.7 Conclusion

Growing sorghum organically is a shining example of economic prosperity, sustainability, and good health. It ensures a cleaner, greener environment by removing synthetic contaminants, fostering biodiversity, protecting water supplies, and enhancing soil health. Sorghum's deep roots make it a potent ally in the battle against climate change because they improve soil stability, reduce erosion, and increase carbon sequestration. Organic sorghum is a superfood in terms of nutrition, full of dietary fiber, vital minerals, and antioxidants. Its high phenolic content promotes gut and heart health, lowers oxidative stress, and boosts immunity. A tasty and healthful substitute for processed grains, this gluten-free, low-glycemic grain is perfect for people with diabetes and gluten intolerance. By increasing market value, decreasing reliance on pricey synthetic inputs, and improving long-term financial security, organic sorghum gives farmers more economic leverage. Profitable prospects are guaranteed by the rising demand for organic products worldwide, while sustainable farming practices like crop rotation and composting enhance soil fertility and production even more. Another significant benefit is climate resilience; organic sorghum is a game-changer for areas that are vulnerable to drought since it can withstand extreme weather conditions with little water. Its versatility guarantees a future in which people and the environment prosper by promoting sustainable agriculture methods and ensuring food security.

References

Abbas SY, Fareed H, Zyara A (2024) The nutritional value of germinated sorghum. Humanit Nat Sci J 5(6):274–296. https://doi.org/10.53796/hnsj56/16

Abobatta WF, Fouad FW (2024) Sustainable agricultural development. In: Advances in environmental engineering and green technologies book series, pp 1–27. https://doi.org/10.4018/979-8-3693-4240-4.ch001

Alavi S, Mazumdar SD, Taylor JRN (2018) Modern convenient sorghum and millet food, beverage and animal feed products, and their technologies. In: Sorghum and millets: chemistry, technology, and nutritional attributes. Elsevier, pp 293–329. https://doi.org/10.1016/B978-0-12-811527-5.00010-1

Anjum S, Hanamaratti NG, Bandiwaddar TT, Shaila HM, Chattanavar SN (2024) Screening sorghum breeding lines for identification of resistance sources for multiple fungal diseases. J Sci Res Rep 30(12):263–270. https://doi.org/10.9734/jsrr/2024/v30i122671

Aragaw G, Terefe H (2024) Response of sorghum genotypes to anthracnose (Colletotrichum sublineolum) resistance under field conditions in eastern Ethiopia. PLoS One 19(12):e0316016. https://doi.org/10.1371/journal.pone.0316016

Aparna MC, Baskaran R, Ezhilarasi V, Kalpana R, Boominathan P, Jagadeeswaran R (2024) Advances and trends in weed management: a comprehensive review. Plant Sci Today 11(sp4). https://doi.org/10.14719/pst.5141

Babajani A, Muehlberger S, Feuerbacher A, Wieck C (2023) Drivers and challenges of large-scale conversion policies to organic and agro-chemical free agriculture in South Asia. Int J Agric Sustain 21. https://doi.org/10.1080/14735903.2023.2262372

Bhall R, Sharma BL, Yadav S, Kumar A, Arvind A (2024) Innovative strategies for controlling sorghum shoot fly: seed treatments and botanical insecticides. J Adv Biol Biotechnol 27(10):661–668. https://doi.org/10.9734/jabb/2024/v27i101487

Bhaumik S, Rajeev R, Kumar S, Kumari A (2024) Organic farming a pathway to achieve sustainable agriculture development: a comprehensive review. Int J Environ Climate Chang 14(11):419–435. https://doi.org/10.9734/ijecc/2024/v14i114557

Bradicich PA, Faris AM, Gordy JW, Brewer MJ (2024) Natural increases in parasitoid and predator abundances and a shift in species dominance point to improved suppression of the sorghum aphid since its invasion into North America. Insects 15(12):958. https://doi.org/10.3390/insects15120958

Chavan NYN, Koshti NR (2023) Constraints analysis in production and marketing of organic vegetables using Garrett's ranking technique. Asian J Agric Ext Econ Sociol 41(8):229–233. https://doi.org/10.9734/ajaees/2023/v41i82001

Cuevas HE, Peiris KHS, Bean SR (2023) Assessment of grain protein in tropical sorghum accessions from the NPGS germplasm collection. Agronomy 13(5):1330. https://doi.org/10.3390/agronomy13051330

D'Almeida CTS, Silva Oliveira MEA, da Silva Lima LR, Carvalho CWP, Mameri H, 1Ferreira MSL (2025) Sorghum. In: Millet: crops for climate resilience and for food and nutritional security, pp 239–269. https://doi.org/10.1007/978-981-95-1256-0

Dahlberg J (2019) The role of sorghum in renewables and biofuels. In: Sorghum: methods and protocols, pp 269–277. https://doi.org/10.1007/978-1-4939-9039-9_19

Dar KA, Singh L, Saad AA, Rai U, Tanwar H, Khatoon A (2024) Climate smart agriculture: innovating sustainable practices for a changing climate. J Sci Res Rep 30(11):568–576. https://doi.org/10.9734/jsrr/2024/v30i112585

Desta KT, Choi Y, Shin M-J, Yoon H, Wang X-H, Lee Y-J, Yi J-Y, Jeon Y, Lee S (2023) Comprehensive evaluation of nutritional components, bioactive metabolites, and antioxidant activities in diverse sorghum (Sorghum bicolor (L.) Moench) landraces. Food Res Int 173(Pt 2):113390. https://doi.org/10.1016/j.foodres.2023.113390

Dhar AK, Kumari BM, Kavithamani D, Boopathi NM, Periasamy M (2024) Understanding the advances in sorghum grain quality improvement: an overview. Plant Sci Today. https://doi.org/10.14719/pst.3527

El Mahdi N, Lemée L, Blancart Remaury Q, Eloy L, Nhiri N, Lakhssassi N, Cacciola F, Nhiri M (2024) Potential of sorghum seeds in alleviating hyperglycemia, oxidative stress, and glycation damage. Molecules 29(15):3445. https://doi.org/10.3390/molecules29153445

Gallo TCB, Beaumal V, Houinsou-Houssou B, Viau M, Ribourg-Birault L, Sotin H, Bonicel J, Boire A, Queiroz VAV, Mameri H, Guyot S, Riaublanc A, Nicoletti VR, Berton-Carabin CC (2024) Sorghum protein ingredients: production, compositional variability and enhancement of aqueous dispersibility through homogenization. Fut Foods. https://doi.org/10.1016/j.fufo.2024.100323

Getachew F, Bayabil HK, Hoogenboom G, Kiker GA, Yu Z, Li Y (2023) Development of climate-smart sorghum ideotype for climate resilience in Ethiopia. Field Crop Res 109135. https://doi.org/10.1016/j.fcr.2023.109135

Gidi M (2023) Sorghum as a model crop for drought stress tolerance. Adv Biosci Bioeng. https://doi.org/10.11648/j.abb.20231103.14

Hamidi NH, Ahmed OH, Omar L, Ch'ng HY, Johan PD, Paramisparam P, Musah AA, Jalloh MB (2023) Co-application of inorganic fertilizers with charcoal and sago bark ash to improve soil nitrogen availability, uptake, use efficiency, and dry matter production of sorghum cultivated on acid soils. Sustainability 15(1):827. https://doi.org/10.3390/su15010827

Hu W-B, Zhou L, Chen J (2022) Conversion sweet sorghum biomass to produce value-added products. Biotechnol Biofuels Bioprod 15(1). https://doi.org/10.1186/s13068-022-02170-6

Isha I, Gautam P, Pareek N, Chandra S, Paul J, Singh A (2024) Establishment and evaluation of soil fertility gradient on the basis of sorghum fodder yield, nutrient uptake and post-harvest soil test values. Int J Plant Soil Sci 36(12):38–46. https://doi.org/10.9734/ijpss/2024/v36i125181

Jaćimović S, Kiprovski B, Ristivojević P, Dimić D, Nakarada Đ, Dojčinović B, Sikora V, Teslic N, Pantelić N (2023) Chemical composition, antioxidant potential, and nutritional evaluation of cultivated sorghum grains: a combined experimental, theoretical, and multivariate analysis. Antioxidants 12(8). https://doi.org/10.3390/antiox12081485

Jiménez CE, Hernández IZ, Aguilar FB, Jiménez JR, Guevara-Hernández F (2025) Evaluación de una enmienda orgánica sobre variables agronómicas y rendimiento de Sorghum bicolor en La Frailesca, Chiapas, México. Agriscientia 41(2):113–122. https://doi.org/10.31047/1668.298x.v41.n2.42951

Khalid S, Qadeer S, Rahman A, Zamir MSI, Shafiq S, Iftikhar A (2024) Biological control agents for pest management and advancements in crop development. In: Advances in environmental engineering and green technologies book series, pp 203–226. https://doi.org/10.4018/979-8-3693-3061-6.ch009

Khadim MD, Wesal AB, Peezhand AW (2024) Overview of the impact of compost on bulk density, aggregate consistency and cation exchange capacity of soils and its consequential effect on crop productivity. Cognizance J 4(6):344–359. https://doi.org/10.47760/cognizance.2024.v04i06.021

Kirton CC (2024) Enhancing climate resilience in sorghum production. Int J Agric Nutr. https://doi.org/10.33545/26646064.2024.v6.i1b.144

Kisinyo PO, Opala PA, Gudu S (2019) Response of sorghum (Sorghum bicolor (L.) Munch) and chemical characteristics of soil to organic and inorganic fertilizers on Kenyan lower midlands acid soil. Int J Plant Soil Sci:1–8. https://doi.org/10.9734/IJPSS/2019/V28I130099

Kumar A, Juyal R, Nautiyal M (2024a) The economics of organic farming: financial aspects and market trends. https://doi.org/10.20944/preprints202412.1150.v2

Kumar KMH, Kumari S (2024) Sweet sorghum supply chain for bio-ethanol production. In: Advances in logistics, operations, and management science book series, pp 261–274. https://doi.org/10.4018/979-8-3693-4330-2.ch010

Kumar R, Kaur P, Shergill R (2024b) An integrated approach to weed control in sorghum and pearl millet. In: Handbook of integrated weed management for major field crops, pp 62–70. https://doi.org/10.2174/97898153050981240100009

Kydyrbaeva EO, Байдыбекова СК, Syzdykbayeva N (2024) Food storage: objectives, scientific principles. Problemy Agrorynka 4:187–196. https://doi.org/10.46666/2024-4.2708-9991.17

Liman L, Tantalo S, Muhtarudin M, Penta W, Priscilla K, Ramadhanu R, Wijaya AK, Adhianto K (2023) Effect of substitution of corn (Zea mays) fodder with sorghum (Sorghum bicolor var. Numbu) fodder on nutrient digestibility and daily weight gain of beef cattle. J Agric Sci. https://doi.org/10.4038/jas.v18i3.9449

Loewen SA, Maxwell BD (2024) Site-specific weed management on organic grain farms using variable rate seeding and data driven simulation. Weed Res 65(1). https://doi.org/10.1111/wre.12669

Majhi A (2024) Impact of organic farming in economic development in agriculture. Indian Sci J Res Eng Manag 08(11):1–8. https://doi.org/10.55041/ijsrem38387

Malhi SS, Brandt SA (2014) Potential of management practices and amendments to prevent nutrient deficiencies under organic cropping. https://harvest.usask.ca/handle/10388/8980

Mansoor M, Padmaja B, Bindhu GN, Ramprakash T (2024) Effect of integrated weed management practices on growth and yield of Rabi sorghum (Sorghum bicolor. L). J Exp Agric Int 46(11):615–621. https://doi.org/10.9734/jeai/2024/v46i113083

Matzrafi M, Blank L, Lati R (2024) Selection and adaptation to weed management methods: implications for non-chemical and integrated weed management approaches. Pest Manag Sci. https://doi.org/10.1002/ps.8463

McGinnis MJ, Webster MJ (2024) Current and potential future uses of sorghum to increase nutrient density for human foods. J Food Sci. https://doi.org/10.1111/1750-3841.17571

Mohamed HI, Fawzi EM, Basit A, Lone R, Sofy MR (2022) Sorghum: nutritional factors, bioactive compounds, pharmaceutical and application in food systems: a review. Phyton 91(7):1303–1325. https://doi.org/10.32604/phyton.2022.020642

Mwamahonje A, Mdindikasi Z, Mchau D, Mwenda ET, Sanga DN, Garcia-Oliveira AL, Ojiewo CO (2024) Advances in sorghum improvement for climate resilience in the global arid and semi-arid tropics: a review. Agronomy 14(12):3025. https://doi.org/10.3390/agronomy14123025

Nasidi M, Agu RC, Deeni YY, Walker GM (2016) Utilization of whole sorghum crop residues for bioethanol production. J Inst Brew 122(2):268–277. https://doi.org/10.1002/JIB.324

Okosun OO, Allen KC, Glover JP, Reddy GVP (2021) Biology, ecology, and management of key sorghum insect pests. J Integr Pest Manag 12(1). https://doi.org/10.1093/JIPM/PMAA027

Ola A, Dodd IC, Quinton J (2015) Can we manipulate root system architecture to control soil erosion? Soil 1(2):603–612. https://doi.org/10.5194/SOIL-1-603-2015

Pal AD, Gandhi R (2024) Determination of proximate composition, phytochemical content and gut-health-promoting efficacy of developed sorghum-based functional beverage. J Sci Res 16(2):537–550. https://doi.org/10.3329/jsr.v16i2.68262

Panday P, Pant SC, Singh P, Yadav H (2024) Agricultural warehousing. In: Advances in logistics, operations, and management science book series, pp 217–238. https://doi.org/10.4018/979-8-3693-4330-2.ch008

Patel KM, Desai LJ, Patel P, Patel VK (2024) Nitrogen management through organics on the basis of soil test values (STV) in sorghum (Sorghum bicolor)–lucerne (Medicago sativa) cropping system. Indian J Agron 69(4):404–409. https://doi.org/10.59797/ija.v69i4.5543

Patra PC, Jaswal A, Fatima I (2025) Enhancing food security through sustainable agriculture: investigating the allelopathic effects of sorghum on weed management in field pea (Pisum sativum var. arvense). Nat Environ Poll Technol 24(S1):273–283. https://doi.org/10.46488/nept.2024.v24is1.020

Pontieri P, Troisi J, Calcagnile M, Bean SR, Tilley M, Aramouni FM, Boffa A, Pepe G, Campiglia P, Del Giudice F, Chessa A, Smolensky D, Aletta M, Alifano P, Del Giudice L (2022) Chemical composition, fatty acid and mineral content of food-grade white, red and black sorghum varieties grown in the Mediterranean environment. Foods 11(3):436. https://doi.org/10.3390/foods11030436

Prom LK, Ahn E, Cuevas HE, Liu J, Isakeit T, Magill C (2024) Association and interrelationship among agronomic traits and fungal diseases of sorghum, anthracnose and grain mold. Crops 4(4):651–666. https://doi.org/10.3390/crops4040045

Rajeshkumar S, Anand GS, Narayanan SL, Subramanian E, Mini ML (2024) Red sorghum [Sorghum bicolor L. Moench]: A biochemical, nutritional hub and key food supplement - a review. Plant Sci Today 11(sp4). https://doi.org/10.14719/pst.5738

Rani S, Johar V (2024) Organic farming: a way forward for sustainable development. In: Futuristic trends in agriculture engineering & food sciences, pp 66–81. https://doi.org/10.58532/v3bcagp1ch4

Rezaee N, Hone E, Sohrabi HR, Johnson S, Liu Z, Chatur P, Gunzburg S, Martins RN, Fernando WMADB (2024) Sorghum grain polyphenolic extracts demonstrate neuroprotective effects related to Alzheimer's disease in cellular assays. Foods 13(11):1716. https://doi.org/10.3390/foods13111716

Rugwete N, Mutibvu T, Halimani T (2024) Inclusion of sorghum in ruminant diets towards methane emission mitigation and improved meat quality: a review. J Dairying Foods Home Sci. https://doi.org/10.18805/ajdfr.drf-333

Saini S, Meena HR (2024) Organic farming and marketing challenges: a case study of Haryana. ShodhKosh J Vis Per Arts 5(4). https://doi.org/10.29121/shodhkosh.v5.i4.2024.2436

Sandhiya V, Balakrishnan TR (2024) Constraints faced by the farmers in adoption of recommended sorghum practices by Atma in Dindigul District of Tamil Nadu, India. Int J Curr Microbiol Appl Sci 13(11):137–141. https://doi.org/10.20546/ijcmas.2024.1311.016

Sathiyabama N, Sathiya Bama K, Jayashree R, Kaleeswari RK, Ganesan KN, Kalpana R, Anandham R (2024) Nutrient alchemy: optimizing multicut fodder sorghum for yield, quality and environmental balance. Plant Sci Today. https://doi.org/10.14719/pst.4809

Sharawat V (2025) Influence of integrated nutrient management on sustaining legume crop productivity and reducing environmental impact: a review. J Food Legum 38(1):14–25. https://doi.org/10.59797/jfl.v38i1.1548

Sharma S, Bipana KC (2024) Sustainable agriculture and its practices: a review. Turk J Agric Food Sci Technol 12(12):2631–2639. https://doi.org/10.24925/turjaf.v12i12.2631-2639.6622

Shrestha S, Beneton K, Abit MJM, Shrestha S, Dar A (2024) Perspective chapter: management of weeds in organic farming system—special focus on organic vegetable farms of the USA. IntechOpen. https://doi.org/10.5772/intechopen.1004309

Singh S, Devi NB, Divakaran D, Kumar A, Singh S, Tyagi G, Nath A, Kumar A (2024) Soil fertility management: role of organic amendments and bio-fertilizers: a review. Int J Res Agron 7(12):766–772. https://doi.org/10.33545/2618060x.2024.v7.i12j.2253

Slanev K (2024) Prospects for sorghum utilization as human food. Rasteniev″dni Nauki 61(6):46–51. https://doi.org/10.61308/gsle3638

Soma DM, Kiba DI, Ewusi-Mensah N, Gnankambary Z, Lompo F, Sedogo MP, Abaidoo RC (2018) Changes in sorghum production, soil P forms and P use efficiency following long-term application of manure, compost and straw in a Ferric Lixisol. Acta Agric Scand Sect B Soil Plant Sci 68(5):401–411. https://doi.org/10.1080/09064710.2017.1414871

Stefoska-Needham A (2024) Sorghum and health: an overview of potential protective health effects. J Food Sci. https://doi.org/10.1111/1750-3841.16978

Takola E, Korell L, Beckmann M, Bonfanti J, Reitz T, Tamburini G (2025) Does sustainable agriculture promote biodiversity and yield? A second-order meta-analysis. https://doi.org/10.32942/x25620

Tanwar R, Panghal A, Chaudhary G, Kumari A, Chhikara N (2023) Nutritional, phytochemical and functional potential of sorghum: a review. Food Chem Adv. https://doi.org/10.1016/j.focha.2023.100501

Tirkey RS (2024) The role of organic farming for climate change mitigation and sustainable development: a review. Int J Multidiscip Res 6(4). https://doi.org/10.36948/ijfmr.2024.v06i04.26493

Tiwari AK (2024) Insect pests in agriculture identifying and overcoming challenges through IPM. Arch Curr Res Int 24(3):124–130. https://doi.org/10.9734/acri/2024/v24i3651

Tonni NH, Sarker UK, Ahmed M, Upama SA, Rahman MM, Uddin MR (2024) Relative benefits of aqueous extract of sorghum residues with herbicide on weed control and yield of wheat varieties. J Bangladesh Agric Univ 22(3):326–334. https://doi.org/10.3329/jbau.v22i3.76405

Tsehaye Y, Menamo TM, Abay F, Tadesse TT, Bantte K (2024) Multi-locus genome-wide association study for grain yield and drought tolerance indices in sorghum accessions. Plant Genome e20505. https://doi.org/10.1002/tpg2.20505

Uyi O, Toews MD (2024) Melanaphis sorghi: a review and synthesis of its control options 10 years post detection of a new invasive haplotype in The United States of America. CAB Rev. https://doi.org/10.1079/cabireviews.2024.0046

Varma NR, Wadatkar H, Salve R, Kumar TV (2024) Advancing sustainable agriculture: a comprehensive review of organic farming practices and environmental impact. J Exp Agric Int 46(7):695–703. https://doi.org/10.9734/jeai/2024/v46i72623

Wiryawan DM, Ariestanti CA, Prasetyaningsih A (2024) Sorghum (Sorghum bicolor L. Moench var. bioguma) cookies (SoKis): source of antioxidant and prebiotic. Agritech 44(4):289. https://doi.org/10.22146/agritech.87733

Yang W, Zhao Y, Zhang R, Wang L, Yang Y, Xia S, Jia Y, Qu Z (2025) Greenhouse gas emission, carbon footprint and the ecosystem carbon budget of sorghum production: replacing chemical fertilizer with partial manure in saline soils. Elsevier. https://doi.org/10.2139/ssrn.5090285

Zhao T, Kubota H, Hernandez-Ramirez G (2024) Contrasting soil organic carbon concentrations and mass storage between conventional farming and organic farming: a meta-analysis. Sustainability 16(24):11260. https://doi.org/10.3390/su162411260

Chapter 12
Optimizing Millet Nutrition and Yield Through Sustainable Fertilization and Soil Nutrition Practices

Kannan Pandian, Mohamed Roshan Abu Firnass Mustaffa, Surya Teja Varanasi, Manimaran Gajendiran, Abishek John Prince, and Mahalingam Govindaraj

Abstract Millets, which are adopted to abiotic stresses, are an important source of food and nutrition in dry and semi-arid regions. They are still underestimated and generally viewed as food for underprivileged groups, limiting their widespread adoption despite their high-nutritional profile and climate readiness. However, soil health has deteriorated due to a continuous reliance on chemical fertilizers without organic inputs, resulting in multi-nutrient deficiencies and lower millet yields. To improve the sustainability of millet production, balanced fertilization is required to enhance nutrient management. Soil organic carbon (SOC), nutrient buffering capacity, and overall soil health can all benefit from integrated nutrient management (INM), which includes organic amendments and biofertilizers. Climate-smart approaches, including nanofertilizers, site-specific nutrient management (SSNM), and precision fertilization, all aim to increase nutrient use efficiency while minimizing negative environmental impact. In the meantime, incorrect assumptions about millet's low nutrient requirements, insufficient extension services, and dangerous fertilizer supply chains impede the implementation of balanced fertilization. Despite improvements, challenges remain regarding developing millet-specific fertilizer, optimizing nutrient applications in different agro-ecological zones, and evaluating long-term effects on soil microorganisms and carbon sequestration. Future studies should include digital tools such as artificial intelligence and remote sensing to

K. Pandian (✉)
Centre for Water and Geospatial Studies, Tamil Nadu Agricultural University, Coimbatore, Tamil Nadu, India
e-mail: kannan.p@tnau.ac.in

Mohamed Roshan Abu Firnass Mustaffa · S. T. Varanasi · M. Gajendiran · A. J. Prince
Department of Soil Science and Agricultural Chemistry, Tamil Nadu Agricultural University, Coimbatore, Tamil Nadu, India

M. Govindaraj
Alliance of Bioversity International and the International Centre for Tropical Agriculture (CIAT), Cali, Colombia

M. Govindaraj (ed.), *Breeding Biofortified Millets for Nutritional Security*, Sustainability Sciences in Asia and Africa,
https://doi.org/10.1007/978-981-95-9827-4_12

ensure correct nutrition regulation. As a result, this chapter covers various fertilization techniques, challenges, and potential future directions, emphasizing their importance for sustainable agriculture, food security, and climate resilience.

Keywords Crop nutrition • Fertilizers • Millets • Nutritional security • Sustainability

12.1 Introduction

Malnutrition remains a global challenge, with 1.9 billion adults overweight or obese and 462 million undernourished (WHO 2018). Including nutrient-dense millet in meals provides a sustainable way to address hidden hunger and food security. Often referred to as "Nutri-cereals," millets belong to the Poaceae family and have been grown for 8000 years in the mountains in central China (Nainwal et al. 2018). Africa leads in millet cultivation, accounting for 59% of the total area and 55% of global output, followed by Asia with 38% of the area and 42% of production. In 2024, global millet production reached 31.1 million tonnes (USDA), with India being the largest producer, followed by Niger, China, and Mali. Recognizing their importance, the United Nations General Assembly (UNGA) declared 2023 as the "International Year of Millets" to promote their benefits and encourage consumption (Harish et al. 2024).

India is a leading producer of millets, cultivating diverse species such as pearl millet (*Cenchrus americanus*), finger millet (*Eleusine coracana*), kodo millet (*Paspalum scrobiculatum*), foxtail millet (*Setaria italica*), proso millet (*Panicum miliaceum*), barnyard millet (*Echinochloa esculenta*), little millet (*Panicum sumatrense*), teff (*Eragrostis tef*), guinea millet (*Brachiaria deflexa*), fonio (*Digitaria exilis*), browntop millet (*Urochloa ramosa*), and Job's tears (*Coix lacryma-jobi*), as displayed in Table 12.1 (FAO 2023). Among these, pearl millet and sorghum are widely grown; others are categorized as "minor millets," or "small millets." From a nutritional standpoint, in every 100 g, millets provide 7.3–12.5 g of protein, 60.9–75.7 g of carbohydrates, 1.1–5.8 g of fat, and 1.2–14.7 g of crude fiber, surpassing traditional cereals. They are mineral-rich, containing 1.6–4.7 g of minerals, 14–344 mg of calcium, 121–296 mg of phosphorus, 2.2–17.47 mg of iron, and 3.01–60.6 mg of zinc (Dayal et al. 2023). Millets contain essential amino acids, particularly sulfur-rich ones, along with beneficial phytochemicals like phytic acid. Being gluten-free, they are ideal for individuals with coeliac disease and offer significant health benefits for those with diabetes and heart disease (Coulibaly et al. 2011). Their high nutritional value supports food security and promotes better health outcomes (Fig. 12.1). As climate-resilient crops, millets contribute to multiple UN Sustainable Development Goals (SDGs). Their ability to thrive in diverse agro-climatic conditions with minimal inputs enhances food security (SDG 2) and supports sustainable farming under drought and heat stress (SDG 13). Their rich nutritional profile improves health (SDG 3), while millet-based agriculture drives

Table 12.1 Classification of millets and its nutritional benefits

Common name	Scientific name	Origin	Nutritional benefits
Pearl millet	*Pennisetum glaucum*	West Africa	Gluten-free, antioxidant-rich, full of iron, magnesium, phosphorus, copper, iron, and protein. Helps control cardiovascular diseases and diabetes.
Finger millet	*Eleusine coracana*	East Africa	Rich in thiamin, copper, magnesium, phosphorus, selenium, and iron. With a low glycemic index, it supports bone health and diabetes management.
Foxtail millet	*Setaria italica*	China	Rich in thiamin, pantothenic acid, copper, magnesium, and phosphorus, while also providing iron, niacin, vitamin B6, and zinc.
Proso millet	*Panicum miliaceum*	Asia	Gluten-free, rich in vitamin B, magnesium, and phosphorus. Aids in lowering cholesterol and improving digestion.
Barnyard millet	*Echinochloa frumentacea*	India	High in fiber, iron, and calcium; low glycemic index. It helps with weight loss and managing blood sugar levels.
Little millet	*Panicum sumatrense*	India	Rich in B vitamins, minerals like potassium and zinc, and dietary fiber. Supports skin health and improves metabolism.
Kodo millet	*Paspalum scrobiculatum*	India	High in fiber, protein, and antioxidants. It helps reduce obesity and diabetes.
Sorghum	*Sorghum bicolor*	Africa	Gluten-free, rich in antioxidants, fiber, and protein. Supports heart health, aids digestion, and helps manage diabetes.
Teff	*Eragrostis tef*	Ethiopia	High in calcium, iron, and resistant starch. Gluten-free, beneficial for bone health, and managing anemia.
White Fonio	*Digitaria exilis*	West Africa	Rich in amino acids, iron, and zinc. Gluten-free, supports immune function and energy production.

rural economic growth, reducing poverty (SDG 1). Promoting millets is vital for food security, climate resilience, and economic stability (Pradhan et al. 2021).

Despite their agronomic and nutritional benefits, millet cultivation has declined over the past decades due to lower productivity and poor economic returns. With 91% of millet farming in rainfed areas, nutrient and moisture deficits limit yields to three to four times below potential yield (Mubeena et al. 2019). While chemical fertilizers boost yields, excessive use depletes soil health, reduces long-term production, and drives farming expenses. Enhancing millet profitability requires balanced fertilization using strategic interventions while preserving soil sustainability (Bana et al. 2013). Integrated Nutrient Management (INM), combining organic and inorganic fertilizers, ensures a steady nutrient supply, improves soil biodiversity, and sustains long-term fertility for resilient millet production. Furthermore, advanced fertilization strategies such as nano fertilizers, slow-release fertilizers, foliar application, and artificial intelligence-driven precision farming play a vital role in enhancing millet production. Understanding millet-specific nutrient

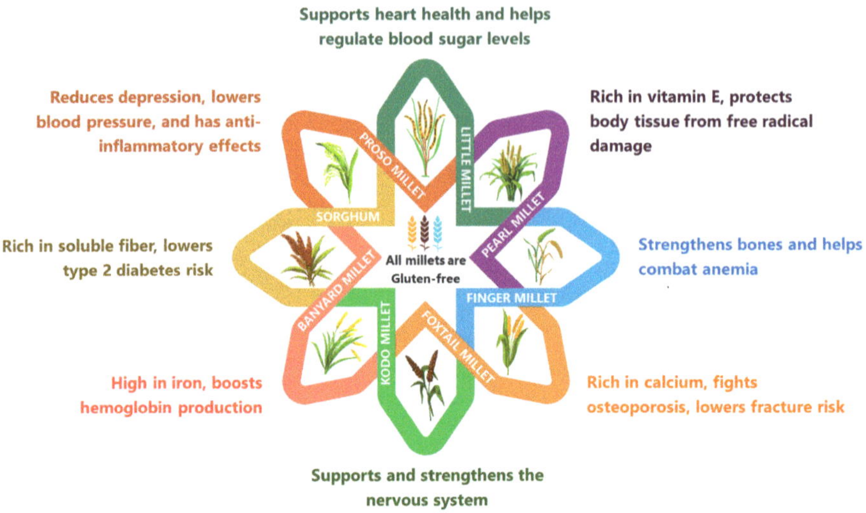

Fig. 12.1 Types of millets and its nutritional value

requirements through soil testing and precision agriculture optimizes fertilizer use, minimizing waste while maximizing yields (Babele et al. 2022). Balanced fertilization supplies essential macronutrients (N, P, K) and micronutrients (Fe, Zn, B), promoting optimal growth and sustainability. Integrating these approaches enhances food security and climate resilience with minimal environmental impact. This chapter examines balanced fertilization's role in millet production, exploring advanced fertilization techniques, their effects on soil health, and modern technologies for optimizing nutrient use. It provides insights into improving millet yields, ensuring sustainability, and enhancing farmers' economic viability.

12.2 Types of Fertilizers for Sustainable Millet Production

Fertilizer recommendations for millet crops depend on nutrient uptake patterns, soil conditions, and agronomic practices. Sorghum (*Sorghum bicolor*) typically requires 80–100 kg ha^{-1} nitrogen, 40–50 kg ha^{-1} phosphorus (P_2O_5), and 40 kg ha^{-1} potassium (K_2O), with nitrogen applied in two to three splits (Cothren et al. 2000). Pearl millet (*Pennisetum glaucum*) responds well to 60–100:40–50:40–50 kg ha^{-1}, with side-dressing around 25–30 days after sowing (DAS). Finger millet (*Eleusine coracana*) needs 40–60:20:20 kg ha^{-1}, with nitrogen applied in two splits (Prakash et al. 2008). Other small millets, such as foxtail, barnyard, kodo, little, and proso millet, require 40:20:00 kg ha^{-1}, as potassium has shown limited yield benefits. In rainfed conditions, fertilizer rates are reduced by 50% due to moisture constraints (Rana et al. 2012). However, millet-specific precision nutrient management remains underexplored, highlighting the need for further research.

12.2.1 Inorganic Fertilizers (Conventional Fertilizers)

Inorganic fertilizers play a crucial role in millet farming by providing readily available nutrients that enhance plant growth and yield potential (Raundal et al. 2017). Ammonium nitrate (33–34% N) offers quick-release nitrogen with reduced volatilization losses, while urea (46% N) supports vegetative growth and chlorophyll synthesis. Phosphorus fertilizers such as diammonium phosphate (DAP, 18-46-0), single super phosphate (SSP, 16% P_2O_5), and triple super phosphate (TSP, 44–48% P_2O_5) are vital for root development and early establishment, especially in phosphorus-deficient soils. Potassium fertilizers like sulphate of potash (SOP, 50% K_2O) and muriate of potash (MOP, 60% K_2O) enhance stress tolerance, grain filling, and disease resistance, making them essential for millets in dry regions (Damalas and Koutroubas 2024). Customized formulations based on soil test recommendations and NPK combinations (e.g., 20-20-20, 15-15-15) help to further promote balanced fertilization.

Raundal et al. (2017) reported the highest little millet yield (1.34 t ha^{-1} grain, 1.61 t ha^{-1} stover) with 150% RDF, which was on par with 125% RDF and superior to 100% RDF (10:20:00 of N: P2O5:K2O kg ha^{-1}). Mubeena et al. (2019) found that 100% RDF (60:30:20 kg ha^{-1}) significantly increased foxtail millet yield (2.14 t ha^{-1}), with 125% RDF yielding similar results due to better nitrogen and phosphorus availability. Efficient application methods, including basal placement, top-dressing, foliar sprays, and fertigation, improve nutrient uptake. However, over-reliance on inorganic fertilizers can lead to soil acidification, nutrient imbalances, and reduced microbial activity (Damalas and Koutroubas 2024). Integrating inorganic fertilizers with organic amendments is essential for sustaining soil health and long-term millet productivity.

12.2.2 Organic Fertilizers

Sustainable millet production relies on organic manures, which enhance soil fertility, stimulate microbial activity, and promote long-term soil health compared to inorganic fertilizers (Li et al. 2025). The addition of farmyard manure (FYM), compost, green manures, vermicompost, and agricultural residues enhances soil organic matter, improving structure, moisture retention, and nutrient availability. Sivagamy et al. (2024) reported that combining 12.5 t ha^{-1} FYM with 6.5 t ha^{-1} in-situ sunhemp green manuring at 45 days after planting significantly improved finger millet grain (2420 kg ha^{-1}) and straw yields (3549 kg ha^{-1}), with a 3% Panchakavya foliar spray further enhancing productivity. Sun et al. (2022) demonstrated that partially replacing chemical nitrogen fertilizers with organic alternatives improved foxtail millet yield and quality. A 75% replacement enhanced grain appearance (yellow pigment, grain diameter) and taste (amylose, soluble sugar content), while a 50%

replacement improved gel consistency and nutritional quality (crude protein, amino acids).

Similarly, Wang et al. (2024) found that substituting chemical nitrogen with fermented mealworm manure raised nitrate, ammonium, total, and alkaline nitrogen levels (13.59–52.56%, 4.47–18.27%, 4.40–12.09%, and 1.70–32.37%, respectively). A 25% substitution increased grain yield (2.78 t ha^{-1}), whereas a 50% replacement reduced yield (2.30 t ha^{-1}), indicating optimal substitution levels for maximizing productivity. Adapting mulching strategies to local conditions is crucial for sustainable millet production. Li et al. (2025) conducted a meta-analysis of 751 datasets from 58 studies, revealing that mulching increased yield by 30.9% and water-use efficiency (WUE) by 27.3% compared with traditional flat planting. Unlike synthetic fertilizers, which supply instantaneous nutrients but can impair soil quality over time, organic manures release nutrients gradually; therefore, reducing leaching losses and enhancing nutrient-use efficiency. Figure 12.2 depicts the comparison of organic and inorganic fertilizers in enhancing soil properties and crop uptake.

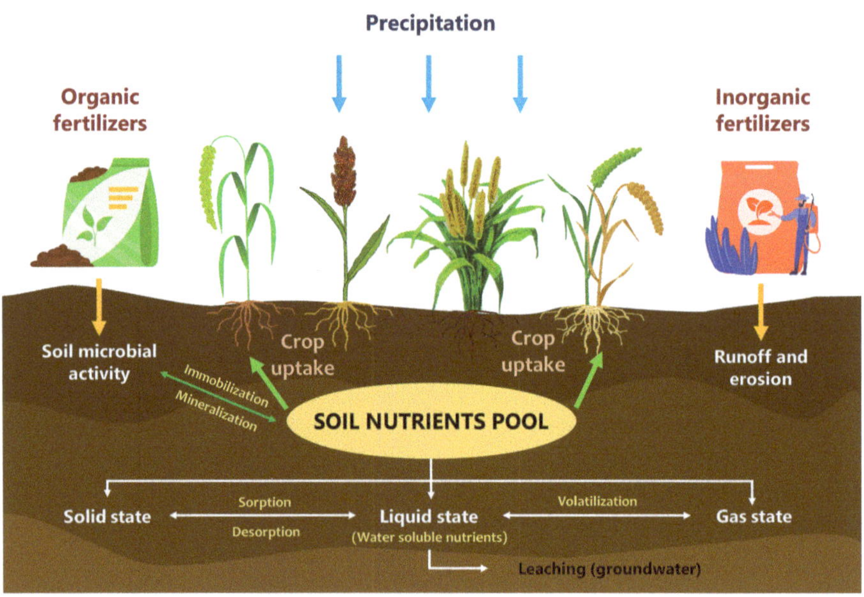

Fig. 12.2 Comparison of organic vs inorganic fertilizers in improving soil health and crop uptake

12.2.3 Biofertilizers

The high costs of chemical fertilizers restrict access for resource-poor farmers in arid regions. As a sustainable alternative, biofertilizers improve soil fertility by harnessing beneficial microbes like *Rhizobium, Azospirillum, Azotobacter,* and *Phosphobacteria,* which enhance nitrogen fixation and phosphorus solubilization (Kennedy et al. 2004). While blue-green algae (BGA) and *Azolla* offer a sustained nitrogen source, mycorrhizal fungus further boosts root growth and nutrient absorption, which makes biofertilizers perfect for millet-based systems. Integrating biofertilizers into intercropping systems, such as pearl millet with mung bean, significantly boosts nitrogen fixation and yield (Monisha et al. 2019). Abdullahi et al. (2014) demonstrated that applying biofertilizers like AMF *Glomus mosseae* and *Azospirillum brasilense* along with 2.5 t ha^{-1} poultry manure improved millet growth, increasing plant height (72.6 cm), tillers (4.1), and dry biomass, while enhancing root colonization (62%).

Ashwani and Rajesh (2017) reported that applying 100% nitrogen through chemical fertilizers combined with *Azospirillum* achieved the highest sorghum grain yield (2.53 t ha^{-1}), stover yield (5.74 t ha^{-1}), ear head length (24.7 cm), test weight (14.38 g), protein content (12.02%), and maximum net returns (Rs. 49,490 ha^{-1}) with a benefit–cost ratio of 2.66, outperforming chemical fertilizers alone. Monisha et al. (2019) observed that 50% RDF + 25% poultry manure N-based +25% biofertilizer maximized finger millet tillers, panicle length (17.8 cm), and yield (2385 kg ha^{-1}). This suggests that biofertilizers, when combined with organic manure, enhance productivity in low-input systems while reducing manure requirements. Liquid microbial consortia (LMC) formulations, with high colony counts (10^9 CFU mL^{-1}) and extended shelf life, provide a more effective alternative to traditional biofertilizers, ensuring better adaptation to diverse climates (Patil et al. 2023). Applied through seed treatment, soil inoculation, foliar sprays, or fertigation, biofertilizers improve microbial colonization, boosting nutrient use efficiency and stress tolerance. Given millet's adaptability to marginal soils, integrating biofertilizers supports soil health, stabilizes yields, and strengthens food security in resource-limited regions.

12.2.4 Integrated Nutrient Management (INM) for Sustainable Millet Production

INM enhances soil fertility and millet productivity by combining organic, inorganic, and biological inputs (Sharma et al. 2024). Organic amendments like FYM, compost, and green manure improve soil structure and microbial activity, while inorganic fertilizers provide essential macronutrients for immediate uptake. Biofertilizers, including nitrogen-fixing bacteria, P-solubilizing bacteria, and mycorrhizal fungi, enhance nutrient availability and root absorption (Sharma et al.

2016). This synergy improves nutrient-use efficiency, supports sustainable soil health, and boosts millet yields, particularly in low-input systems. Figure 12.3 shows the mechanism of different sources of fertilizers utilized for INM in enhancing soil health and efficient nutrient uptake in millets. Issoufa et al. (2019) found that millet glume-derived compost (MGD-Compost) increased millet grain yield by 59–91% over control, while its integration with NP fertilizer improved yields by 36–82% compared to NP alone.

Sharma et al. (2016) reported that soil test crop response (STCR)-based fertilization with 10 t FYM ha^{-1} achieved the highest grain (3.62–6.13 t ha^{-1}) and stover yields (7.0–8.01 t ha^{-1}), increasing pearl millet–wheat system yields by over 100% under integrated nutrient management. Rangaiah et al. (2024) found that NPK requirements for producing per kg of kodo millet grain were 0.049, 0.0047, and 0.035 kg under STCR NPK alone, slightly increasing with FYM addition. STCR NPK + FYM achieved a higher yield (1710 kg ha^{-1}) for a targeted yield of 1700 kg ha^{-1}. Similarly, Mohamed et al. (2023) reported that producing 1 tonne of

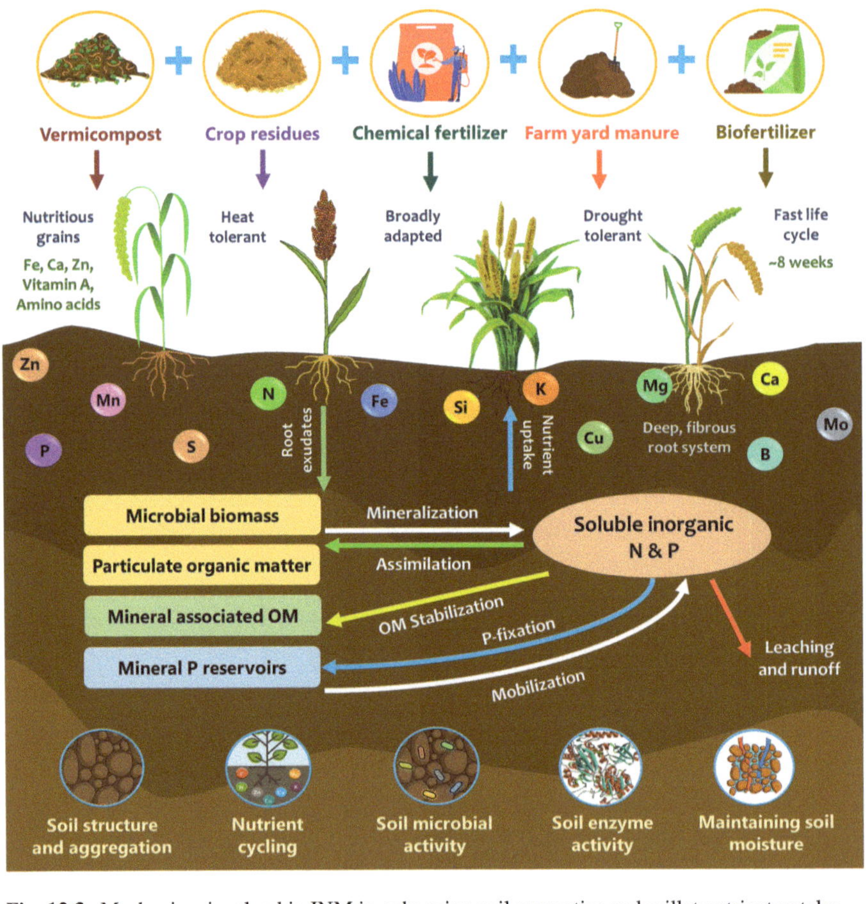

Fig. 12.3 Mechanism involved in INM in enhancing soil properties and millet nutrient uptake

finger millet required 48 kg N, 23 kg P, and 44 kg K, with varying soil, fertilizer, and FYM contributions. Zhou et al. (2025) found that applying 60 kg·ha^{-1} N, 40 kg·ha^{-1} P_2O_5, and 2000 kg·ha^{-1} organic matter increased foxtail millet dry matter by 15.04–33.7%, improved water use efficiency (WUE) by 8.73–35.11%, and enhanced yield by 9.23–35.17%, ensuring stable millet production.

Sheoran et al. (2024) studied the long-term effects of organic and nitrogen fertilization on pearl millet under a millet-wheat system. Applying FYM in both seasons increased grain N (1.99%), P (0.17%), and K (0.37%), with 15 t FYM ha^{-1} boosting protein (13.56%), stover N (32.76%), and root length (29.83%) over control. Combining 15 t FYM ha^{-1} with 120 kg N ha^{-1} enhanced seedling length by 30.54% over N alone, proving superior for quality and yield. Sankar et al. (2011) conducted a long-term study (1984–2008) and found that the application of FYM at 10 t ha^{-1} + 100% NPK achieved the highest yield (3167 kg ha^{-1}, 22.7% variation) and improved soil organic carbon (0.45%), N (204 kg ha^{-1}), P (69 kg ha^{-1}), and K (107 kg ha^{-1}), proving superior to 100% NPK alone.

Sharma et al. (2024) assessed the long-term (10 years) impact of STCR–IPNS (integrated plant nutrition system) on soil nutrient status and found that continuous manure and fertilizer application increased soil organic carbon by 78% (0–15 cm) and 42% (15–30 cm) over the control. INM in millet production improves nutrient cycling, microbial interactions, and nutrient efficiency (Sheoran et al. 2024). Organic matter enhances microbial biomass, enzyme activity, and soil structure, improving water retention. Biofertilizers aid nitrogen fixation and phosphate solubilization, while inorganic fertilizers complement slow-releasing organics, minimizing nutrient losses. This synergy boosts root growth, stress resilience, and grain yields, sustaining long-term soil health and optimizing millet productivity in an eco-friendly manner.

12.2.5 Foliar Application of Nano-Fertilizers

Nano-fertilizers offer a cutting-edge approach to plant nutrition, enhancing nutrient use efficiency, reducing losses, and boosting crop productivity. Their nanoscale formulation ensures controlled and targeted nutrient release, optimizing plant uptake while minimizing environmental impact (Bhardwaj et al. 2014). Nano-NPK fertilizers supply essential macronutrients in nano-form, reducing leaching losses, while nano-urea minimizes volatilization and denitrification for sustained nitrogen availability (Udapudi et al. 2024). Nano-micronutrient fertilizers (Zn, Fe, Cu, Mn) enhance enzymatic activities, and nano-biofertilizers incorporate beneficial microbes for soil health. Applied via foliar spray, soil incorporation, or seed priming, nano-fertilizers optimize nutrient delivery across crop growth stages (Shree et al. 2024). In drought-prone regions where soil nutrient absorption is limited, foliar spraying ensures rapid nitrogen uptake, bypassing soil-related constraints like fixation and leaching. This technique is particularly beneficial for millet, often grown in nutrient-deficient soils, as it quickly corrects deficiencies, enhancing plant vigor and yield (Senthamil and Tamilmounika 2022). Micronutrients such as zinc, iron, boron, manganese, and copper are crucial for stress tolerance, chlorophyll synthesis, and reproductive development, with foliar application ensuring immediate availability. Proper dilution, spray timing, and uniform coverage maximize nutrient uptake while preventing leaf burn (Sharma et al. 2022).

Udapudi et al. (2024) demonstrated that applying 75% RDN with nano-urea seed treatment and foliar spray at 90 DAS significantly increased number of leaves (36.73 $hill^{-1}$), leaf area (982.90 cm^2 $hill^{-1}$), and leaf area index (3.28), absolute (0.74 g day^{-1}), relative (0.029 g g^{-1} day^{-1}), and crop growth rate (24.80 g m^2 day^{-1}), also boosted grain yield (3397 kg ha^{-1}) of finger millet and NPK uptake (86.6, 25.9, and 55.4 kg ha^{-1}). Meena et al. (2024) reported that 75% RDN with two nano-urea sprays (4 ml L^{-1}) at 25 and 45 DAS significantly increased pearl millet height (200.8 cm), dry matter (266.1 g), tillers (24.3), ear head length (28.5 cm), grain yield (2285 kg ha^{-1}), and net returns (Rs. 63,300 ha^{-1}) over 100% RDN alone.

Shree et al. (2024) found that applying 75% RDF + vermicompost (5 t ha^{-1}) + nano-urea sprays (30 and 45 DAS) increased plant height (136.6 cm), leaf area index (4.15), crop growth rate (6.4 g^{-1} m^2 d^{-1}), chlorophyll index (50.8), grain yield (27.2 q ha^{-1}), and stover yield (73.1 q ha^{-1}) of pearl millet over nano-DAP sprays. Similarly, Sneha et al. (2024) reported that 100% RDN + foliar nano nitrogen and zinc achieved the highest finger millet grain yield (3453 kg ha^{-1}) and straw yield (5048 kg ha^{-1}), comparable to 75% RDN + nano fertilizers, highlighting improved nitrogen use efficiency with reduced nitrogen input. Arya et al. (2022) reported that 50% RDN basal +0.5% nano-urea foliar spray at panicle initiation and booting stages significantly enhanced productive tillers (3.79), earhead length (26.82 cm), grains per earhead (2462), grain yield (3314 kg ha^{-1}), and stover yield (5136 kg ha^{-1}), matching the effect of 50% RDN basal +25% RDN split application. Sharma et al. (2022) found that 100% RDN + two nano-urea sprays (4 ml L^{-1}) at 30 and 45 DAS improved pearl millet height (166.6 cm), dry matter (854 g m^{-2}), chlorophyll (2.69 mg g^{-1}), and nutrient content over a single spray. Nanoparticles (NPs) enhance millet resilience to abiotic stresses like drought, salinity, heat, and nutrient deficiencies by regulating physiological and biochemical processes. It enhances nutrient use efficiency, reduce environmental pollution, and improve plant stress tolerance with lower application rates than conventional fertilizers (Senthamil and Tamilmounika 2022). They minimize waste and optimize nutrient uptake, boosting crop growth. However, challenges include potential toxicity, production complexities, and nanoparticle accumulation concerns. Despite this, they offer a sustainable solution for addressing nutrient deficiencies. Figure 12.4 displays the mechanism of foliar application of nano fertilizers and its effective translocation in plants.

12.2.6 Advanced Fertilization Technologies in Millets

Advanced technologies in millet fertilization enhance efficiency, sustainability, and productivity. Precision application, sensor-based fertilization, fertigation, micro-dosing, controlled-release fertilizers, and AI-driven decision-making optimize

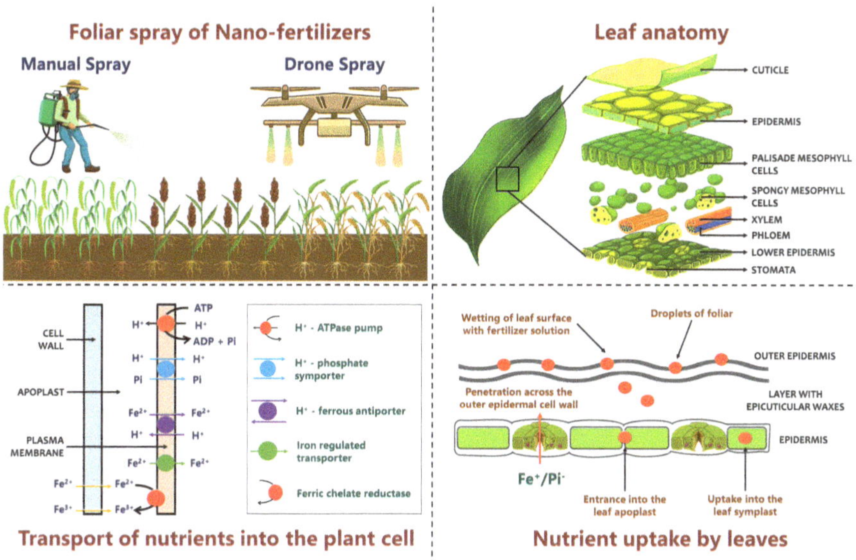

Fig. 12.4 Mechanism behind the translocation of nano fertilizers in plants via foliar application

nutrient delivery while reducing waste and environmental impact (Thind and Dhillon 2021). Ibrahim et al. (2016) found that fertilizer microdosing (2 g DAP hill^{-1}, 6 g NPK hill^{-1}) achieved 86% and 79% of grain yields compared to 200 kg NPK ha^{-1} broadcasting in 2013 and 2014. Microdosing increased lateral root length density by 72% and 40%, improving soil nutrient uptake and millet yield. Precision fertilizer application uses GPS-guided variable rate technology (VRT) and remote sensing for site-specific nutrient management (Gandah et al. 2000). Soil health mapping and Decision Support Systems (DSS) provide customized fertilization plans, though frequent soil testing and technical expertise are required (Thind and Dhillon 2021).

Wang et al. (2024) investigated selenium fertilizer rate and placement and found that optimal selenium enrichment was achieved with 100.65–120.15 kg ha^{-1} fertilizer, placed 10.28–11.76 cm deep and 6.74–7.29 cm horizontally for foxtail millet. Zhao et al. (2025) studied nitrogen and foliar molybdenum fertilization on millet variety Changnong 47 and found that the highest yield (5869 kg ha^{-1}) occurred with 0.3% Mo, a 13% increase over the control. Kaleri et al. (2024) examined zinc and nitrogen effects on millet variety YBS-98, finding that 180 kg N ha^{-1} + 20% Zn resulted in a plant height of 222.98 cm, stem diameter of 0.91 cm, leaf area per tiller of 1780.81 cm^2, and green fodder yield of 77.91 t ha^{-1}. Sensor-based fertilization, using chlorophyll meters and optical sensors, detects nutrient deficiencies in real-time, improving nitrogen use efficiency but requiring calibration and expertise. Fertigation delivers nutrients through irrigation, ensuring uniform uptake and water conservation, but needs proper maintenance (Kumar et al. 2024). Controlled-release fertilizers enhance plant uptake and reduce leaching losses, improving millet yield and stress tolerance, though adoption is limited by high costs and environmental concerns. AI and IoT analyze real-time soil and crop data, automating fertilizer recommendations for precision but requiring infrastructure and connectivity. Mobile apps integrate soil health, weather, and AI insights for location-specific guidance, improving decision-making, but needing digital literacy (Thind and Dhillon 2021). Advancing these technologies in millet fertilization enhances sustainability, fertilizer efficiency, and crop resilience across diverse agro-climatic conditions. Table 12.2 shows the influence of different nutrient management practices on millet yield.

12.2.7 Biofortification in Millets

Plant development and human nutrition depend heavily on micronutrients, especially iron. Although the main source of iron for humans is soil, it can also be found in foods derived from plants and animals. Nonetheless, data indicate that soil iron concentration and bioavailability have decreased over time, which has an impact on food's nutritional value. Millets have more iron than cereals, which are an important dietary staple but have better nutritional value. (Siwela et al. 2020). Millets are one of the first cereals to be grown for human consumption and have great potential to help with micronutrient shortages. A low-cost method of increasing the availability of micronutrients in staple crops and lowering the chance of nutritional deficiencies is biofortification. This entails using agronomic techniques and genetic improvement to increase the nutrient content and bioavailability in edible plant portions (Jadhav et al. 2024). To increase the micronutrient content of millets, a variety of biofortification strategies have been used, such as traditional breeding, molecular methods, micronutrient fertilization, nanotechnology, and microbial interventions (Fig. 12.5). By increasing the efficiency of nutrient absorption, these tactics seek to improve nutritional outcomes both in biofortified (to maintain or increase nutrition in low-fertility soil) and non-biofortified varieties. Nanja Reddy et al. (2025) found that foliar application of 0.5% $ZnSO_4{\cdot}7H_2O$ at flowering and 20 DAF increased grain-Zn to 40 μg g^{-1}. Soil application of $ZnSO_4{\cdot}7H_2O$ (12.5 kg ha^{-1}) enhanced grain yield from 444 to 504 g m^{-2}. Enhancing yield and stress resilience in pearl millet requires conventional breeding and advanced tools like genomic selection, speed breeding, genome editing, and precision phenotyping. Integrating iron and zinc traits into breeding programs is crucial for nutrient-rich hybrids (Yadav et al. 2024). For example, Govindaraj et al. (2019) found significant genetic variability in pearl millet (30–140 mg kg^{-1} Fe, 20–90 mg kg^{-1} Zn), enabling the development of high-yielding, nutrient-rich biofortifies cultivars, including the open-pollinated Dhanashakti and hybrids ICMH 1202, ICMH 1203, and ICMH 1301, with 70–75 mg kg^{-1} Fe and 35–40 mg kg^{-1} Zn.

Samtiya et al. (2023) reported that ICP–OES mineral analysis revealed high iron content (87.79 mg kg^{-1} in HHB-311 and 84.26 mg kg^{-1} in Dhanashakti) and zinc content (55.05 mg kg^{-1} in HHB-311 and 52.43 mg kg^{-1} in Dhanashakti),

Table 12.2 Influence of different nutrient management practices on millet yield

Common name	Nature of fertilization	Source of fertilization	Application rate	Yield increases over control	References
Foxtail millet	Balanced fertilization	NPK	50:30:30 kg ha^{-1}	Increased grain yield by 23.6%	Shahzadi et al. (2025)
Sorghum	INM	Crop residues and Inorganic fertilizers	Rapeseed green manure + Sorghum straw return with 70% of chemical fertilization	Higher grain yield by 10.49%	Wang et al. (2025)
Pearl millet	Organic fertilizer	Compost	Compost at 5 t ha^{-1}	Increased yield by 44% than control	Assane et al. (2024)
Foxtail millet	Organic + Inorganic	Fermented mealworm feces, Urea, DAP, and Potassium chloride	25% of chemical fertilizers have been replaced with organic fertilizers	Higher grain yield of 65.88%	Wang et al. (2024)
Kodo millet	IPNS	Inorganic fertilizers + FYM	30:30:7.5 kg ha^{-1} of N, P_2O_5, and K_2O + 12.5 t ha^{-1} of FYM	Higher grain yield by 65%	Rangaiah et al. (2024)
Finger millet	INM	NPK + Poultry manure	75% RDF (45–30-22.5 NPK kg ha^{-1}) + 3.0 t ha^{-1} poultry manure	Higher grain yield (2590 kg ha^{-1})	Sreenivasulu et al. (2023)
Pearl millet	Foliar application of nano fertilizers	IFFCO Nano N and Zn liquid fertilizers	Nano-N + nano-Zn with 100% RDF (150:75:75)	Higher grain yield (57.1–65.4%)	Upadhyay et al. (2023)
Pearl millet	Foliar application of nano fertilizers	Nano-urea	Nano-urea (2 ml L^{-1})	Increased grain yield by 8.96%	Khan et al. (2023)
Barnyard millet	IPNS	Inorganic fertilizers + FYM	60:30:40 kg ha^{-1} of N, P_2O_5, and K_2O along with 12.5 t ha^{-1} of FYM	Increased Grain yield by 27.73%	Selvam et al. (2022)
Foxtail millet	Biofertilizers	*Azospirillum* seed inoculation	75% nitrogen from urea +25% from poultry manure + Azospirillum seed inoculation	Higher grain yield (2.31 t ha^{-1})	Marwein et al. (2019)
Little millet	Organic + Inorganic	Urea, DAP, MOP, and Vermicompost	100% RDF + Vermicompost @ 2 t ha^{-1}	Higher grain yield by 21%	Rani et al. (2017)
Teff	Organic + Inorganic	Complex fertilizer (23–10-5NPS), and FYM	100% RDF + 5 t ha^{-1} of FYM. (RDF-40:60:0 NPK ha^{-1})	Higher grain yield by 42%	Mebratu et al. (2017)

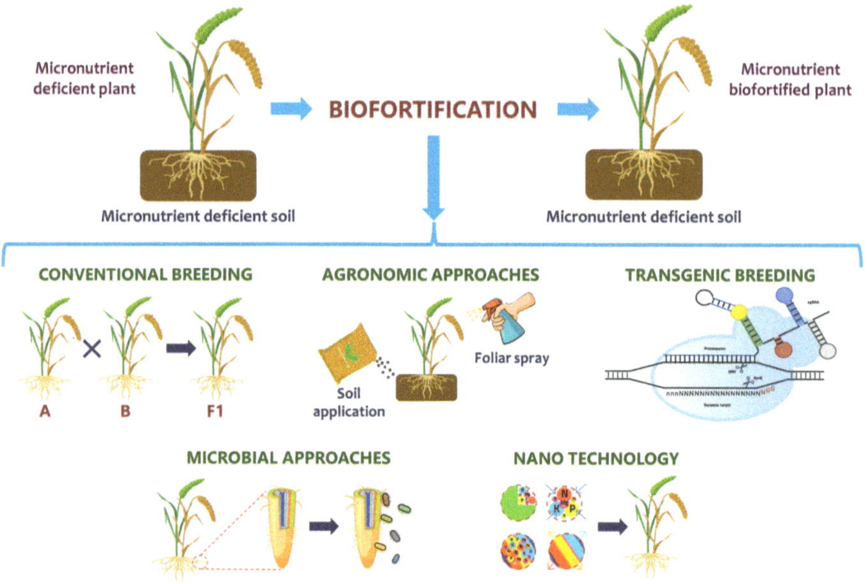

Fig. 12.5 Various biofortification approaches for enhancing micronutrient content in millets

highlighting the nutritional benefits of these biofortified pearl millet varieties. Similarly, Jadhav et al. (2024) reported that AHB-1200 had higher fat content, while Dhanshakti contained more protein (11.46%) and carbohydrates (74.15%). Dhanshakti also had elevated iron (84.43 mg kg^{-1}) and calcium (918.2 mg kg^{-1}), making it a valuable biofortified millet for addressing malnutrition and micronutrient deficiencies. Srinivas et al. (2022) evaluated *Streptomyces albus* (CAI-24, KAI-27) and *S. griseus* (MMA-32) for plant growth promotion and nutrient enhancement in pearl millet under greenhouse and field conditions. Biofortified hybrid ICMH-1201 treated with these strains showed significant increases in grain yield (13–23% in glasshouse, 9–12% in field) and nutrient content (Fe: 40–53%, Zn: 10–15%, Ca: 11–29%) over the control, whereas non-biofortified PA-9444 showed no such gains. KAI-27 and MMA-32 enhanced shoot, root, and grain yield, while CAI-24 and MMA-32 improved Fe content (up to 49%).

Govindaraj et al. (2020) analyzed the nutrient profile, diversity, and interrelationships of grain minerals (Ca, K, Mg, Na, P, S, Cu, Fe, Mn, Zn, Mo, Ni) in 122 pearl millet hybrids and open-pollinated varieties in India. The cultivars showed two- to four-fold variation in mineral content, with significant positive correlations among minerals. A strong Fe–Zn association ($r = 0.71$, $P < 0.01$) indicated potential for simultaneous improvement without compromising other nutrients. Cluster I, with ICTP 8203, Ajeet 38, Sanjivani 222, PAC 903, and 86 M 86, had the highest Fe (56 mg kg^{-1}) and Zn (49 mg kg^{-1}). Moreover, Govindaraj et al. (2022) assessed micronutrient (Fe, Zn, Mn, Na) and macronutrient (P, K, Ca, Mg) variability in 14 pearl millet trials involving hybrids, inbreds, and germplasm. Significant genetic variation was observed (Ca: 4.2–40.0 mg 100 g^{-1}, Fe: 24–145 mg kg^{-1}, Zn: 22–96 mg kg^{-1}, Na: 3.0–63 mg kg^{-1}). Fe and Zn were highly correlated ($r = 0.58–0.81$; $p < 0.01$) and significantly associated with Ca ($r = 0.26–0.61$; $p < 0.05$) and Mn ($r = 0.24–0.50$; $p < 0.05$), supporting joint selection for Fe, Zn, and Ca. Therefore, biofortification, through crop breeding, genetic modification, and mineral or biofertilizer application, is a promising strategy to combat iron deficiency and improve global nutritional security. Table 12.3 outlines various nutrient management techniques in millets, highlighting their role in enhancing nutrient bioavailability and overall nutritional security.

12.3 Role of Balanced Fertilization in Enhancing Millet Productivity

Balanced fertilization optimizes millet productivity, enhances nutrient use efficiency, and maintains soil health (Kumar et al. 2024). It ensures the right mix of macronutrients (N, P, K) and micronutrients (Zn, Fe, Mn) based on soil and crop needs, improving yield, quality, and climate resilience. Nutrient deficiencies disrupt physiological processes, reducing yield and grain quality. A proper nutrient balance enhances uptake and assimilation, minimizing losses through leaching, volatilization, or fixation, supports microbial enzymatic activity, enhancing organic matter decomposition and nutrient cycling (Sharma et al. 2024). Duan et al. (2024) assessed balanced NPK fertilization on foxtail millet productivity and found that the 27-14-10 NPK treatment achieved the highest yield (8487.58 kg hm^{-2}), a 26.53% increase over the control. Moreover, soil enzyme activities (urease, sucrase, glutaminase) increased by 26.67%, 36.27%, and 33.33%, respectively. Lu et al. (2024) found the optimal nitrogen rate for high-yield millet varieties was 141.3 kg ha^{-1}. Nitrogen-efficient varieties yielded 5607.3–5637.2 kg ha^{-1} at 151.5 kg ha^{-1} N, while nitrogen-inefficient ones produced 4749.8–4847.7 kg ha^{-1} at 134.5 kg ha^{-1} N. Excessive nitrogen (>360 kg ha^{-1}) increased NH_4^+-N and NO_3^--N, raising environmental risks.

Guan et al. (2022) conducted a two-year study to determine the optimal N:P:K ratio for foxtail millet, and the highest yield (3927.0 kg ha^{-1}) was achieved with 112.5 kg N, 90.0 kg P, and 150.0 kg K ha^{-1}, a 68.8% increase over the control. Balanced fertilization reduces greenhouse gas emissions, minimizes nutrient losses, and supports sustainable soil management (Patil et al. 2023). Efficient nitrogen management using nitrification inhibitors, slow-release fertilizers, and legume intercropping lowers nitrous oxide emissions, while compost and biochar enhance carbon sequestration (Kumar et al. 2024). Conservation practices like minimum tillage and residue retention mitigate CO_2 emissions. Split fertilizer applications, deep nutrient placement, and precision phosphorus management reduce leaching, runoff, and volatilization (Duan et al. 2024). Site-specific nutrient management, customized fertilizers, and silicon application counteract soil acidification and micronutrient imbalances. Integrating these strategies enhances productivity, promotes

Table 12.3 Influence of different nutrient management practices on nutrient uptake in millets

Common name	Nature of fertilization	Source of fertilization	Application rate	Nutrients uptake	References
Kodo millet	IPNS	Inorganic fertilizers + FYM	30:30:7.5 kg ha^{-1} of N, P_2O_5, and K_2O + 12.5 t ha^{-1} of FYM	Total NPK uptake increased by 69.1, 28.2, and 92.0 kg ha^{-1}, respectively	Rangaiah et al. (2024)
Finger millet	Biofortification	Endophytes and $ZnCO_3$	3 mL of Endophytic isolates: 2 fungal and 2 bacterial isolates along with $ZnCO_3$ as Zn source	Endophytes increased the Zn content in grains by 12.1–18.8% compared to the control	Chaudhary et al. (2023)
Pearl millet	Foliar application of nano fertilizers	IFFCO Nano N and Zn liquid fertilizers	Nano-N + nano-Zn with 100% RDF (150:75:75)	Increased Cu and Zn uptake of 4603–4386 and 1144–1195 mg ha^{-1}, respectively	Upadhyay et al. (2023)
Foxtail millet	Biofortification	Enriched organic manures	Enriched compost with Zn and Fe application + foliar spray of panchagavya (4210 kg ha^{-1})	Increased Zn concentration in grain (13.9 mg kg^{-1}) and straw (19.1 mg kg^{-1})	Krupashree et al. (2022)
Pearl millet	Foliar application of nano fertilizers	IFFCO Nano N Urea and Inorganic fertilizers	RDF + Two sprays of nano urea at 4 ml L^{-1} water at 30 and 45 DAS	Nutrient uptake is 22.73% of N, 4.89% of P, and 23.58% of K	Sharma et al. (2022)
Finger millet	Biofortification	ZnSO4 and Zn-EDTA	Foliar sprays of 0.2% $ZnSO_4$ and Zn-EDTA twice at 30 and 60 days after sowing	Total Zn and Fe uptake increased by 149 g ha^{-1} and 1497 g ha^{-1} with $ZnSO_4$ spray, while Zn-EDTA spray resulted in a higher uptake of 279 g ha^{-1} and 1862 g ha^{-1}, respectively.	Kumar and Pandey (2020)
Sorghum	Foliar application of nano fertilizers	Nano N fertilizer and inorganic fertilizers	2.5 times reduction of RDF through nano N fertilizer	Increased NPK uptake of 3.06, 0.56, and 2.23 g pot$^{-1,}$ respectively	Rani et al. (2019)
Little millet	Organic + Inorganic	Urea, DAP, MOP, and Vermicompost	100% RDF + Vermicompost @ 2 t ha^{-1}	Increased Zn (240 g ha^{-1}) and Fe uptake (945 g ha^{-1})	Rani et al. (2017)

climate-smart agriculture, and ensures food security while minimizing environmental impact.

12.4 Constraints in the Adoption of Balanced Fertilization for Enhancing Millet Production

Balanced fertilization is essential for maximizing millet yields, improving soil health, and ensuring sustainable production. However, several constraints hinder its widespread adoption, especially in regions where millets are traditionally grown under low-input conditions. These challenges include socioeconomic barriers, limited awareness, and input accessibility issues (Srivastava et al. 2023). Addressing these constraints through targeted interventions can enhance the adoption of balanced fertilization and improve millet productivity.

12.4.1 Socioeconomic Barriers

Many smallholder farmers cultivating millets in rainfed and marginal lands struggle with financial limitations, making it difficult to afford balanced fertilization. The high cost of customized fertilizers, organic amendments, and biofertilizers discourages investment in improved nutrient management practices (Kishore et al. 2021). Compared with staple cereals such as rice and wheat, millets receive less government support in terms of subsidies and incentives for balanced fertilization (Srivastava et al. 2023). This limits farmers' motivation to invest in nutrient management strategies. Some fertilization techniques, such as split applications, deep placement, and organic matter incorporation, are labor-intensive and time-consuming. Smallholder farmers with limited labor resources often avoid these practices. The absence of a well-developed millet value chain reduces incentives for farmers to adopt advanced fertilization strategies (Rouamba et al. 2021). Addressing these barriers requires targeted policies to strengthen millet markets, provide financial support for fertilization, and promote sustainable intensification programs to enhance productivity and resilience.

12.4.2 Lack of Awareness and Knowledge

Limited awareness of balanced fertilization hinders its adoption in millet farming. Many farmers rely on traditional practices, believing millets need minimal fertilization. Unlike input-intensive crops such as maize and wheat, millets have historically been grown with low external inputs, reinforcing this misconception (Maitra et al.

2022). Weak extension services and limited scientific advisory support further contribute to poor nutrient management. Farmers often lack access to soil testing and site-specific recommendations, leading to imbalanced nitrogen (N), phosphorus (P), and potassium (K) applications while neglecting essential micronutrients like sulfur (S), zinc (Zn), and silicon (Si) (Huang and Karimanzira 2018). Biofertilizers, such as nitrogen-fixing and phosphate-solubilizing microbes, are underutilized due to limited awareness and availability. Strengthening extension programs, integrating digital advisory platforms, and providing farmer training on balanced fertilization are crucial for improving millet productivity and resilience.

12.4.3 Limited Access to Inputs

The availability and affordability of quality fertilizers, biofertilizers, and soil amendments remain major challenges in millet-growing regions. Fertilizer companies prioritize major cereals over millets, limiting the production and distribution of millet-specific fertilizers (Gudadhe et al. 2023). Smallholder farmers struggle to access the right fertilizers at the right time, hindering balanced fertilization. Weak input supply chains in remote and arid areas further restrict access to essential fertilizers. Organic amendments, such as compost and fortified biofertilizers, are costly and require proper storage, making them less accessible to smallholder farmers (Kumar et al. 2024). High transportation costs, poor storage, and price fluctuations exacerbate the issue. Additionally, subsidies favoring rice and wheat discourage millet farmers from investing in balanced fertilization, leading to reliance on cheaper but imbalanced options. Strengthening millet-specific fertilizer supply chains, promoting local biofertilizer production, and enhancing government support can improve nutrient management in millet-based farming systems.

12.5 Strategies to Promote Balanced Fertilization

Through targeted subsidies for biofertilizers, micronutrient-enriched formulations, and customized fertilizers, governments can include balanced fertilization into agricultural plans. Fertilizer use efficiency will be improved by promoting soil health cards, site-specific nutrient management, and digital advising platforms. For broad adoption, it is essential to fortify millet value chains and guarantee fair access to fertilizer. Extension organizations, non-governmental organizations, and research institutes can enhance information transmission through farmer training, field demonstrations, and mobile advising services. Real-time nutrition management will be further optimized by digital technologies like smartphone apps and AI-driven decision support systems (Maitra et al. 2022). Financial incentives such as subsidies, credit schemes, and rewards for sustainable practices can encourage balanced fertilization. Supporting organic amendments, biofertilizers, and slow-release fertilizers through targeted funding fosters eco-friendly nutrient management (Kishore et al. 2021). Community-led distribution and cooperative fertilizer banks increase accessibility and affordability, while group procurement lowers expenses and fosters information exchange. Adoption is further increased by bolstering regional compost and biofertilizer production through farmer cooperatives. By establishing minimum support prices (MSPs) for millets, investments in fertilization are encouraged, and economic security is provided. Sustainability will be improved by stepping up research on climate-smart nutrient management and fertilizers tailored to millet. Balanced fertilization may greatly increase millet yields, enhance soil health, and create resilient farming systems by combining supply chain enhancements, farmer education, and governmental support.

12.6 Research Priorities Prospects

Millets are highly nutritious, climate-resilient crops that play a significant role in ensuring food security, especially in arid and semi-arid regions. However, the sustainable production of nutrient-dense millets requires the adoption of balanced fertilization techniques that enhance soil fertility, optimize nutrient use, and improve grain quality while minimizing environmental impact. Key future directions to achieve these goals are as follows.

(a) *Precision Nutrient Management*

- Soil Testing and Plant Tissue Analysis: Implement systematic soil and plant tissue analysis to determine the exact nutrient requirements of millet crops.
- Site-Specific Nutrient Management (SSNM): Develop and promote SSNM strategies to optimize fertilizer use based on local soil conditions and crop needs.
- Decision-Support Tools: Leverage technology such as mobile applications, artificial intelligence (AI), and machine learning for real-time recommendations on nutrient application.

(b) *Integrated Soil Fertility Management (ISFM)*

- Organic and Inorganic Fertilizer Integration: Encourage a combination of farmyard manure, compost, green manure, and biofertilizers along with balanced chemical fertilizers.
- Microbial Biofertilizers: Use nitrogen-fixing bacteria (Rhizobium, Azospirillum) and phosphate-solubilizing microbes to improve nutrient uptake efficiency.
- Crop Rotation and Intercropping: Implement millet-legume cropping systems to naturally enhance soil nitrogen levels and reduce dependency on synthetic fertilizers.

(c) *Customized Fertilizer Formulations*

- Millet-Specific Fertilizer Blends: Develop fertilizer formulations enriched with essential micronutrients such as zinc, iron, and calcium to enhance the nutritional quality of grains (agronomic biofortification strategy).
- Slow-Release and Nano-Fertilizers: Promote the use of advanced fertilizer technologies for better nutrient efficiency and reduced leaching losses.
- Foliar Sprays for Micronutrient Fortification: Encourage the adoption of foliar nutrient sprays to improve grain fortification and overall crop health.

(d) *Organic and Regenerative Farming Approaches*

- Organic Soil Amendments: Expand the use of biochar, vermicompost, and compost to improve soil organic matter and microbial activity.
- Conservation Agriculture: Adopt practices such as no-till farming, cover cropping, and minimal soil disturbance to maintain soil structure and fertility.
- Agroecological Farming: Promote biodiversity-friendly millet farming systems that integrate natural ecosystem services.

(e) *Water-Efficient Nutrient Application*

- Fertigation Techniques: Utilize fertigation (applying fertilizers through irrigation) to enhance nutrient absorption and minimize wastage.
- Micro-Irrigation Systems: Implement drip and sprinkler irrigation systems to improve water-use efficiency and reduce nutrient runoff.
- Climate-Resilient Millet Varieties: Develop and promote drought-resistant millet varieties that require fewer inputs and withstand extreme weather conditions.

(f) *Policy and Farmer Support Programs*

- Government Incentives: Introduce subsidy programs and incentives for farmers practicing balanced fertilization and organic nutrient management.
- Capacity Building and Training: Conduct farmer training programs on sustainable fertilization techniques and soil health management.
- Research and Extension Services: Strengthen agricultural research institutions and extension programs to disseminate knowledge on millet fertilization and productivity improvement.

(g) *Climate-Smart Fertilization Strategies*

- Carbon Sequestration Techniques: Incorporate biochar and organic amendments to enhance soil carbon content and improve soil fertility.
- Microbial Inoculants: Promote microbial-based biofertilizers to increase nutrient use efficiency and reduce greenhouse gas emissions.
- Adaptive Fertilization Schedules: Develop flexible fertilization schedules aligned with changing climate conditions to optimize nutrient application.

Therefore, sustainable millet production with balanced fertilization techniques requires a holistic approach integrating precision agriculture, organic soil management, water-efficient strategies, and farmer-centric policies. By adopting these future-oriented practices, farmers can enhance millet productivity, improve grain nutrition, and contribute to environmental sustainability. Continued research, government support, and technological innovations will be key to ensuring the long-term viability of millet cultivation in the face of climate change and resource constraints.

12.7 Conclusions

To increase millet production and guarantee long-term agricultural sustainability, balanced fertilization is essential. Optimizing nutrient management is crucial to overcoming production restrictions and enhancing yield stability because millets are mostly farmed on rainfed and marginal soils. Grain quality, environmental degradation, and nutrient usage efficiency can all be improved by combining mineral fertilizers with organic amendments, biofertilizers, and site-specific nutrient management. Despite its advantages, several obstacles prevent widespread use. Economic limitations, restricted access to high-quality fertilizers, inadequate extension services, and a lack of knowledge about nutrient management are common problems faced by smallholder farmers. A multifaceted strategy including financial assistance, technology breakthroughs, regulatory interventions, and capacity-building programs is needed to address these obstacles. Sustainable millet farming can be supported by promoting intercropping, crop rotation, and agroforestry systems, which can improve soil fertility and lessen nutrient depletion. Research on biostimulants and soil microbial interactions can enhance plant health and nutrient availability even more. Promoting sustainable millet production would require a comprehensive strategy that combines legislative assistance, farmer-centric interventions, and scientific innovation. In millet-based agricultural systems, balanced fertilization can improve essential nutrition, environmental sustainability, increase economic growth, and fortify food security by tackling present issues and seizing new opportunities.

Acknowledgments The authors acknowledge the Tamil Nadu Agricultural University, Coimbatore, for providing library and web source facilities for this book chapter.

Competing Interests The authors have no conflict of interest.

References

Abdullahi R, Sheriff HH, Buba A (2014) Effect of biofertilizer and organic manure on growth and nutrients content of pearl millet. ARPN J Agric Biol Sci 9(10):351–355

Arya GR, Manivannan V, Marimuthu S, Sritharan N (2022) Effect of foliar application of nano-urea on yield attributes and yield of pearl millet (*Pennisetum glaucum* L.). Int J Plant Soil Sci 34(21):502–507. https://doi.org/10.9734/IJPSS/2022/v34i2131293

Ashwani S, Rajesh S (2017) Effect of seed bed and integrated nitrogen management on growth and yield of sorghum (*Sorghum bicolor* L.). Int J Curr Microbiol App Sci 6(12):401–407

Assane MM, Magagi AS, Bodo BS, Mahamane A, Marin B, Morvan X, Guillaneuf A, Fronteau G (2024) Compost, a natural fertilizer to improve pearl millet [*Pennisetum glaucum* (L.) R. br.] growth and its productivity in the Sahel (South-Eastern Niger): an experiment to convince small producers to manage their soil sustainably. Int J Biol Chem Sci 18(3):864–881. https://doi.org/10.4314/ijbcs.v18i3.11

Babele PK, Kudapa H, Singh Y, Varshney RK, Kumar A (2022) Mainstreaming orphan millets for advancing climate smart agriculture to secure nutrition and health. Front Plant Sci 13:902536

Bana RS, Rana KS, Dass A, Choudhary AK, Pooniya V, Vyas AK, Kaur R, Sepat S, Rana DS (2013) A manual on dryland farming and watershed management. IARI, New Delhi, p 104

Bhardwaj D, Ansari MW, Sahoo RK, Tuteja N (2014) Biofertilizers function as key player in sustainable agriculture by improving soil fertility, plant tolerance and crop productivity. Microb Cell Factories 13. https://doi.org/10.1186/1475-2859-13-66

Chaudhary R, Kumar V, Gupta S, Naik B, Prasad R, Mishra S et al (2023) Finger millet (*Eleusine coracana*) plant–endophyte dynamics: plant growth, nutrient uptake, and zinc biofortification. Microorganisms 11(4):973

Cothren JT, Matocha JE, Clark LE (2000) Integrated crop management for sorghum. In: Smith CW, Federiksen RA (eds) Sorghum: origin, history, technology, and production. Wiley, pp 409–441

Coulibaly A, Kouakou B, Chen J (2011) Phytic acid in cereal grains: structure, healthy or harmful ways to reduce phytic acid in cereal grains and their effects on nutritional quality. Am J Plant Nutr Fertil Technol 1(1):1–22

Damalas CA, Koutroubas SD (2024) Potassium supply for improvement of cereals growth under drought: a review. Agron J 116(6):3368–3382

Dayal P, Prajapati SK, Sow S, Ranjan S (2023) Millets: nutritional profile and health benefits as nutri cereals. In: Soil and crop management practices for sustainable agriculture. Elite Publishing House, pp 106–118

Duan Y, Wang C, Li L, Han R, Shen X, Han G, Wang J, Nie M, Zhou X, Du H, Yuan X (2024) Effect of compound fertilizer on foxtail millet productivity and soil environment. Plants 13(22):3167. https://doi.org/10.3390/plants13223167

FAO (2023) Unleashing the potential of millets—international year of millets 2023. Background paper, Rome. https://doi.org/10.4060/cc7484en

Gandah M, Stein A, Brouwer J, Bouma J (2000) Dynamics of spatial variability of millet growth and yields at three sites in Niger, West Africa and implications for precision agriculture research. Agric Syst 63(2):123–140

Govindaraj M, Kanatti A, Rai KN, Pfeiffer WH, Shivade H (2022) Association of grain iron and zinc content with other nutrients in pearl millet germplasm, breeding lines, and hybrids. Front Nutr 8:746625. https://doi.org/10.3389/fnut.2021.746625

Govindaraj M, Rai KN, Cherian B, Pfeiffer WH, Kanatti A, Shivade H (2019) Breeding biofortified pearl millet varieties and hybrids to enhance millet markets for human nutrition. Agriculture 9(5):106. https://doi.org/10.3390/agriculture9050106

Govindaraj M, Yadav OP, Rajpurohit BS, Kanatti A, Rai KN, Dwivedi SL (2020) Genetic variability, diversity and interrelationship for twelve grain minerals in 122 commercial pearl millet cultivars in India. Agric Res 9:516–525. https://doi.org/10.1007/s40003-020-00470-7

Guan R, Pan H, He W, Sun M, Wang H, Cui X, Lou Y, Zhuge Y (2022) Fertilizer recommendation for foxtail millet based on yield response and nutrient accumulation. J Plant Nutr 45(3):332–345. https://doi.org/10.1080/01904167.2021.1943679

Gudadhe NN, Raut AA, Bisht K, Bisht K (2023) Assessment of production trends, nutritional benefits, constraints, policy support, processing and value addition of millets for achieving higher productivity and nutritional security. Indian J Fertil 19(10):1020–1031

Harish MS, Bhuker A, Chauhan BS (2024) Millet production, challenges, and opportunities in the Asia-pacific region: a comprehensive review. Front Sustain Food Syst 8:1386469. https://doi.org/10.3389/fsufs.2024.1386469

Huang Z, Karimanzira TTP (2018) Investigating key factors influencing farming decisions based on soil testing and fertilizer recommendation facilities (STFRF)—a case study on rural Bangladesh. Sustainability 10(11):4331

Ibrahim A, Abaidoo RC, Fatondji D, Opoku A (2016) Determinants of fertilizer microdosing-induced yield increment of pearl millet on an acid sandy soil. Exp Agric 52(4):562–578. https://doi.org/10.1017/S0014479715000241

Issoufa BB, Ibrahim A, Abaidoo RC, Ewusi-Mensah N (2019) Combined use of millet glume-derived compost and mineral fertilizer enhances soil microbial biomass and pearl millet yields in a low-input millet cropping system in Niger. Arch Agron Soil Sci 65(8):1107–1119. https://doi.org/10.1080/03650340.2018.1554247

Jadhav V, Kshirsagar RB, Raut V, Agarkar BS, Kadam GR (2024) Advanced physico-chemical and nutritional characterization of biofortified pearl millet varieties: a comparative evaluation of Dhanshakti and AHB-1200. Int J Adv Biochem Res 8(9S):1291–1295. https://doi.org/10.33545/26174693.2024.v8.i9Sp.2338

Kaleri AA, Lund MM, Manzoor D, Jamali AR, Channa GM, Arif M, ur Rehman M, Khan SZ, Fazal S (2024) Integrated impact of inorganic fertilizers (nitrogen+ zinc) growth and yield of millet (*Pennisetum glacum* L.). Int J Agric Innov Cutting Edge Res 2(4):61–67

Kennedy IR, Choudhury ATMA, Kecskes ML (2004) Non-symbiotic bacterial diazotrophs in crop-farming systems: can their potential for plant growth promotion be better exploited? Soil Biol Biochem 36(8):1229–1244

Khan R, Chandra S, Nagar R, Khatana RS, Nagar V (2023) Response of pearl millet [*Pennisetum glaucum* (L.) R. Br.] to foliar application of nano nitrogen fertilizer (nano urea). J Progress Agric 14(1):69–73

Kishore A, Alvi M, Krupnik TJ (2021) Development of balanced nutrient management innovations in South Asia: perspectives from Bangladesh, India, Nepal, and Sri Lanka. Glob Food Sec 28:100464

Krupashree R, Rao S, Desai BK, Ananda N, Bhat SN (2022) Zinc and Iron fortification through enriched organics and foliar nutrition on growth, yield and economics of foxtail millet (*Setaria italica*). Environ Conserv J 23(3):62–70

Kumar S, Pandey G (2020) Biofortification of pulses and legumes to enhance nutrition. Heliyon 6(3):03682

Kumar P, Kumawat P, Prakash R, Singh H, Kumari V (2024) Integrated nutrient management approaches in pearl millet: a comprehensive review. Int J Educ Manag Stud 14(3):335–339

Li H, Geng J, Liu Z, Ao H, Wang Z, Xue Q (2025) Mulching improves the yield and water use efficiency of millet in Northern China: a meta-analysis. Agriculture 15(4):397. https://doi.org/10.3390/agriculture15040397

Lu Y, Zhao Y, Xia X, Liu M, Wei Z, Wang J, Liu J, Cui J, Li S (2024) Evaluation of nitrogen fertilizer supply and soil nitrate thresholds for high yields of foxtail millet. Agriculture 14(10):1711. https://doi.org/10.3390/agriculture14101711

Maitra S, Praharaj S, Hossain A, Patro TSSK, Pramanick B, Shankar T, Pudake RN, Gitari HI, Palai JB, Sairam M, Sagar L (2022) Small millets: the next-generation smart crops in the modern era of climate change. In: Omics of climate resilient small millets. Springer, Singapore, pp 1–25

Marwein BS, Singh R, Chhetri P (2019) Effect of integrated nitrogen management on yield and economics of foxtail millet genotypes. Int J Curr Microbiol Appl Sci 8(08):2543–2546. https://doi.org/10.20546/ijcmas.2019.808.295

Mebratu Y, Raghavaiah CV, Ashagre H (2017) Productivity, nutrient removal, soil nutrient dynamics and economics of tef (*Eragrostis tef* (Zucc.) trotter) production in relation to genotypes and integrated nutrient management on Vertisols of mid highlands of Oromia region of Ethiopia, East Africa. Int J Life Sci 5(1):21–34

Meena KR, Sharma OP, Dhaker DL, Sain RS, Yadav A, Bhamboo K (2024) Effect of nitrogen levels and foliar application of nano-urea on productivity of pearl millet [*Pennisetum glaucum* (L.) R. Br. emend Stuntz]. Int J Res Agron 7(10):255–258. https://doi.org/10.33545/2618060X.2024.v7.i10Sd.1764

Mohamed BA, Santhi R, Maragatham S, Gopalakrishnan M, Ravikesavan R, Geetha P (2023) Sustaining soil fertility and yield by inductive cum targeted yield model-based fertilizer prescriptions for finger millet (*Eleusine coracana* L.) in Alfisols of Tamil Nadu, Southern India. Agric Sci Digest 500:1–7. https://doi.org/10.18805/ag.D-5762

Monisha V, Rathinaswamy A, Mahendran PP, Kumutha K (2019) Influence of integrated nutrient management on growth attributes and yield of foxtail millet in red soil. Int J Chem Stud 7(3):3536–3539

Mubeena P, Halepyati AS, Chittapur BM (2019) Effect of date of sowing and nutrient management on nutrient uptake and yield of foxtail millet (*Setaria italica* L.). Int J Bio-Resour Stress Manag 10(1):92–95

Nainwal K, Verma O, Reena. (2018) Conservation of minor millets for sustaining agricultural biodiversity and nutritional security. J Pharmacogn Phytochem 1:1576–1580

Nanja Reddy YA, Jayalakshmi P, Narayana Reddy AB, Udaya Kumar M, Ashok EG, Gowda MVC (2025) Agronomic biofortification of finger millet for zinc, and its influence on yield and partitioning of zinc, iron, and calcium in the plant. J Plant Nutr 48(1):23–38. https://doi.org/10.1080/01904167.2024.2377822

Patil PB, Goudar G, Preethi K, Rao JS, Acharya R (2023) Millets: empowering the society with nutrient-rich superfoods to achieve sustainable development goals. J Drug Res Ayurvedic Sci 8(Suppl 1):S100–S114

Pradhan P, Sapkota TB, Kropp JP (2021) Why food systems transformation is crucial for achieving the SDGs. Ruralia 21(55):10

Prakash V, Gupta A, Srivastava AK (2008) Millet production techonology in North West Himalayas. Indian Farm 58:7–12

Rana KS, Kumar D, Bana RS (2012) Agronomic research on pearlmillet (*Pennisetum glaucum* L.). Indian J Agron 57(3rd IAC: Special issue):45–51

Rangaiah KM, Nagaraju B, Kasturappa G, Kadappa BP, Narayanaswamy UKS, Sab MSH, Veerabadraiah GG, Srivastava S, Dey P (2024) Optimizing nutrient management for Kodo millet (*Paspalum scrobiculatum* L.) in the Alfisols of Southern India through modeling soil, plant, and fertilizer interactions. Sci Rep 14(1):31852. https://doi.org/10.1038/s41598-024-83265-y

Rani B, Zalawadia NM, Buha D, Rushang K (2019) Effect of different levels of chemical and nano nitrogenous fertilizers on content and uptake of N, P, K by sorghum crop cv. Gundari. J Pharmacogn Phytochem 8(5):454–458

Rani YS, Triveni U, Patro TSSK (2017) Integrated nutrient management for enhancing the soil health, yield and quality of little millet (*Panicum sumatrense*). Int J Bio-resour Stress Manag 8(1):26–32. https://doi.org/10.23910/IJBSM/2017.8.1.1672

Raundal PU, Pawar PP, Musamade AM, Mahajan MS, Desale SB (2017) Response of little millet varieties to different levels of fertilizers under rainfed condition. A peer-reviewed multidisciplinary. Int J 18

Rouamba A, Shimelis H, Drabo I, Laing M, Gangashetty P, Mathew I, Mrema E, Shayanowako AIT (2021) Constraints to pearl millet (*Pennisetum glaucum*) production and farmers' approaches to striga hermonthica management in Burkina Faso. Sustainability 13(15):8460

Samtiya M, Chandratre GA, Dhewa T, Badgujar PC, Sirohi R, Kumar A, Kumar A (2023) A comparative study on comprehensive nutritional profiling of indigenous non-bio-fortified and bio-fortified varieties and bio-fortified hybrids of pearl millets. J Food Sci Technol 60(3):1065–1076. https://doi.org/10.1007/s13197-022-05452-x

Sankar GM, Sharma KL, Dhanapal GN, Shankar MA, Mishra PK, Venkateswarlu B, Grace JK (2011) Influence of soil and fertilizer nutrients on sustainability of rainfed finger millet yield and soil fertility in semi-arid Alfisols. Commun Soil Sci Plant Anal 42(12):1462–1483. https://doi.org/10.1080/00103624.2011.577863

Selvam R, Santhi R, Maragatham S, Chandrasekhar CN, Ganapathi PS (2022) Effect of initial soil fertility and integrated plant nutrition system on yield and NPK uptake by barnyard millet. Int J Plant Soil Sci 34(4):40–47

Senthamil E, Tamilmounika R (2022) Recent approaches for foliar nutrition in small millets. Compend Agric Allied Sci 2:41–52

Shahzadi H, Khan S, Al-Hashimi A, Basra SM, Mehmood K, Wahid MA, Nawaz M, Irshad S, Haseeb A, Rais A, Gul S (2025) Preliminary study on unlocking growth and yield potential of USDA foxtail millet (*Setaria italica* L.) lines with NPK fertilization. BMC Plant Biol 25(1):22. https://doi.org/10.1186/s12870-024-05960-2

Sharma J, Goyal V, Bhardwaj KK, Duhan D, Dahiya R, Yadav R, Ranjan S, Dey P (2024) Ten year impact of soil test crop response based integrated plant nutrient supply on nutrient use efficiency and productivity for sustainability of pearl millet-wheat cropping system. J Soil Sci Plant Nutr 24(2):3131–3149. https://doi.org/10.1007/s42729-024-01739-4

Sharma SK, Sharma PK, Mandeewal RL, Sharma V, Chaudhary R, Pandey R, Gupta S (2022) Effect of foliar application of nano-urea under different nitrogen levels on growth and nutrient content of pearl millet (*Pennisetum glaucum* L.). Int J Plant Soil Sci 34(20):149–155. https://doi.org/10.9734/IJPSS/2022/v34i2031138

Sharma VK, Pandey RN, Kumar S, Chobhe KA, Chandra S (2016) Soil test crop response based fertilizer recommendations under integrated nutrient management for higher productivity of pearl millet (*Pennisetum glaucum*) and (*Triticum aestivum*) wheat under long term experiment. Indian J Agric Sci 86(8):1076–1081. https://doi.org/10.56093/ijas.v86i8.60561

Sheoran S, Prakash D, Raj D, Mor VS, Yadav PK, Gupta RK, Alamri S, Siddiqui MH (2024) Long-term organic and N fertilization influence the quality and productivity of pearl millet under pearl millet-wheat sequence in north India. Sci Rep 14(1):19503. https://doi.org/10.1038/s41598-024-70009-1

Shree K, Augustine R, Manuel RI, Balaganesh B, Kumar D (2024) Effect of soil and foliar applications on growth and productivity of pearl millet (*Pennisetum glaucum* L.). Asian J Soil Sci Plant Nutr 10(2):236–241. https://doi.org/10.9734/AJSSPN/2024/v10i2280

Sivagamy K, Parasuraman P, Karunakaran V, Ananthi K, Rajesh M, Balaji T, Harisudan C, Sheikh T, Vijayashanthi VA, Yogameenakshi P, Arul Prasad S (2024) Effect of organic manures and foliar application of different organic sources of nutrients on irrigated finger millet in the semi-arid region of southern India. Appl Ecol Environ Res 22(2):10.15666/aeer/2202_11491158

Siwela M, Pillay K, Govender L, Lottering S, Mudau FN, Modi AT, Mabhaudhi T (2020) Biofortified crops for combating hidden hunger in South Africa: availability, acceptability, micronutrient retention and bioavailability. Foods 9(6):815

Sneha M, Thimmegowda M, Hanumanthappa D (2024) Effect of nitrogen levels and method of fertilizer application on growth and productivity of finger millet. Mysore J Agric Sci 58(1)

Sreenivasulu K, Rao MS, Jyothi PA, Rao SG, Rao AU (2023) Growth, yield attributes and yield as influenced by crop establishment and nutrient management in fingermillet (*Eleusine coracana* L.). Andhra Agric J 70(4):431–436

Srinivas V, Naresh N, Pratyusha S, Ankati S, Govindaraj M, Gopalakrishnan S (2022) Exploring plant growth-promoting Streptomyces spp. for yield and nutrition traits in pearl millet hybrids. Crop Pasture Sci 73(5):484–493. https://doi.org/10.1071/CP21438

Srivastava P, Sangeetha C, Baskar P, Mondal K, Bharti SD, Singh BV, Agnihotri N (2023) Unleashing the potential of millets promoting nutritious grains as vital cereal staples during the international year of millets: a review. Int J Plant Soil Sci 35(18):1860–1871

Sun MJ, Chao Y, He W, Kang XR, Yang QG, Wang H, Pan H, Lou YH, Zhuge YP (2022) Changes in foxtail millet (*Setaria italica* L.) yield, quality, and soil microbiome after replacing chemical nitrogen fertilizers with organic fertilizers. Sustainability 14(24):16412. https://doi.org/10.3390/su142416412

Thind TS, Dhillon BS (2021) Sustainable development of agriculture for National Food and Nutritional Security and Farmers' livelihood. Agric Res J 58(3)

Udapudi P, Pushpa K, Sukanya TS, Krishnamurthy R (2024) Alterations in the nutrient status of soil and enzymatic activity of finger millet cultivated soil accompanying to nano urea application. Environ Ecol 42(2A):716–725. https://doi.org/10.60151/envec/VGQA5833

Upadhyay PK, Dey A, Singh VK, Dwivedi BS, Singh T, Rajanna GA, Babu S, Rathore SS, Singh RK, Shekhawat K, Rangot M (2023) Conjoint application of nano-urea with conventional fertilizers: an energy efficient and environmentally robust approach for sustainable crop production. PLoS One 18(7):e0284009. https://doi.org/10.1371/journal.pone.0284009
Wang J, Han G, Duan Y, Han R, Shen X, Wang C, Zhao L, Nie M, Du H, Yuan X, Dong S (2024) Effects of different organic fertilizer substitutions for chemical nitrogen fertilizer on soil fertility and nitrogen use efficiency of foxtail millet. Agronomy 14(4):866. https://doi.org/10.3390/agronomy14040866
Wang R, Wang C, Liu T, Chen Y, Liu B, Xiao J et al (2025) Effects of different organic materials and reduced nitrogen fertilizer application on sorghum yield and soil nutrients. Sci Rep 15(1):6914
WHO (2018) Global strategy for women's, children's, and adolescent's health 2016–2030 and the 2030 Agenda for sustainable development
Yadav OP, Gupta SK, Govindaraj M, Singh DV, Verma A, Sharma R, Mahala RS, Srivastava SK, Birthal PS (2024) Strategies for enhancing productivity, resilience, nutritional quality, and consumption of pearl millet [Pennisetum glaucum (L.) R. Br.] for food and nutritional security in India. Crop Sci 64(5):2485–2503. https://doi.org/10.1002/csc2.21346
Zhao Y, Wang G, Liu M, Xue H, Zhao P, Han B, Fan H, Wang R, Wang L, Guo E, Zhang A (2025) Study on improving yield and antioxidant enzyme activities in millet by rationing molybdenum and nitrogen. Front Plant Sci 15:1524347. https://doi.org/10.3389/fpls.2024.1524347
Zhou T, Liu Q, Yang S, Wang Z, Zhang P, Wang X, Ji X, Zhang X (2025) Optimized fertilization patterns increase foxtail millet biomass on the distribution and transformation in Loess Plateau of China. PLoS One 20(2):e0318199. https://doi.org/10.1371/journal.pone.0318199

MIX
Papier | Fördert gute Waldnutzung
FSC® C083411

Kolibri 360 GmbH
Ferdinand-Jühlke-Straße 7
99095 Erfurt, Deutschland
produktsicherheit@kolibri360.de